LIFELINES OF OUR SOCIETY

Infrastructures Series

Edited by Paul N. Edwards and Janet Vertesi

A list of books in the series appears at the back of the book.

LIFELINES OF OUR SOCIETY

A GLOBAL HISTORY OF INFRASTRUCTURE

DIRK VAN LAAK

FOREWORD BY PAUL N. EDWARDS

TRANSLATED BY ERIK BUTLER

THE MIT PRESS   CAMBRIDGE, MASSACHUSETTS   LONDON, ENGLAND

Originally published as *Alles im Fluss. Die Lebensadern unserer Gesellschaft—Geschichte und Zukunft der Infrastruktur*. © 2018 S. Fischer Verlag GmbH, Frankfurt am Main

The translation of this work was funded by Geisteswissenschaften International—Translation Funding for Humanities and Social Sciences from Germany, a joint initiative of the Fritz Thyssen Foundation, the German Federal Foreign Office, the colleting society VG WORT and the German Publishers & Booksellers Association.

The MIT Press would like to thank the anonymous peer reviewers who provided comments on drafts of this book. The generous work of academic experts is essential for establishing the authority and quality of our publications. We acknowledge with gratitude the contributions of these otherwise uncredited readers.

This book was set in Stone Serif by Westchester Publishing Services. Printed and bound in the United States of America.

Library of Congress Cataloging-in-Publication Data

Names: Laak, Dirk van, author. | Butler, Erik, 1971– translator.
Title: Lifelines of our society : a global history of infrastructure / Dirk van Laak, Paul N. Edwards, Erik Butler ; foreword by Paul N. Edwards ; translated by Erik Butler.
Other titles: Alles im Fluss. English
Description: Cambridge, Massachusetts : The MIT Press, [2023] | Includes bibliographical references and index.
Identifiers: LCCN 2022050330 (print) | LCCN 2022050331 (ebook) | ISBN 9780262546386 (paperback) | ISBN 9780262375962 (epub) | ISBN 9780262375955 (pdf)
Subjects: LCSH: Infrastructure (Economics)—History.
Classification: LCC HC79.C3 L3313 2023 (print) | LCC HC79.C3 (ebook) | DDC 338.9—dc23/eng/20221107
LC record available at https://lccn.loc.gov/2022050330
LC ebook record available at https://lccn.loc.gov/2022050331

10 9 8 7 6 5 4 3 2 1

CONTENTS

FOREWORD vii
Paul N. Edwards
PREFACE TO THE ENGLISH-LANGUAGE EDITION xi
ACKNOWLEDGMENTS xv

INTRODUCTION: AS LONG AS IT KEEPS WORKING . . . 1

I THE CLASSICAL ERA OF INFRASTRUCTURE

1 PUBLIC WORKS: THE NINETEENTH CENTURY 21
2 LIFELINES OF COMMUNITY: THE EARLY TWENTIETH CENTURY 47
3 THE MEASURE OF MODERNITY: THE LATE TWENTIETH CENTURY 79

II NODES OF INFRASTRUCTURAL DEBATE

4 PUBLIC, PRIVATE, OR PARTNERSHIP? HOW INFRASTRUCTURE IS ORGANIZED 113
5 PRESTIGE PROJECTS: SYMBOLIC VALUE AND FAILURE 129
6 CRUMBLING BRIDGES: THE LIFE CYCLES OF INFRASTRUCTURES 143
7 ACHILLES' HEELS: THE VULNERABILITY OF LARGE-SCALE NETWORKS 157

8 THE COLD PERSONA: USERS AND OPERATORS OF INFRASTRUCTURES 173

CONCLUSION: DOWN THE DRAIN? 189

NOTES 207

BIBLIOGRAPHY 259

INDEX 311

FOREWORD

Paul N. Edwards

Sometime around 2020, the sum of all human-made things on earth attained the staggering mass of 1.2 trillion tons, exceeding that of the entire biosphere. All the more jaw-dropping is the speed of this dubious achievement: 120 years earlier, at the dawn of the twentieth century, human-made mass was just 3.5 billion tons.[1] Meanwhile, by 2020 humanity's total energy consumption, including solar energy converted in crops and forestry as well as fossil fuels, nuclear energy, and renewables, reached some twenty-seven terawatts—nearly one-third of the ninety terawatts of biochemical energy produced in the biosphere by photosynthesis.[2]

Much of this energy-hungry, unimaginably colossal construction consists of infrastructure. Roads. Railways and subways. Ports, airports, and spaceports. Container ships and oil tankers. Trucks. Pipelines. Mines, oilfields, and refineries. Electric grids, hydroelectric dams, and power plants. Skyscrapers. Hospitals. Industrial agriculture and monoculture forestry. Communication cables, data centers, and Internet hubs. Sewers, reservoirs, and municipal water pipes. And millions upon millions of other objects, devices, and systems built by human beings.

This extraordinary book is about the heavy weight of infrastructure: not its mass per se, but the histories and stories of its growth, its human purposes, its governance (or lack thereof), and the enormous inertia it can acquire. Infrastructure is built to last, and much of it endures longer than a human lifetime, longer than most governments and corporations. The infrastructures of modernity represent a kind of second nature. They set the background conditions of life for generations born into them, who are committed (willingly, knowingly, or not) to their ongoing repair and maintenance. Path dependence—the "inertia of the installed base"—rules.[3] Infrastructure, including its human components, resists change unless it offers clear advantages for many or profits for a few.

The Roman Empire famously built roads throughout Europe and the Mediterranean basin, many of them still in use today. The ancient Chinese had their Great Wall, for a time the only human structure visible from space. Unlike these constructions, designed mainly for ruling elites and their armies, modern infrastructures typically have broader public purposes. *Public works*, a common phrase of the nineteenth and early twentieth centuries, defines the initial period of van Laak's study, when governments first took on provision of fundamental systems and services such as roads, sewers, and canals. This was also the period of the first and second Industrial Revolutions, powered by steam and fueled by coal, whose transport from mines to "coketowns," factories, and cities shaped the outline of canals and railways across England, Europe, and the United States. With the advent of the telegraph, communication cables followed those routes. These later became main highways for cars and trucks, linking "motowns" such as Detroit and Los Angeles. Today, Internet backbone lines trace those same paths, first laid down many decades earlier.[4] Open to most, but not all, infrastructures with public purpose underlie the daily lives of billions.

This is not one story, but many, a myriad of ventures by governments, corporations, and multinationals. In every case, infrastructure confronted tensions between public purposes and private profit, symbolic and practical value, initial investment and long-term maintenance, open systems and security against human threats and natural disasters. Although the main threads of his story come from Europe and the United States, van Laak does not neglect to canvass the colonial empires that supplied much of the material and energy for Western infrastructure—often facilitated by extractivist systems, such as railways, built in the colonies by native peoples under duress, with mainly negative consequences for their bodies, cultures, and lifeways. Not only in the colonies but also in Europe and the United States, those enrolled in infrastructure projects often labored under adverse conditions ranging from dangerous and poorly paid to conscription, forced labor, and slavery. Countless miners, railway workers, bridge builders, and others sacrificed lives, limbs, and health in service of public infrastructure, and such unheralded sacrifices continue to this day. Yet infrastructures also improved the quality of life for vast numbers of people (including most readers of this book), in terms of mobility, communication, health, longevity, and standard of living.

To tell these stories, van Laak weaves together an astonishing quantity and variety of historical scholarship from the last several decades, along with numerous sources in multiple languages. Yet this is no dry, academic tome; the writing in van Laak's tour de force is brisk, lively, and entertaining. The book brims with stories and examples. It eschews the drab, tedious language of the academic literature review, yet nearly every sentence has a footnote for those seeking deeper understanding. As scholarship, this book is generous and comprehensive; as a reading experience, the book invites wonder, engagement, and reflection.

During the Cold War, many feared that a US-Soviet nuclear exchange might "bomb us back to the Stone Age." With all the easily accessible minerals and fossil fuels mined out, went this line of thought, humanity would never get out of it again. Today, it's not just ores and fuels; we are even running out of sand for concrete and glass production. Yet merely to survive the coming centuries of climate change—a direct consequence of fossil-fueled infrastructure—whole new infrastructures are needed. Resilience is today's new watchword, revising the taken-for-granted infrastructural tolerances of a bygone age when temperature, weather, and sea levels could be presumed stable; forests, fish, and other resources seemed inexhaustible; and buried raw materials lay waiting for discovery. One can seriously wonder whether re-infrastructuring the world for genuine sustainability is even remotely plausible.

In his final chapter, van Laak examines the condition of infrastructure users today, who often confront (and desire) impersonal digital interfaces requiring little or no human interaction. Even more strangely, such interaction as does occur happens with tech support workers on the other side of the planet who cannot themselves access the systems and services they support. Viewing modern history through van Laak's lens of infrastructure restores a large measure of humanity and a depth of understanding to this increasingly anonymous world. I wish you, reader, the same deep satisfaction this book has brought to me.

PREFACE TO THE ENGLISH-LANGUAGE EDITION

Since this book's initial publication, in 2018, events have offered many occasions to reflect on infrastructures and their history:

- In the spring of 2021, the container ship *Ever Given* was stuck in the Suez Canal for six days. This unintentional blockade triggered a traffic jam in the global transport system, driving home just how much international commercial logistics depend on the smooth operation of many parts. In broader perspective, the insight was hardly new: shortly after opening in 1869, the Suez Canal quickly became the "aorta of the British Empire."
- The same year, in the wake of the COVID-19 pandemic, the US government inaugurated the largest infrastructural program since the New Deal. But, as had been the case for decades, policy makers disagreed about how, exactly, to understand the scope of projects. Does *infrastructure* refer only to transportation routes, or does the term include social and cultural initiatives?
- Russian aggression in Ukraine in early 2022 demonstrated, once again, the key importance of seizing and destroying infrastructure in modern warfare. Airports, bridges, energy supply systems, and lines of digital communication all provided targets. Civilian arrangements and institutions become critical infrastructures in times of trouble: they make society as a whole vulnerable to blackmail and represent Achilles' heels of states.

Today, there is more talk than ever of infrastructure, which has come to mean the very foundations of life in society. The ubiquity of the term attests to just how much collective well-being relies on planning, concessions, and the availability of resources, technology, and services. Our day-to-day activities presuppose that various offerings and options will be available. Such provisions, however, are precarious and always at risk of being interrupted or vanishing. Accordingly, interest groups are constantly calling for infrastructures to be maintained and, indeed, expanded. The slightest disturbance in the artificial environment they have created—which now counts as normal—activates unconscious fears.

The idea for this book was hatched about ten years ago. At the time, there was no way to foresee the infrastructural turn in the humanities and social sciences, which, in the last few years, has been booming. *Infrastructure* is used to group all manner of cultural processes and developments. To a great extent, the lack of a clear definition is what makes it so appealing; a broad margin exists for relatively free and loose associations by critics and commentators.

Infrastructures promise to open up spaces and save time. Levels of development and the number of facilities and services inspire a sense of being either modern and up-to-date or backward and stuck in the past. Connection to infrastructures represents a decisive criterion for quality of life. Being excluded from them is reckoned a catastrophe. For all that, increased reliance also brings inherent ambiguities to light. Although infrastructures make possible the expansion of horizons and the circulation of goods both material and immaterial, the same conditions can be exploited to exercise social control and restrict access to vital resources.[1]

When discussion concerns infrastructures, participants usually have the future in mind, specifically prospects for maintaining or improving the common weal. The operative rhetoric tends to suggest that such provisions have always existed. In historical perspective, however, it is clear that technological developments and political measures emerged over time and now are interwoven to such an extent that the whole is difficult to survey. Over the last two hundred years, we will see, the material and organizational arrangements of transport, communication, supply, and disposal have shaped the rhythms and patterns of life itself.

Infrastructures serve to divide and distribute labor, whether performed by human or machine, and they draw on resources that are available only in limited quantity. They make it possible for many people to participate in social and political decision-making and to enjoy relative prosperity and luxury. They create opportunities. By the same token, however, they give rise to dependency and debt. Moreover, infrastructures represent potential liabilities in ecological terms; disrepair indicates the fragility of social conditions on the whole. In this respect, infrastructures and their history record the seismic shifts of the Anthropocene.[2]

This study explores the subterranean social relations that are built into services and institutions.[3] Accordingly, the account offered of schemes of development, their architects, and the people who use them often stands at odds with standard histories. Only in part does the matter concern geographic borders or the particularities of national cultures. In fact, I argue, infrastructures are the concrete basis for colonization and globalization.[4] They collapse and reorganize space and synchronize and standardize time.[5]

The point is not to define infrastructures once and for all or to explain them in light of any single theory. Nor does this book consider their impact in terms of a privileged category—say, power, economics, or social dynamics. Instead, the goal is to examine infrastructures as parts of a history that has been shaped not by exceptional individuals so much as by the interplay of varied forces and interests. The merit of such an approach has been made clear, in recent years, by numerous case studies with a specific focus and well-defined field of investigation.[6]

The following pages also take up the formative effect on the people who use infrastructures, once they are up and running, observed by researchers time and again. Travelers, for example, adapt to timetables and fee schedules, and they make certain codes of expected conduct their own. In other words, infrastructures are not just a matter of technology, engineering, and organization but also include lived, day-to-day practices.[7] Such processes receive due attention here.

Infrastructures fundamentally concern expectations of the present and the future. They keep things flowing. They exist in a positive sense to the extent they are used—that is, inasmuch as they have a history. Although a concept of Western origin, infrastructure represents a global phenomenon; therefore, examples in this book come from all over the world. Detailed empirical studies are now being conducted almost everywhere. The history of infrastructure is an established discipline. In 1999, the sociologist Susan Leigh Star self-depreciatingly remarked that looking at infrastructures means studying "boring things."[8] The intervening decades have proved the very opposite.

Dirk van Laak, August 2022

ACKNOWLEDGMENTS

The thought and reflections that inform this book evolved over the course of many years. I owe the idea of writing a cultural and historical analysis of infrastructures in everyday life to my teacher and adviser Lutz Niethammer. In the early 1990s, the topic represented new and almost exotic terrain; since then, it has come to stand at the center of historical and cultural studies. It was my good fortune that the Volkswagen Foundation looked with favor on my proposal to draw up a balance sheet of research to date; I am extremely grateful to the foundation for the Opus Magnum grant, which enabled me to escape academic routine for a year. Colleagues at Giessen and the administration of the Justus Liebig University also supported my efforts. In particular, criticism and comments offered by Benjamin Brendel, Jürgen Dinkel, Sabine Dworog, Denise Eisenbeiser, Tanja Hommen, Jeannette van Laak, and Kai Nowak helped craft a much better and more readable text. Thanks for further improvements are due to Maria Paula Diogo, Axel Dossmann, Alexander Eckermann, Nikola Medenwald, and Johan Schot. Participants at colloquia in Berlin, Copenhagen, Erlangen, Frankfurt am Main, Heidelberg, Munich, Regensburg, and Zurich offered valuable feedback. I am happy to share credit; of course, responsibility for any shortcomings is mine alone.

Dirk van Laak, Leipzig, January 2018

INTRODUCTION: AS LONG AS IT KEEPS WORKING . . .

It's a familiar sight: people constantly checking their smartphones. At social gatherings and on roadways, at the movies or during sporting events, it seems these devices are never set aside, much less turned off. People today are chronically online, and that's evidently how they like it. But their faces often reflect the stress of being overwhelmed.

The smartphone grants the power to survey, log into, and swipe away the whole world. It symbolizes everything that intellectual enlightenment, industrial revolution, and political liberalism have achieved since the eighteenth century: independence from assigned social space and station, access to global streams of data, and membership in a global community. The smartphone has become the remote control of a networked society and has given rise to comprehensive customer choice, up-to-date and informed citizenship, and free and self-determined personhood.

At the same time, the system behind this device makes people dependent on networked technologies and everything they offer. By requiring us to communicate and make decisions incessantly, the smartphone epitomizes more than two centuries of developments in a field that, somewhat blandly, is called *infrastructure*.

BEYOND EVERYDAY LIFE

When still a student, I listened in on a conversation between scholars visiting from the German Democratic Republic. One detail has really stuck in my mind: the East Germans, who were in the West for the first time, did not marvel at the consumer offerings. They were ready for that. What astonished them was the seamlessness with which West German *infrastructures* operated. Not long afterward, the GDR collapsed. I largely failed to register what had happened, even though I was studying the history of municipal utilities. My research on long-term structures such as railway lines, power cables, and water supply had distracted me from political events of the day. In retrospect, I recognize a hidden logic at work in my oblivion.

Attention to infrastructures reveals any number of paradoxes. This clumsy word suggests that matters are better left to the experts. Provisions for communications, travel, the delivery of goods, and waste removal free us from worrying about where the electricity or water that we use comes from or how people and commodities get from one place to another. Yet infrastructures constitute a constant object of discussion and debate. Who should build and maintain them? How are they to be protected? And why are they all in such bad shape? Only when they stop working—as occurred in the GDR—does the suspicion arise that operations might not be as straightforward as they appear.

Why do most people pay so little mind to what affects their day-to-day lives? What makes the *conditio sine qua non* of every modern society[1] seem so profane and trifling? Even though the health and wealth of a given society are reflected by its infrastructure, the category itself has come to connote a shortcoming—so much so that the historian Paul Edwards and colleagues have referred to it as an "all-encompassing solution and an omnipresent problem, indispensable yet unsatisfactory, always already there yet always an unfinished work in progress."[2]

It's deemed proof positive of backwardness when telephone lines are not working or running water is not available. Infrastructures promise better and more comfortable living conditions. They require constantly connecting with new, more efficient ways of managing transport, communication, and energy. As a rule, advocates couch matters in terms of practical necessity—alongside utopian promises of a brighter future.

This book does not propose a systematic, comprehensive overview of the economic, political, and legal dimensions of infrastructure—much less its technological aspects—even though these factors all play a role. Instead, it sets out to show that the growth of infrastructure networks has not been steady and has varied in keeping with time and place. The task is to situate and examine infrastructures historically—in particular, their influence on (post)modern culture, everyday life, and cultural practices. To do so requires a shift of perspective away from what normally counts as history. It also means taking a step back from the easy conscience of the welfare state—the calm outlook that infrastructures tend to promote.[3]

Members of my generation, who grew up during the 1960s and achieved political consciousness during the 1970s, expected that infrastructural provisions would continue to expand and improve. Today, however, it seems that infrastructure has come to represent a problem. Wealthy countries

0.1 Artistic renderings of energy and traffic flows are rare. Gustav Wunderwald's 1927 depiction of a Spandau underpass exemplifies the style of New Objectivity (*Neue Sachlichkeit*). The painter died of water poisoning in 1945. © akg-images

face the challenge of setting up new systems (at the moment, digital ones) while maintaining, or sometimes reducing, existing ones. Elsewhere, the task is to provide a minimum of utilities, waste disposal, and arrangements for communication and transportation, or simply to attain a level similar to what has long existed in richer lands. In the main, talk of development for the last 150 to 200 years has revolved around better or worse infrastructures.

By the same token, infrastructures have become the way to measure so-called modernity—that is, levels of connectedness to a global economy and society. Infrastructures set the standard. It would seem there can never be enough of them: someone else always has it better and enjoys the benefits of more comprehensive networks, faster travel, and fuller access to information.

Since the eighteenth century, infrastructures have brought human communities up to date, integrating groups of people with others in spatial, social, and cultural terms and offering members new possibilities for living their lives. Yet infrastructures exercise control, instill conformity, and shape conduct in ways that are not always welcome. Order and good government are assessed in terms of the scope and performance of infrastructures; indeed, these terms are often used as synonyms. The consequence is an environment where routines are observed intuitively and a kind of second, artificial nature predominates. It is impossible to overestimate the importance of the infrastructural model of modernity in defining culture. Regimes of supply, disposal, communication, transportation, and energy guide the habits and behavior of those who employ them. Thus, they constitute an autonomous and independent factor in recent developments; their complexity and impact represent a real marvel.

Nowhere are human designs and technology tied so closely to everyday life as in the realm of infrastructure, which, more and more often, finds mention as a matter of course alongside the weather or stock market. Instead of just alerting people to traffic jams, the media offers reports on overall trends—say, the mounting flow of vehicles from day to day and the implications of this development. It's essential to stay on top of construction, strikes, interruptions in production or delivery, and anything else that might require a change of routine. Instances of dysfunction occasion debate about who is responsible for arranging and financing it all, as well as ways to prevent similar inconveniences. Infrastructural matters have long since left the niche sector of nostalgic railroad enthusiasts and stamp collectors; they now stand front and center in discussions ranging from getting to work on time to setting national and international policy.

It is plain that we are growing increasingly aware how much we depend on infrastructures. It is just as clear that our ability to understand how they work is falling behind. The ready availability of anonymous services has compounded our sense of vulnerability, and digital technology has increased real and perceived dependencies. Presumably, the process will continue for the foreseeable future. The vital importance of infrastructures looms into view when they stop working smoothly, fall into a state of disrepair, are hobbled by strikes or attacks, or when prices rise. A low smartphone battery is all it takes for many people to start getting nervous.

The greater attention being paid to infrastructures points to something else, too. Evidently, the expansion of networks of circulation has peaked.[4] This may seem surprising in light of many developing countries working hard to establish infrastructural systems in the first place. No one would deny them the right—indeed, the need—to do so. Everyone should be invested in providing as many people as possible with clean water and hygienic conditions. That said, examining such provisions and their transformations over time, in the nineteenth and twentieth centuries, reveals how much their expansion presumes continuous economic growth, free markets, and the domination of nature. These axioms of industrialized Western societies are being called into question for good reason.

One of the main questions addressed in this book is how infrastructures have changed everyday life by forging new routines, standards of conduct, and expectations. Emphasis is placed, then, on transformations of what, at a given period of history, has counted as normal. One feature of infrastructural development has long been—and will continue to be—that it relieves us of thinking too much about the past, present, and future. Infrastructures are supposed to spare us the day-to-day difficulties of securing material existence, open room for creative pursuits, and enable people to broaden their horizons. They bring about *flows* that we join as need arises—by turning a tap, boarding a train, or logging onto the Internet. Infrastructure as a whole might be defined as everything stable that enables human beings, goods, and ideas to circulate. Accordingly, our focus is on these flows' material dimension: tracks, roadways, pipes, and other conduits down or through which things pass.

Our subject does not include heroic visionaries, epochal shifts, or festive anniversaries. It concerns what the architectural historian Siegfried Giedion calls "anonymous history."[5] Infrastructural planners orient themselves on ordinary human beings and recurring needs in the modern world. From the outset, then, we should note that the infrastructural networks surrounding us did not result from a coherent program so much as they emerged in response to different, and competing, interests. They represent the material substrate of social constellations, the crystallized state of conditions at a given moment. Such arrangements, however, are constantly being negotiated and modified. By foregrounding continuity, which lies below the threshold of ordinary awareness, we see the degree to which infrastructures are subject to change over time.

The account offered here treats secular processes of the modern age. No limits have been placed on extension in space, but in temporal terms our discussion bears mainly on the last two hundred years. Nor does this study claim to be exhaustive. Vantage points and examples have a subjective tint; alternating between illustrative cases and global assessments, the material intersects with other fields of inquiry at every turn. This review of events from the eighteenth century to the present aims to disclose and incorporate a new register of meaning. To date, histories written on a grand scale have hardly treated infrastructure, even though it represents a phenomenon independent from industrialization and urbanization, the emergence of bureaucracy, and the rise of the welfare state. In part I, the first three chapters, which discuss the most essential phases of development in the nineteenth and twentieth centuries, rectify this omission and identify the principal features of processes that have occurred everywhere else in the world at different times and places.

Analysis occurs on urban and national, European and North Atlantic, and global levels; each of these registers contains elements of the others. Restricting the discussion to a specific region or example would fail to do justice to the expansive nature of infrastructures, which open new worlds in a literal sense. If a western European, North Atlantic perspective predominates, it is because more research exists in this area. Even so, the big picture is planetary; by definition, infrastructure promotes circulation and growth worldwide.

Since the 1970s, the classical period of building material networks has faded away in so-called developed countries; the era of universal access to standardized services offered by a monopolistic provider is drawing to a close.[6] But elsewhere the modern infrastructure ideal is only starting to gather speed. Chapters in part II concentrate more systematically on discussions of infrastructure that have come to the fore in the last two hundred years: organization, financing, visibility, symbolic value, evolution, decline, and vulnerability—as well as the expectations placed on users and operators. Finally, the most salient issues in contemporary debate are taken up alongside visions of the future.

Our conclusions can be only provisional. In this regard, the book mirrors infrastructures themselves: it represents a formation dependent on the conditions of the moment.[7] Whatever infrastructures provide is developed but ultimately depurposed; at some point, it becomes obsolete. Everything we consider normal and self-evident is subject to rapid change. Infrastructures represent the precondition for networked activity on a global scale that determines the course of everyday life; yet at times they threaten to fall apart. When I first took up the matter in 1989, the scope of the next transformation—the digital revolution—could scarcely be foreseen. It is important to be ready for a new turn of events; a look at what has happened to date can help us anticipate coming developments.

WHERE TO BEGIN?

On March 1, 1952, the *Frankfurter Allgemeine Zeitung* reported that a "bizarre word" had surfaced at a NATO (North Atlantic Treaty Organization) conference in Lisbon; although "baffling to German ears," technical specialists, government agencies, and organs of the press eagerly adopted it. In keeping with editorial standards, the reputable publication took care to set the offending term apart in special type, recommending, moreover, "always translate the expression *infrastructure* into German." After all, the article continued, nothing encourages thoughtlessness so much as the "promiscuous use of incomprehensible words."[8]

The ambiguity of the term later drove parties allergic to jargon to despair. Their ranks included Bernard S. Katz. In a 1989 column, the advertising executive remarked that *infrastructure* had become inescapable—it was just too tempting for people who thought that obscure language hinted at importance to pass up.[9]

Some twenty-five years later, according to the journalist Till Briegleb, infrastructure had "dissolved into an all-purpose metaphor for just about any kind of system."[10] Indeed, for some time now, it has enjoyed one boom after another in broader and broader contexts, providing a synonym for conditions required not just for technological changes but also for social or cultural processes of any kind.[11] *Infrastructure* shines with the glow of necessity. Presumably, and for all its clumsiness, it owes its prominence to counting as a concept, writ large. This notion, we will see, emerged in western Europe and the United States over the course of the nineteenth century, took on greater dimensions in the twentieth, and finally came into its own in the 1950s and 1960s.[12] Now, it dominates the entire globe—even if, as we will also see, not everyone is happy about it.

In recent decades, more and more voices have been calling for infrastructure to be viewed in historical perspective. Using the lens of social psychology, Harald Welzer notes how it shapes ways of thinking: "Institutional infrastructures regulate growth; the material ones manifest it, and mental infrastructures translate it into lifeworlds, equipping the inhabitants of growth societies with the associated biographies and notions of self."[13] The practical course he recommends—abandoning the idea of perpetual growth—is one of the most radical interpretations to have been proposed.

But couldn't it be said that the concept applies to the ancient and medieval worlds? How else could one describe the wells, streets, canals, city walls, granaries, and arsenals of premodern times? Indeed, since the nineteenth century, the Roman Empire has provided the standard reference point for how government established networks of supply, transportation, and communication. Rome gave posterity the first paved highway between regions, the Via Appia (312 BC); aqueducts as far away as southern France; and the *limites*, or border system, in German territories. Time and again, scholars have drawn attention to the elaborate system providing Rome with water and the impressive Cloaca Maxima, which disposed of waste, in addition to Rome's street design and efficient postal system.

Officials traveling along the *cursus publicus* benefited from a highly developed logistical system that included relay and rest stations, teams of draft animals, and replacement vehicles.[14]

To this day, knowing about these institutions forms an integral part of the cultural repertoire of the educated classes in Europe. *Opera publica* was a phrase in ancient Rome; in modern languages it became *obras públicas, réseaux publics, public works*, and *öffentliche Arbeiten*. Now that Europe has come to be just one global province among others, the field of reference includes great achievements elsewhere, not as familiar as Europe's but imposing a belated modesty on them: the canal networks of ancient China; Japanese roadways; irrigation systems in Egypt, India, and Central America; and the Silk Road, for example.[15]

To be sure, the term *infrastructure* can be applied to all the preceding. Still, I prefer to restrict its use to developments over the last two centuries. Our operative definition does not depend on a particular set of technologies, all of which had forerunners of one kind or another. Rather, it concerns the circulation of commodities, people, and ideas, as well as efforts to provide services to, and enable communication among, all citizens. Such goals arose in tandem with the Enlightenment, political revolution in the United States and France, industrialization, and the ascent of a liberal economic order and mass society.

Earlier projects of building roads, exchanging information, and providing welfare services fall outside the scope of inquiry because we are concerned with systems on which the majority of a given population relies from day to day. Infrastructures are not just available; they are actively used. This definition allows us to compare different epochs and regions while acknowledging the particularity of arrangements in a given time and place.

Where should we start, then? There are several options. The classic academic approach would begin with ideas and tie them to pioneers, pathbreakers, and prophets. In that case, we would have to take up early modern utopian schemes for creating harmonious and just societies that meet the needs of all people. For the most part, such visions of a stable future without material need or social marginalization evoke a somewhat boring land of milk and honey.

Alternatively, we could emphasize practical politics. This would mean starting with the emergence of the modern state, policies of mercantilism and cameralism, and plans for developing clearly defined territories. In *The Wealth of Nations* (1776), Adam Smith, the father of modern economics, assigned the "sovereign or commonwealth" the duty

> of erecting and maintaining those public institutions and those public works, which, though they may be in the highest degree advantageous to a great society, are, however, of such a nature that the profit could never repay the expense to any individual or small number of individuals, and which it therefore cannot be expected that any individual or small number of individuals should erect or maintain.[16]

On the basis of observations along these lines, generations of political scientists have made such "works" their business; in Germany, the term *öffentliche Anstalten* was used for the first time just two years after Smith's book was published.[17]

Then again, we could take for the point of departure the occasion when the mentality shifted: the notion—formulated in the eighteenth century at the latest—that progress and civilizational advances are the dividends paid by circulation, the free movement of human beings, goods, and ideas.[18] In this framework, sedentary ways of life closed off from interaction with other groups were regarded as backwardness, a condition to be overcome. Conversely, the web, or network, became a positive image for human communities enjoying the benefits of both centralization and decentralization.[19] The picture implies conflicts between infrastructures, which are expansive by nature, and economic realms sealed off by national borders—the problem of the "isolated state" examined by Johann von Thünen (1826). Since then, disagreement has prevailed between those who argue along protectionist lines and those advocating free trade; to this day, infrastructures are set up or curtailed in keeping with decision-makers' preference one way or the other.

Another starting point could be changes from an environmental perspective. The use of fossil fuels for industrial machinery represented a more intensive exploitation of natural resources to meet human needs. Greater activity of this kind had been long anticipated, but means for doing so were limited.[20] Well before the dawn of the machine age, measures were taken to make nature more calculable—for example, dams and levees in marshlands and coastal regions.[21] Johann Wolfgang von Goethe made a point of including such developments at the end of the second part of *Faust*; the effectiveness of machinery suggested that natural conditions could be changed at will.

We could also begin with human needs, first and foremost the wish for material security and comfort. In this case, it would be impossible to go back too far. That said, the amenities provided by the forerunners of today's infrastructures were enjoyed by a vanishingly small number of people. As a rule, a given service—say, the opportunity to communicate readily—was the privilege of the nobility or an elite and performed by specialized personnel. It was not until the nineteenth century that broader segments of the population enjoyed the same benefits.

Our history of infrastructures could also focus on material provisions for transport and communication. If so, the fourteenth-century shift from the "time of cathedrals" to the faster tempo of trade described by the historian Jacques Le Goff would represent a starting point.[22] Alternatively, we could look at streets, conveyances, and postal relays from the sixteenth century on; relevant institutions would include optical telegraphs already long in use. This framework would highlight tensions between military and civilian infrastructures, the often-improvised defensive arrangements made in times of crisis and long-term provisions under conditions of peace.

Yet another option would be to place the story of great men front and center. This approach is common in accounts of technological advances, especially among writers belonging to the fields in question. Often with a defiant undertone, they point to heroes whose contributions to general well-being have not been paid due respect. Indeed, it is difficult to overestimate the impact of work by the likes of George Stephenson, Isambard Kingdom Brunel, Heinrich von Stephan, Thomas Alva Edison, or Nikola Tesla. These individuals have had encomiastic volumes written about their lives and inspired many enthusiastic amateurs; the length and scope of their Wikipedia entries are no accident.

Were we to expand the view beyond individual pioneers, the emergence of classes of professionals and cultures of expertise represents a major index: military and civil engineers, planners and administrators, bankers and entrepreneurs, inventors and developers, construction and mining concerns, and political officials at all levels. But with few exceptions, infrastructures are not linked with the names of the parties who designed and set them up. They almost always have come into being through complex, cumulative networking and coordination efforts.

In sum, the most advisable course is to view infrastructure as the result of negotiation and compromise.[23] The topic stands at the heart of social existence, and many disciplines have made important contributions. History tends to factor into the discussion in the negative—when disorganization or obsolescence prevails because institutions simply grew (instead of being planned from the ground up). Indeed, the sphere of infrastructure disavows history—and politics, for that matter—because it claims to serve the common good; the pretense of impartiality is a defining characteristic. Thus, the book does not look for when a given practice or institution originated so much as the point at which it demonstrated incontestable significance. The *longue durée* is more important than the short term, and quotidian routines tell more than stories of innovation, diffusion, and change; what has escaped notice often proves to be more revealing than what is readily apparent.

SPHERES OF DEFINITION

Of course, innumerable definitions have been proposed. For Paul Edwards, mentioned earlier, *modernity* means living in and through infrastructures, which constitute its very essence.[24] Other scholars understand infrastructures as "a symbol of the collective *we*."[25] Infrastructures have been said to be the "matter" of the social sphere or to amount to its "condition of functioning."[26] By turns, such "infrastructuralism" has been described as an arrangement, dispositive, rhizome, or consolidation.[27] In concrete technical terms, this translates to more laconic observations—for instance, "Modern civilization is unthinkable . . . without working sewers."[28]

Even if many questions remain unresolved, this book confers order on how they are framed. Since the eighteenth century, thinkers and scholars have sought to shed light on the requirements for functioning markets, which cannot exist without state measures. Adam Smith and others who followed his lead are wrongly said to have demanded that governments refrain from economic intervention.[29] In fact, many liberal economists called for state action in emergencies, and they acknowledged that certain economic spheres defy the rules of supply and demand.

Since the civil engineer and economist Jules Dupuit introduced them to France in 1844, *travaux publics* have represented an integral component of governmental policy.[30] The corresponding terms in English-speaking and German-speaking lands are *public works* and *öffentliche Arbeiten*. In the nineteenth century, ministries and departments were founded expressly to oversee the building of national roadways and communications networks. In a parallel development, the science of logistics attended to the distribution of wares, circulation of human beings, and diffusion of ideas. To this day, theories developed in this framework are the cornerstone of statewide and, more specifically, regional infrastructure policies. The goal is to activate social development and promote economic growth. Until the late twentieth century, the latter meant increased productivity, and commercial enterprises routinely called for suitable provisions to be put in place to fuel business in new areas.

As the nineteenth century drew to a close, a second field of infrastructural policy emerged. Policies assigned tasks to the state and municipalities that hitherto had been performed by extended families or by conscientious employers. To secure the loyalty of specialized workers, many companies in the mining industry—for instance, Krupp, Thyssen, and Cockerill—built accommodations and founded institutions to ensure the well-being of employees. Later, many of the same arrangements were taken over by insurance plans and the welfare state (*Vorsorgestaat*).[31] Modern nation-states created systems for traffic and transport, water supply, and waste disposal—that is, fixed assets on a massive scale.

In the period between the two world wars, an umbrella term for such institutions was coined: *public utilities*. Expanding infrastructures would catalyze overall economic vitality.[32] Later, in 1953, the Estonian economist Ragnar Nurske promoted the idea that the economies of "underdeveloped" regions would flourish once their stability had been secured by infrastructural "scaffolding."[33]

Since the early 1960s, a general agreement has prevailed that infrastructures provide the basic operations required for economic growth, integration into markets, and steady supply streams. In 1966, the economist Reimut Jochimsen elaborated a provisional "theory of infrastructure."[34] Even if, in the intervening decades, no consensus has emerged about everything it includes, infrastructure represents the main sector for public investment and planning, alongside military spending and measures for ensuring the social safety net.

The vast sums involved make infrastructure a political item of the first order. Projects both demonstrate power and serve common interests, however one defines them. Canals, roads, postal systems, and communications networks promote social interaction and unification.[35] Only through infrastructure did Breton peasants turn into Frenchmen or Catalan fishermen become Spanish.[36] In Europe, telegraph wires and power cables bound a patchwork of regions into nations.[37] Administrative and technological measures achieved legitimacy inasmuch as they created prosperity at home and averted danger from abroad.[38]

To this very day, as times and systems undergo further change, the political theater of groundbreaking ceremonies, cutting ribbons, or pushing a button for the first time has survived. Such demonstrations of confidence in the future, the sound organization of communal life, and good government blur the boundaries between the realms of politics, bureaucracy, and civil society.[39] "This is collective power," Michael Mann observes, "'power through' society, coordinating social life through state infrastructures."[40] But such power also depends on popular consent.[41] The sociologist Trutz von Trotha calls it *Organisationsmacht*, or "organizational power":

> [Its] productivity—the fact that it makes a difference and changes things—exerts persuasive forces. This [principle] holds for trees planted, buildings, roads, bridges, and railroad tracks, in which organizational power creates *accomplished facts* [*vollendete Tatsachen*]; their *finality* signifies its unconditional claim.[42]

Examples of organizational power include undertakings in societies where survival depends on collective effort—when the elements threaten settlements or must be controlled for agricultural purposes (to grow grain or rice, for example). Max Weber and Karl August Wittfogel used the terms "bureaucratic water-engineering state" (*wasserbaubürokratischer Beamtenstaat*) and "hydraulic society" some hundred years ago. Historians might debate whether such projects determined the shape of social systems and government to the extent that the authors implied, but it is clear that they conferred, stored, and distributed power in both real and symbolic terms.[43]

In complex modern societies, power over transport, communication, and supply simply cannot be exercised directly from the top down.

Moreover, processes of control, integration, or exclusion can be turned against those who would wield them to others' disadvantage. The significance of this is evident after examining twentieth-century projects of state socialism. As the historians Jens Ivo Engels and Gerrit Jasper Schenk put it, infrastructures "capacitate" (*ermächtigen*). That said, they do so not only for politicians, planners, and operators but also for users.[44] The field they constitute abounds in indirect and technocratic modes of intervention, as well as spaces to reverse the flow of power.[45]

The urbanist Keller Easterling has observed that infrastructures are producing some of the most enduring changes in our globalizing world. Because the systems operating in cities emerge more rapidly than administrative structures for regulating them, they qualify as "extrastatecraft."[46] Consulting firms active across the globe, multinational corporations, and bodies such as the International Organization for Standardization are constantly working to create uniform conditions that hold universally.[47]

In non-Western societies, of course, infrastructure meets with fundamentally different ways of dealing with power—as the ups and downs of developmental policy have shown. For many decades, efforts have been made to transfer a concept shaped by the West to "underdeveloped" countries without considering cultural or historical factors. Planners of the unreconstructed sort still view putting basic infrastructures in place as a panacea for "backward" and "primitive" conditions.[48] When the formula fails to deliver, the set response is that the societies in question do not understand the specific "language of infrastructure."[49]

Far more often, however, the problem lies elsewhere. Infrastructures represent a kind of fetish, and in many cases they have been instituted because they radiate a certain kind of "enchantment."[50] Thus, following World War II, many roadways were built in Balkan countries even though the private ownership of automobiles was practically nonexistent.[51] It seemed that the mere symbolism of traffic and circulation would be enough to consolidate, spatially and socially, regions only loosely connected until this point.[52]

Accordingly, infrastructures have often been defined as social programming. As early as 1950, the business journalist Ferdinand Fried described the matter fairly precisely:

One enters into an immense network of mutual contractual entanglements and thereby . . . into a vast scheme of organization, without which modern existence would lapse into anarchy. . . . In this way, modern man finds himself in a magic circle from which he can no longer escape. All things influence and exaggerate each other into furious necessity, driving people together into ever-larger communities of shared purpose.[53]

For their part, sociologists have described such communities of purpose in terms of "structuration" (Anthony Giddens) and society as a "megamachine" (Lewis Mumford).[54] By encoding human action, habitus, and knowledge, infrastructures form the basis for cultural practices of arranging life, communicating; and exchanging information. Moreover, they enable these aspects of existence to be controlled.[55] Thus, when coin-operated water meters were introduced into homes, a certain civic and consumer morality came along with water, and users were trained to become "calculating subjects."[56]

Large-scale technological systems, the very heart and soul of infrastructures, have generated their own branches of research[57] that examine how political and social decisions are inscribed in them and their long-term effects. The question, Do artifacts have politics? demands a more nuanced response than the one offered in 1980 by the philosopher of technology Langdon Winner, who claimed that New York bridges had been built so low to ensure that only automobiles driven by whites could use roadways, and people of color traveling in buses could not. The allegation turned out to be a myth.[58]

The Spanish sociologist Manuel Castells's phrase for infrastructures, "spaces of flow," refers to the permeable boundaries that prevail between technology, nature, and urban culture. According to another contemporary sociologist, Elisabeth Heidenreich, the expectation that flexibility will be available at any given time represents "one of the most far-reaching aspects of modernity."[59] Under these conditions, the natural world turns into a raw material, or fuel; in a word, it is commodified. Infrastructures liquefy both social relations and interactions with the natural world. Because they are built to last and require complex organization, they set the course for society over the long term, establishing path dependencies.

Anthropologists, ethnologists, and cultural theorists have long noted that humankind enhances its comparatively limited physical strength by technical means. Along the lines described by the sociologist Arnold Gehlen, tools and instruments act as prostheses to supplement "deficient human being." Amplifying the scope and reach of bodily organs, they are ways to increase convenience, secure control, and broaden horizons. The philosopher Hans Blumenberg has argued that technological progress aims to gain more time—that is, to extend the feeling of life.[60] The sociologist Wolfgang Essbach, in contrast, stresses the increasing reliance on "second nature," or the "bioartificial symbiosis"[61] that modern humans live in with their surroundings.

In the early twentieth century, Georg Simmel analyzed the sensory overload that is part and parcel of metropolitan life. Urbanites must learn to filter out excess stimuli so as not to break down and fall victim to a fit of nerves—a fashionable disease of the day.[62] A perceived "annihilation of space and time," which the cultural historian Wolfgang Schivelbusch has discussed in relation to the railroad system, was a phenomenon remarked time and again by diagnosticians of the modern condition. More recently, the claim has been advanced that postmodern culture and life are characterized by constant acceleration, or even a "speed virus."[63] As a rule, blame attaches to innovations in the fields of communication and movement, which overload systems of traffic and flow.

How is it that millions of people meet the everyday challenges of roadways and get from one place to another? How do they deal with vast quantities of information pouring in from every side? According to the sociologist Susan Leigh Star, the planning behind large-scale networks does not need explaining so much as their practical operation does, which is constantly being modified. Paradoxically, she observes, people tend to prefer the familiar and routine to alternatives that are more functional or less expensive.[64] Her findings confirm the extent to which infrastructures and the habits they engender stabilize ordinary life.

The field of media studies has examined today's networked society with an eye to the environments and lifeworlds that infrastructures arrange, or "dispose."[65] Often, scholars remark, media are invented and developed in a military context. Cannons, artillery, radio technology, aviation, satellites, teletechnologies, and the Internet activate space by breaking down boundaries.[66] Even in a civilian setting, such dynamism fosters forms of behavior that depart from previous norms.[67]

The ethnological perspective, which analyzes relationships between self and other, has shed light on industrialized societies and the infrastructures on which they depend. Thus, in 1996, the anthropologist Marc Augé was asked whether he could conduct his research without traveling halfway around the world; after all, different cultures exist side by side in London, Paris, and Los Angeles.[68]

If social relations are increasingly mediated by technology, then it is necessary to analyze infrastructural power, too: for instance, exclusion from networks or the inequitable distribution of services.[69] Especially in the booming megacities of the Global South, infrastructures vary in telling ways. According to the architectural historian Swati Chattopadhyay, places such as Lagos and metropolises in East Asia exist in a realm beyond urban planning. The concept of infrastructure, which comes from the West, needs to be rethought to account for conditions elsewhere; subaltern populations engage with systems along wholly different lines, displaying both creativity and stubborn resistance.[70] Day-to-day life in these places is not determined by reliable exchange and connections; instead, unpredictability sets the course.[71] For the uprooted and mobile inhabitants of "infracities" (for instance, the townships of Johannesburg or Cape Town, where development is piecemeal), the real infrastructure is made up of flexible chains of acquaintance and kinship.[72] In other words, everything comes full circle: in overpopulated areas in a state of semipermanent emergency, human beings, not large-scale technical systems, ensure survival.[73]

In sum, more and more disciplines have turned their attention to infrastructures in recent years because these systems link highly varied, and even opposing, sectors of society to one another. Infrastructures are the material expression of social and political negotiation; they structure communities and shape individual character and life. Drawing on the approaches outlined in the preceding, the following chapters seek to portray, as vividly as possible, the diverse and surprising forms of what otherwise remains impersonal and nameless. On the one hand, infrastructures stand for stability, continuity, and new possibilities and opportunities. On the other, they harbor the potential for rupture, danger, and new burdens. Infrastructures reflect social and cultural processes; whether they remain invisible or stand in plain view, they have become a defining feature of modern life.[74]

I

THE CLASSICAL ERA OF INFRASTRUCTURE

1

PUBLIC WORKS: THE NINETEENTH CENTURY

CANALS

In the last days of World War II, the British writer Robert Fordyce Aickman, known for his supernatural fiction, discovered a book by his colleague Lionel T. C. Rolt. *Narrow Boat* offers an atmospheric account of a trip Rolt and his bride took down the waterways of central England in 1939. At that point, the canal system had fallen into disuse and all but returned to nature. Aickman felt inspired and suggested to Rolt that the nearly seven thousand kilometers of navigable waterways be revived to serve ends opposite to the commercial purposes for which they had originally been built.

On February 15, 1946, the two writers founded the Inland Waterways Association to promote opening up the channels of England to tourism. Aickman's travel guide, *Know Your Waterways*, enjoyed multiple printings,[1] and the example set a precedent. To this day, a trip by boat in England or Ireland remains a prime attraction for vacationers, one of the most agreeable ways to retreat from the hustle and bustle of modern life. Few people who go on holiday know that reclaiming waterways marked the beginning of an era of industrial and infrastructural history.

The beginnings of these canals go back to the Romans, who undertook strategically useful excavations in the foreign lands they occupied from

CE 43 on. In medieval and early modern times, more artificial courses connecting rivers and lakes were made. Plans existed for building more still, but costs proved prohibitive. Later on, all across Europe, absolutist rulers and planners in their employ embraced the idea of bettering the gifts of nature and demonstrating the power of human ingenuity to impose order, especially through idealized city and garden layouts. Central to this ambition was mastering the elemental force of water.[2]

The construction of the Canal du Midi, completed in 1681 under Louis XIV (and subsequently continued as the Canal des Deux Mers, linking the Atlantic to the Mediterranean), was particularly sensational; none other than Lionel T. C. Rolt would later chronicle its history.[3] Over and above increased economic efficiency, the canal offered a way for the sovereign to display his might and mark out the space he governed.[4] In England, a comparably symbolic date is 1761, when a canal was opened that enabled Francis Egerton, third Duke of Bridgewater, to ferry coal from mines in Worsley to Manchester for the foundries and factories there; this was one of the first routes created to meet the needs of the Industrial Revolution, which was just beginning.

Other canals soon followed, yielding a broad network of navigable waterways throughout central England. The system enabled the cheap and reliable circulation of raw materials—coal, in particular, but also foodstuffs and wares for industrial and domestic use. For some time, operations relied on horses, which pulled the narrow ships along at a relatively leisurely pace. Despite the sluggish, biogenic rhythm, canals evened out disparities between regions rich in raw materials and agricultural areas. Economic experts—cameralists and mercantilists—and politicians seized on this feature when advocating further construction.[5]

At first, the importance of the channels was not plain to all, but ultimately they came to be recognized as a major step toward a worldwide economy: energy sources could now be transported wherever needed. The age of mineral and fossil fuels had dawned, and the era of low-yield energy sources, such as wood, began to fade. Only through this development did the idea of continued growth become possible.[6]

The eighteenth century also witnessed the rise of a class of experts who advised political leaders and pushed for the development of canals and everything else that, much later in history, would come to be known as infrastructure. In effect, they put themselves forward as capable of steering society as a whole;[7] only by striking a balance between abundance and scarcity in different regions could producers of raw material, manufacturers, specialized labor forces, and urban and industrial centers develop properly. Even remote areas of a country could be valorized, and markets benefited from new levels of production and exchange. Over the course of time, something like a national economy emerged, as well as the rudiments of an infrastructure state—that is, government in which power is defined through canals, roadways, and other human-made structures.[8]

The Bridgewater Canal made do without locks, but water bridges were needed to navigate intersecting rivers. Here a further goal of modern infrastructures was in evidence: to be as independent as possible from the landscape. Historically, such efforts have often failed because of natural conditions, economic and strategic factors, and issues of land ownership. All the same, road and civil engineers have held fast to the goal of ensuring the most efficient routing possible. In countries with a centralized government—France and Spain, for example—development reflected the political order, with transport lines arranged in a star pattern centered on capital cities.[9]

The English age of canals reached its peak at the end of the eighteenth century. For the most part, waterways were built on private initiative, with a view to increasing the value of adjacent land. By the mid-nineteenth century, nearly twenty-five thousand barges were in use.[10] Other countries developed closed canal systems for transport and controlling water levels. In the Netherlands, a specialized corps of engineers formed early on, because the sea was always encroaching on dry land.[11] Specialized knowledge also has a long tradition outside Europe; hydraulic engineering was practiced in Central America, the Middle East, India, and China. It is no coincidence, then, that one of the first ministries for public works to be established in the early nineteenth century was in Egypt. Ever since the Old Kingdom in ancient times, such projects had shaped the landscape.

A defining trait of the late eighteenth and early nineteenth centuries in Europe was the search for new economic opportunities—along with efforts to guard against hazards, which were no longer accepted as passively as in the past.[12] Rivers had long represented essential lines of transportation and supply; projects to make them more suited to human use continued to the same ends. Johann Gottfried Tulla, Carl Friedrich von Wiebeking, and other hydro engineers viewed modifications of the Rhine or Danube as correctives to the landscape. In their eyes, God had deliberately posed challenges for human beings to overcome and set right.[13] Thus, the Rhine, which earlier generations had deemed wild and untamable, was transformed by weirs and locks into a canalized waterway serving industrial, commercial, and urban interests. (By 1975, nine-tenths of its floodplain was gone, having been shortened by a total of 105 kilometers, and the river itself had been given a uniform width.[14])

Settlements and infrastructures built along the banks of the Rhine were designed to display regularity and form a spatial continuum. When flooding did occur, it counted as revenge taken by nature, which now was perceived as a hostile force because consequences proved more serious than ever.[15] But in time, this same nature came to exercise enchantment; during the nineteenth century, and in reaction to human intervention, "Rhine Romanticism" emerged, expressing reverence for the landscape and its wonders. In turn, a domesticated version of poetic sentiments offered tourists another commodity to consume.[16]

Managing waterways represents an ongoing task for government, and international agreements proved necessary early on. The Central Commission for the Navigation of the Rhine was founded in 1831.[17] Because the river holds equal importance for Switzerland, France, Germany, and the Netherlands, it was plain that one country might assert dominance and appropriate it politically, strategically, nationally, culturally, or economically. Hardly any other river has been the object of so much wrangling over the past two centuries.[18]

Rivers often cross political borders. Accordingly, and also with a view to more direct routes, engineers have long sought to create navigable alternatives. Just as the Canal de Deux Mers was intended to spare the French the lengthy journey around the Iberian Peninsula, the Swiss entertained the idea of a Trans-Helvetian Canal that would grant them access to the Mediterranean.[19] Czechoslovakia had designs to reach coasts with links between the Danube, Oder, and Elbe.[20] For decades, Belgium conducted unsuccessful talks with Germany about plans for a Rhine-Maas-Scheldt Canal.[21] In 1895, the Germans charted a triumph with the completion of Kaiser-Wilhelm Canal, which has been known since 1948 as the Kiel Canal. By shortening the route between the North Sea and the Baltic by a full seven hundred kilometers, it not only saved time but also eliminated customs fees formerly paid to Denmark.[22] Even more spectacular achievements were the Suez Canal—which opened in 1869 and was promptly seized by the British—and the Panama Canal, built by the United States. (Alas, the outbreak of World War I eclipsed the latter's opening on August 3, 1914.[23])

Countless other examples could be provided from all over the world. Canals facilitated and accelerated shipping not only on inland waterways but also at sea. Maritime transport—the importance of which is often underestimated (especially today)—did not become a vital infrastructure because of advances in navigation techniques or steam power. Rather, it rose to prominence because of what is now known as global logistics. Up to the nineteenth century, only a few European countries carried out voyages of discovery and exploration; sometimes one power would predominate, sometimes another. For the most part, expeditions were undertaken on the initiative of rulers or private firms—for instance, the British or Dutch East India Company. Particular regions—and, even more, select parties, families, and social groups—amassed great wealth. Over the course of the nineteenth century, young nation-states incorporated the practice into political designs. Hitherto, exchange had been largely confined to luxury goods; now, raw materials and other commodities became an essential part of the overall economy in Europe. Governments and the military calculated the risks and benefits of trade wars (as had been waged against England by Continental powers between 1806 and 1811, under the domination of Napoleon). Expanding merchant fleets usually went hand in hand with building up naval forces. Plans for ports and stations no longer aimed just to reduce freight costs but also followed strategic considerations.[24]

Inland transport and travel developed in parallel to traffic by water. Canal mania peaked in the late eighteenth century; from the 1830s on, it was succeeded by railway mania. Initially used to bring goods to canals, the railway network proved more flexible and came to replace shipping by water in England. In most cases, railroad companies took over waterways, but they often neglected to maintain them. British canals were the first modern infrastructure that, after briefly flourishing, fell into disrepair—until, as noted, a new, recreational use was found for them.

The London Bridge, which had spanned the Thames since 1831, also received a second life. In the 1960s, it became clear that the old structure could not handle the burden of daily automotive traffic, so a new six-lane bridge with heated walkways was built. Without further ado, the American industrialist Robert McCulloch (who had made his fortune manufacturing chainsaws and outboard motors) bought the old bridge. He had it transported to the United States through the Panama Canal and reassembled it at Lake Havasu in Arizona. Rededicated in the presence of the mayor of London in 1971, it has been a tourist magnet ever since.[25]

WESTWARD HO!

Tragedy ensued when William Clark, accompanied by a US marshal, showed up at Mussel Slough in California on May 11, 1880. The Southern Pacific Railroad had sent its representative to evict settlers from land that it nominally owned. In addition to legal paperwork, the new arrivals brought firearms as a precaution. The first farmer to receive a visit enjoyed the support of representatives of the Settlers Land League. When the two parties met, a battle of words broke out that quickly escalated into gunfire. In the end, seven men lay dead, and another was seriously injured.

As was standard practice, the private company had obtained land concessions from the US government along planned routes. Profits from these concessions were to be used to secure investments for constructing the lines. The still-barren area in the San Joaquin Valley, at the heart of California, had been leased to settlers in 1870. The agreement held that if the land could be made arable, they would be able to buy it from the company ten years later for two and a half dollars per acre. However, during the intervening period the property value had risen appreciably, thanks to infrastructural development,[26] and the railroad company now asked for ten times the original price. Having spent years draining the land, cultivating it, and building roads and houses, settlers felt cheated. They appealed for public support and threatened violence if the company stuck to its demands. And violence duly occurred.

The Mussel Slough tragedy is but one of countless conflicts that attended infrastructural development in the nineteenth and twentieth centuries. In this case, the calculations of government and private enterprise acting in concert collided head-on with the interests of residents. To be sure, bloodshed was the exception rather than the rule. All the same, the incident demonstrates that building infrastructures was not simply a matter of engineering and administration.

How the West was won has long been romanticized. In fact, the conquest came from many directions at once, converging on inland territories in ways that gave rise to all manner of disputes—including conflicts where railways and telegraph lines cut through lands inhabited by Native Americans (whom development would "civilize"). Popular culture abounds in accounts of grand-scale plans and clashes. The cast of characters includes explorers, pioneers, builders of roads and railways, and the riders of the legendary Pony Express. Whether the technologies involved were first developed in the United States or not, US society has long defined itself in terms of opening up and laying claim to more space, which represents a matter of birthright. In 1872, the artist John Gast, who immigrated to the United States from Berlin, captured the spirit in *American Progress* (originally, *Westward Ho/Manifest Destiny*). The painting shows animals and Native Americans being driven into the shadows as a bright parade of settlers, stagecoaches, railroads, and ships takes the stage of history. Above them towers the Goddess of Progress, Bible and telegraph cable in hand.

Countless ballads, novels, and films have created a mythology of pushing back the wild frontier. However, critical voices have also spoken up. Thus, the journalist Frank Norris made the incident at Mussel Slough the subject of his 1901 novel, *The Octopus*.[27] As the title indicates, the ruthless machinations of railway companies and corrupt politicians resemble the predations of a many-limbed sea monster—a popular (and enduring) metaphor for large industrial trusts. American entrepreneurs, among whom the Harrimans and Vanderbilts are the most notorious, had counterparts in the likes of Bethel Henry Strousberg and Baron Maurice de Hirsch in Europe. Their immense wealth and the vast power they wielded made railroad barons a constant object of scandal.[28]

The Wild West represents a zone of transition between nature and (technological) civilization. Although shrouded in legend, it is also paradigmatic. Literary and cinematic representations feature three motifs. The first is the displacement of indigenous populations by people of European descent; the second concerns infrastructural development; the third deals

1.1 *American Progress* (1872), by John Gast, depicts the progress attending infrastructures, which dispels "native" darkness.

with discipline and order in pioneer society—both formal legislation and the everyday regimes brought into being through networks of circulation and communication. The processes associated with the rise of infrastructures remain more vivid and tangible in North American culture than almost anywhere else. Even iconic examples in the genre of Western movies focus on conflicts engendered by pushing back frontiers—for example, *Once Upon a Time in the West* (1968).

Expanding lines of transport, fixing national borders, and establishing zones of vigorous economic activity were hardly limited to the United States. However, the drama here proved especially intense: apart from a few long-standing settlements and routes, there was nothing on which the settlers could build. Even lakes and rivers helped development to only a limited extent; shipping routes had to be set up in the first place. Thus, after long years of planning, a 584-kilometer canal between Buffalo and Lake Erie opened in 1825. The Erie Canal consolidated New York's position as the most important port on the East Coast and accelerated settlement in the regions around the Great Lakes. Critics, following the lead of Thomas Jefferson, objected to federal funds being used for this purpose. The attendant questions remain unanswered. Who derives the greatest benefit from initiatives? Who should pay for them? According to what formula? Disagreement about the common good and how to finance projects persists to this day.

Ever since US Treasury Secretary Albert Gallatin proposed, in 1808, to allocate $20 million from the federal budget for internal improvements—and at the very latest since 1815, when Senator Henry Clay set forth the American System (protective tariffs, a national bank, and comprehensive infrastructure measures)—doubters have asked whether such policies make sense and who bears responsibility for expenses. The burden has shifted back and forth between federal and state governments and between municipalities and individuals.[29] Proponents argue for the general welfare of citizens. Opponents counter that measures serve special interests. From the outset, critics warned that using public funds would lead to a bloated bureaucracy and corruption. These debates polarized the US political landscape before the Civil War as much as the issue of slavery. Arguments still deployed were formulated early on, with advocates of public and centrally controlled institutions, on the one hand, and proponents of private, unregulated initiative, on the other. Indeed, the issue gave rise to the formation of economic schools of thought and political parties: Federalists evolved into today's Democrats and Democratic Republicans into today's Republicans.

Routing posed a recurrent problem. What course should a road, railroad line, or canal follow? Which communities would be connected, and which ones left behind? Answers depended on features of the landscape and technical, economic, and political considerations. Where established pathways existed, historical and cultural factors had to be taken into account, too. Who would assume responsibility for planning and financing? In turn, who would be allowed (or forced) to perform work? Construction, especially canals and railroads, was a serious undertaking. Later, dams and highways posed challenges on a similar scale.[30]

One of the first federally financed projects in the United States was the Cumberland Road. Beginning in 1811, it stretched westward from the Potomac in Maryland as a route for settlers; in the 1830s, it became the first road in the United States to be paved and tarred (following techniques pioneered by the Scotsman John Loudon MacAdam). Not long afterward, Congress balked at more funding. President Andrew Jackson told his political rival Henry Clay, in Kentucky, that further expansion would be a state issue; additional federal monies for the project would put other states at a disadvantage and jeopardize a balanced budget. Thus, after a distance of about a thousand kilometers, the Cumberland Road terminated in Illinois, at a town with the evocative name of Vandalia. Today, like the Erie Canal, it enjoys the distinction of being a historic landmark and is maintained as a testament to engineering. In other respects, the so-called National Road has suffered a fate similar to that of the British canals or Route 66 from Chicago to Los Angeles (which came later). It offers the prospect of a nostalgic journey to a supposedly slower or wilder past—dozing like a sleeping beauty waiting for her tourist prince.[31]

Similar debates flared up in the early nineteenth century apropos of river and harbor projects, bridges, and even lighthouses. (The Cape Henry Lighthouse, built in 1791, is the first federal structure in the United States; its likeness is featured on the logo of the American Public Works Historical Society.) Opponents and proponents of federal infrastructure programs fought fiercely. During the twelve years between 1825 and 1837, advocates of nation building through technological interconnection exercised the strongest influence.[32] As the British travel writer Frances Trollope observed in 1832, "There is no point in the national character of the Americans which commands so much respect as the boldness and energy with which public works are undertaken and carried through."[33] Railway construction began at precisely this juncture. It proved even more important than canals and roads for connecting the hinterland to coastal regions, industrialization, and the circulation of raw materials and foodstuffs. In a land so sparsely populated, transportation hubs, railroad stations, and other infrastructural relay points proved vital for settlement. In contrast to Europe, where lines resemble a wheel with the capital in the center, right angles in a grid are characteristic of the United States.[34]

By 1840, opponents of public works had managed to forge a lasting connection between infrastructures, public debt, and special interests.[35] Main lines—for instance, the canals and railroads between Philadelphia and Pittsburgh—had already been built. Over the coming decades, most new arteries of traffic and exchange were developed on private initiative, first serving local and regional interests and only later coming together into a national network corresponding to the vision set out already in the 1780s by the likes of George Washington. In the years following the Civil War, government power proved insufficient to counter railroad companies and their quest for profit, and conflicts like the one in Mussel Slough multiplied.[36] Finally, in 1887, the Interstate Commerce Commission was founded, which became the model for regulating private infrastructure providers.

May 10, 1869, marks a milestone in the age of technological improvement. On this day, the first transcontinental railroad connection was completed at Promontory Hill in Utah, about a hundred kilometers northwest of Salt Lake City; a golden nail connecting tracks crowned the ceremony. Even though track gauges of varying dimensions continued to compete for some time, railway construction in the United States had reached one of its high points,[37] and the international community took notice. Strategic plans worldwide followed the model, with telegraph cables, which served important signaling purposes, running parallel to transcontinental lines. The Suez Canal in Egypt, which opened the same year, provided a further impetus.

The German geographer Ferdinand von Richthofen proposed a railway from Europe to China as early as 1874. The two largest cultural centers on earth should be connected, he contended, because "the results of an influx of western ideas into the heart of China, a consummation against which she has striven so hard, is full of interest to all." In practical terms, this meant that "the enormous amount of manual labour available, and the inexhaustible supplies of coal [would] cause the stream of European manufactures to flow back in its course," providing basic needs more inexpensively than before.[38]

Between these two points lay the dominions of the Russian tsar. In terms of the tasks confronted, Russia occupied a position analogous to its North American counterpart, a world away. By building railways, Russian elites aimed to consolidate national, imperial, and industrialized space while shrinking it in size, so to speak. Doing so would set in motion currents of goods and people and civilize a supposedly backward population. On the whole, schemes borrowed from the West provided the model for a unified empire.[39]

In contrast to the United States, Russia needed to protect territory threatened from without and to quell internal unrest. To this end, the separate Railroad Gendarmerie was founded. The military also sought to have a hand in how lines were routed. Construction began in the European part of the Russian Empire in 1837. Asian regions were incorporated into the network beginning in the 1850s. The 52,000 kilometers of track laid by 1900 were modest compared with the 264,000 kilometers built in the United States ten years earlier. Despite the success charted by the Trans-Caspian Railway (1880) and the Trans-Siberian Railway (1891), many Russians were left out in the cold.[40] The consequences of modernizing an agrarian society were more far reaching and fraught than in North America or western Europe. On the one hand, railroad construction, technical standardization, and attendant planning (schedules, routes, travel guides, and so on) forged unity. Particularly in peripheral regions, forward-thinking civil servants and educated society pushed for more, and more thoroughgoing, changes so they would not be condemned to waste away forever in the "eternal provinces."[41] On the other hand, the process underscored differences between regions and social classes—or brought them to light in the first place. The autocratic tsarist regime lost stability inasmuch as different segments of the population had divergent expectations and experiences.

As contemporary observers noted, the railroad brought order to villages but chaos to urban centers. The tsars felt the consequences of what they had wrought. Availing themselves of new and easier facilities, they could travel throughout their dominions and stage their authority in "modern palaces of transportation," that is, magnificent train stations.[42] This visibility brought increased danger. Two incidents stand out: a terrorist strike on Nicholas I in 1879 and a serious accident near Borki in 1888 from which, miraculously, the family of Alexander III emerged shaken but largely unharmed. Protective measures were strengthened, and regents curtailed their travels.[43] In the balance, railroad construction produced mixed results, fostering challenges to autocratic and centralized government.

Indeed, the utopian promise of modernization outlived the tsars and shaped the history of the Soviet Union.[44]

WORLD PEACE OR WORLD DOMINATION?

From the outset, railways were viewed as agents of historic change. Karl Marx described revolution as the "locomotive of world history."[45] Johann Wolfgang von Goethe's 1828 dictum, made three years after steam-powered trains began traveling between Stockton and Darlington in England, is often quoted in this context: "I have no fear that Germany will fail to be unified; our roads and future railroads will do their part."[46] Patriotically minded economists like Friedrich List, now calling themselves national economists, foresaw a Germany of good neighbors united by infrastructure.[47] In 1833, the newspaper *Sachsenzeitung* declared that "the steam locomotives traveling down railways will likely bring about eternal peace."[48] The 1838 edition of the *Brockhaus*, the leading German encyclopedia of the day, did the same: "History has always directed its course toward this truly divine goal, but it will reach it centuries sooner on the barreling wheels of the railroad."[49]

Henri de Saint-Simon, who inspired early forms of socialism and died the same year as the first railroad journey (1824), had hailed the immense potential of connecting politically separate areas by technical means. Just as modern systems of transport and circulation conquer space, global interconnectedness was expected to facilitate understanding between peoples and, ultimately, eliminate mutual suspicions and animosity.[50] Infrastructures such as canals, railroads, and telegraphy facilitated rational planning and management of not just nature but also human society. Saint-Simonians perceived no danger that greater domination, control, or marginalization would result. Instead, they greeted the prospect of improved general welfare and social peace; the fragmented interests of the world's peoples would give way to solidarity. Accordingly, they conceived projects such as canals in Central America and Egypt and an Iberian waterway to Madrid. A program combining science, technology, and labor would shape social reality.[51] The rhetoric of peace through interconnection thrives to this day; one need only call to mind the rapturous predictions made about the Internet. Railways and telephones, airplanes and automobiles, and radio and television have all been extolled in more or less interchangeable terms.[52]

Of course, dimmer views emphasizing the perils of a networked world have long existed, too. Among credentialed pessimists, military experts stand first and foremost. Even those who dream of superweapons that would guarantee peace once and for all must think along the same lines as their gloomy counterparts.

Telegraphy—a precondition for the rapid circulation of information and news—has long made it plain that technology can serve bellicose or peaceful ends. Messaging at a speed approaching real time evolved from forerunners including the Chappe semaphore, which was widely adopted in France beginning in the 1790s.[53] Between 1830 and 1840, electric telegraphy became an essential technology for railroads, as well as a key means of regulating the flow of goods and trade.[54] Accelerated exchanges fueled consumption and exercised a decisive function in transforming the globe into an integrated economic unit. Over time, prices ceased to be determined locally and were instead shaped by patterns on the world market.[55] By the end of the eighteenth century already, the Hamburg professor Johann Georg Büsch identified the guiding principle for the stock market and the world of finance: "News, arriving only a few hours earlier than usual could be worth sums larger than operating a telegraph line with all the associated costs."[56] Within a few decades, telegraph lines, news agencies, and contacts between stock exchanges had been put in place; gradually, such arrangements assumed global dimensions.[57] Old-fashioned businessmen who relied on a network of agents expressed dismay. Thus, in 1851, the banker James de Rothschild (who still used carrier pigeons for long-distance communication) wrote to his nephews in London that the telegraph, which spread information so quickly, was "ruining affairs."[58] Geographic proximity to sources and a material carrier now were no longer required to exchange data. One can see here the makings of the digital age.[59]

For the rest of the nineteenth century and up until World War I, the big question was who would manage to hold all the strings connecting urban centers, countries, and continents. The goal of peace was not yet achieved. Meanwhile, competition prevailed. "Railroads and telegraphy, steamships and cables," wrote the geopolitically minded journalist Artur Dix in 1901, "are the instruments with which modern *homo sapiens* has opened up and subjugated every corner of the earth. At the same time, they are exemplary tools of political power and the best weapons of modern states in their struggle to carve up the world."[60] In the same context, Dix spoke of the earth as a "little village" (*Dörfchen Erde*) whose inhabitants could all look into their neighbors' windows.[61] Half a century later, the Canadian media theorist Marshall McLuhan would take up the image of a global village in another sense, as a metaphor for empathy and cooperation.[62]

In the rush to divide up the world, advantages did not last for long. In 1895, the geographer Michael Geistbeck noted that "every technological advance . . . becomes common knowledge in a short time."[63] Scientific discoveries had brought new possibilities of expansion and pretexts for colonizing "backward" regions, mainly Africa and southeastern Asia. The imperialism of the day was a matter of not just ideology but also, and especially, technology.[64] Europeans could now calculate the risk of conquering previously impenetrable lands thanks to advances in tropical medicine, better rifles, steam-powered ships, and improved communications.[65] That said, and per Geistbeck's observation, subjugated populations were quick to appropriate the new technology and turn it against their oppressors.

Between 1860 and 1900, a global network of land and sea cables emerged, with centers in London, Paris, Berlin, Vienna, and New York. Locations that had been peripheral until this point—Malta, Suez, Algiers, Aden, Bombay, and Brest, among other places—became nodes of strategic importance.[66] By the turn of the century, the thrill of unification spilled beyond national borders and assumed nearly planetary dimensions.[67] "Spatial distance is today no longer an ocean of time," the *Neues Universum* declared in 1901, "for the power of telegraphy extends to even the most remote corners of the earth. One might almost say that civilization begins with the telegraph wire."[68] But in spite of such enthusiasm, limited access and relatively strict controls prevailed. In contrast to the telephone, only a fraction of the world's population would ever use the telegraph. The enormous cost of laying cable at sea kept costs high. In 1858, after many years of preparation and many unsuccessful efforts, the first transatlantic cable still functioned only in part. Military and business leaders had good reason to pursue improvement but little incentive to popularize the technology or make it more cost effective.[69] As late as 1946, after transmitting his famous "Long Telegram" from Moscow (which proved decisive for policy during the Cold War), the US diplomat George F. Kennan was censured for the enormous expense incurred.[70]

European empires employed the technology as a tool of political control. Shortly after the laying of the first transatlantic cable, Karl Marx observed of India that "unity, imposed by the British sword, will now be strengthened and perpetuated by the electric telegraph."[71] Events would prove him right: just four years later, Sepoy uprisings against British colonial policy were suppressed because the colonizers had the means to gather and relay more complete information.[72] Similarly, Paul Dehn, a German analyst of world affairs, declared in 1904, "Formerly, global empires could not be held together in the long run. Today they are made to cohere by modern means of transportation, railroads, steamships, and telegraphs, which facilitate a tightly centralized administration under the direction a single office that is always informed and equipped to decide even on details."[73] Paris, London, Berlin, Moscow, and Washington were no longer far away from each other or their dependencies. This fact held serious consequences for diplomats, military forces, and administrators abroad, most of whom appreciated the relative freedom accompanying physical distance and had regularly worked around the dirigisme of their governments. Direct contact with political centers meant more to do but fewer independent decisions. Developments challenged established routines, which tended to be based on personal contacts and trustworthy messengers. Accordingly, diplomats took to pointing out the unreliability and high cost of telegraphic communications. Because charges were by character, a terse telegram style emerged, which has made a comeback in SMS and Twitter messages.[74]

In the course of the Crimean War (1853–1856), cables were laid across the Black Sea to London and Paris. Governments could now issue orders to generals at the front and control their activities. (One French commander found an expedient when the "schoolmasterly paternalism" became too much for him: he just cut the wire.[75]) In a parallel development, the British and French citizenry received information through reports from correspondents writing for newspapers. The public now exercised greater influence on political decision-makers.[76] Ever since, a particular dynamic has prevailed—and not just in crisis situations or under autocratic regimes—oscillating between (supposed) facts and efforts to control, manipulate, or suppress them. Unreliability and leaks present an ongoing problem, too. Early networks were incomplete and largely controlled by the British; messages reached their destination by surprisingly circuitous routes, and foreign governments regularly intercepted communications. For this reason (among others), modern intelligence services evolved in tandem with telegraphy and successor technologies. Over the course of the twentieth century, agencies constantly upped the ante through campaigns of deception, encryption, interception, and deciphering. Ultimately, almost no information of real importance managed to be kept secret for any significant duration.[77]

Telegrams could also be instrumentalized to manipulate the enemy. The history of the German Empire offers many examples—for instance, when Otto von Bismarck used the Ems Dispatch in 1870 to stir up conflict with France or when Kaiser Wilhelm II expressed support for the Boer leader Paul Kruger in an 1896 telegram and provoked the British in South Africa. Under Woodrow Wilson, the US government intercepted a dispatch in 1917 from Arthur Zimmermann, state secretary for foreign affairs of the German Empire, to the embassy in Washington. The intelligence it contained, about the Reich's understanding with Mexico, served as a pretext for the United States to declare war.[78]

During World War I, the virtues ascribed to the railroad and telegraph turned out to be vices. Civilizing influence and the promotion of international unity were hardly in evidence. Efforts to achieve standardization—in 1865, the International Telegraph Union had been founded, in 1878 the Universal Postal Union, and in 1884 largely uniform world time (in an agreement at a conference in Washington[79])—had failed to bring positive results. When conflict erupted, adversaries seized on infrastructures as points of vulnerability and cut lifelines. All the same, the myth persisted that direct communication would preserve peace. From the 1960s on, the red telephone hotline between Washington and Moscow represented this hope: in an emergency, the US and Soviet heads of state would pick up the receiver and prevent the world from blowing up.[80]

CITIES IN FLUX

Subterranean networks extend for miles and miles underneath large cities. Pedestrians going about their daily business pay no mind to manhole covers and drains (however artfully designed), much less do they suspect the vast realms below. Since the nineteenth century, the city has epitomized the dialectic between cleanliness and dirt, justice and injustice, the beau monde and the dregs of society, palaces and miserable basement dwellings, heaven and hell on earth.[81] Victor Hugo set *Les Misérables* (1862) in the catacombs of Paris, contrasting them with the life of the French capital above. The sewers are the last refuge for the dispossessed—the excrement of civilization—and works of fiction underscore connections between waste and social status.[82] Films like *The Third Man* (1949) perpetuate the mythology of the bowels of the metropolis with tales of espionage and crime. Harry Lime, who has smuggled and adulterated penicillin for personal gain, meets his end in the Vienna sewers; the power over life leads to death.

The modern-day lower depths represent the opposite of modernity and, with that, its hidden truth. Urban technology and infrastructures are supposed to have prevented the privileged top from consolidating its privilege and leaving the wretched bottom at the mercy of a blind fate. The city is conceived as a living thing. The organism's parts must all be healthy, even if they perform modest functions; otherwise, the body will suffer. In this picture, infrastructures have been—and sometimes still are—assigned the role of veins, arteries, and vessels through which life pulses. The fewer the impediments to circulation, the more robust and healthy the whole will be. The same logic has extended to nations and empires, and sometimes even global markets and world society, but the modern city epitomizes it.

There were antecedents, of course; recall Rome's elaborate arrangements for supplying water and eliminating waste.[83] Articulated roadways, patterns of trade, and market systems are also nothing new. But the expansion and acceleration of industry and trade—and, in some instances, rapid population growth—from the end of the eighteenth century on created unanticipated challenges. Urban settlements were still partly medieval in design and surrounded by walls; now, these erstwhile symbols of security represented an obstacle to progress. As cities grew out and over the centers of old, decision-makers needed to be sufficiently far-sighted to incorporate structures to meet new needs and demands.

European cities did not stand alone at the vanguard of infrastructural development. In 1808, pressure from Napoleonic troops induced the Portuguese king João VI to flee to Rio de Janeiro. The members of the upper crust in Lisbon who accompanied him found their delicate sensibilities offended by the New World and felt unsafe on the streets; the exiled monarch had lighting and sewage systems installed to grant his entourage a "civilized" standard of living.[84] Before long, the city proved so agreeable that the king and his courtiers remained in Brazil even after Napoleon's fall. For a few years, and for the first time, an empire spanning four continents was ruled from a colonial seat.

Many industrial centers in Europe steered a course determined by economic interests. Cities in the English Midlands (Liverpool and Manchester, for example), the German Ruhr region, and parts of northern France developed rapidly because of abundant raw materials. Exploiting resources required housing and services for vast numbers of workers. Facilities were built with mining and industrial facilities in mind; the population itself long remained a secondary consideration. Travelers regularly expressed shock at the sight of urban landscapes crisscrossed by railroads and pipelines.[85] In a few cases, cities elsewhere provided remarkably sophisticated facilities, but Europe and North America set the international standard. Most of the time, concrete circumstances—for instance, disastrous hygienic conditions—combined with competition between rival urban centers to fuel change.

As industrialization and economic growth yielded a self-reinforcing dynamic, urbanization and infrastructures fused into an overall system; distinguishing between cause and effect is difficult because the goal was constant renewal and optimization. On a human level, societies bent on expansion and acceleration present more opportunities for things to go wrong—for instance, when the supply of energy stalls or the efficient elimination of waste fails to occur.[86] Before long, controlling flows inward and outward (which had already rearranged the natural world and landscape) became a matter of vital concern.[87] In the eyes of many, there was too little light and air; cities were congested and overfull. To degrees that varied from one place to another, systems for lighting, supplying water, and disposing sewage were introduced in Europe and North America over the course of the nineteenth century. Doing so meant demolishing walls and fortifications to get rid of narrow, medieval conditions; new houses and streets were designed to facilitate smooth circulation and the ready availability of resources from the surrounding countryside.

In this context, idleness, darkness, and dirt signified obsolescence—an earlier stage of civilization that now had been surpassed. The modern metropolis is obsessed with hygiene, even if diseases are no longer attributed to miasma. The discovery of viruses, bacteria, and other pathogens as bearers of illness, which spread through personal contact, speeded the rise of the "sanitary" or "bacteriological" city.[88] Even before the specific medical causes for diphtheria, cholera, and typhoid were identified, municipal authorities equated gloomy and untidy conditions with poverty and sickness; the former counted as the cause for the latter. In turn, "enlightened" reformers introduced measures that were supposed to overcome social imbalances and help the pauperized proletariat help itself.[89] First and foremost, these steps involved reducing soil moisture, eliminating stagnant water, removing excrement expeditiously, and providing as many households as possible with running water and plumbing; in brief, the goal was to vitalize the urban "bloodstream" and prevent dangerous congestion. The movement to promote public health, which at times took on the character of a veritable crusade, represents an incontestable accomplishment of the liberal middle classes in the nineteenth century. In retrospect, however, it also appears as a canny move to leave standing relations of property and wealth untouched. Getting rid of unsanitary conditions and providing basic services was supposed to resolve the "social question" as a whole.[90]

Providing water and sewage disposal, paving and cleaning streets regularly, canalizing rivers, draining marshlands, and removing refuse and excrement from urban areas set the standard for any city to qualify as modern. In Europe and the United States, a model emerged that formed the basis for active governmental administration later on. First introduced in response to catastrophes (for instance, epidemics, major fires, or other large-scale accidents), policies subsequently aimed at anticipating and managing problematic situations. In the process, they shaped what would come to be known as urbanity: a lifestyle characterized by mobility and communication and an economy based on the division of labor along functional lines and a wide range of leisure activities—in other words, all that private citizens could not provide on their own.

In this, competition between cities and nations played a central role. Modern metropolises sought to house industries and a population that would be enterprising and innovative.[91] In 1904, the progressive author Lincoln Steffens lamented the weak infrastructural development in the United States compared with Europe. Like his contemporary and colleague Frank Norris, he blamed corruption and the lack of governmental oversight for shortcomings.[92] In fact, rivalry among city governments, the financial robustness of municipalities, and local expertise are likely what made health care services and a high standard of living emerge so quickly in Europe.[93] Provisions for social welfare, hygiene, and education were introduced, at least in part, because urban elites feared that rampant poverty would entail too much disquiet from within. Public safety, social order, and economic prosperity were at stake. Initially, epidemic quarantines and sanitary measures benefited only wealthier neighborhoods, but when the advantages of universal services became clear, they came to include poorer parts of town and slums, as well.[94]

Health care experts pointed out the woes of impoverished districts, in particular high rates of infant mortality. The next step was to take action, which often started as campaigns. The practice initiated a pattern that continues to this day; to modernize facilities or enact public health measures, appeals are made to hygienic conditions and general welfare. Ordinances such as the German Gesetz betreffend die Bekämpfung gemeingefährlicher Krankheiten (Law Concerning the Control of Diseases Dangerous to the Public, 1900) gave authorities leverage to intervene in "untenable situations."[95] Under the guidance of professionalized experts, institutions were established to serve the common good, and regulations emerged bearing on fire safety, building codes, medical policy, and so on. Authorities also sought to mobilize public support in order to legitimize political decisions—for instance, by means of surveys (with due interpretation of the statistics they yielded).[96] Still today, the operations of modern cities follow this model.[97]

Medical, technical, and administrative elites played the leading role in developing military and civil infrastructures. Organizations such as the French Corps des Ponts et de Chaussées, the British Institution of Civil Engineers, and the German Verein für öffentliche Gesundheitspflege exchanged practical information on an international level. These professional communities shared a technocratic ideology that was supposed to be impartial and divorced from political calculations, and efforts were evaluated in terms of effectiveness and efficiency; the models of organization developed would be applied to society as a whole.[98] For the most part, experts worked behind the scenes and employed a jargon difficult for outsiders to understand. Most of them—but more in France and Germany than in England and the United States—believed that the state was better suited to building and operating services than private entrepreneurs.[99] They shaped what would come to be called the public sector, in contradistinction to the invisible hand of the market.

As Rudolf Virchow, a physician and prominent advocate of intervention, described the situation in 1884,

> The difference between estates is growing smaller with each passing day. Each new institution—water mains, the sewage system, street lights—inasmuch as it is universal, adds something democratic to conditions. People must do as others do. The one can afford only two burners, and another perhaps a dozen, but everyone uses the same gas, water, and sewers.[100]

Here something like the liberal vision of modern consumer society, with engineering leveling class differences, is already in evidence. Sanitation, in this view, represents an "instrument for purifying the physical and social body of the city."[101] Virchow, like the members of the reform-minded Fabian Society in England and the followers of Saint-Simon in France, belonged to the left-leaning part of the bourgeoisie; although hoping to make social boundaries more permeable, he did not want to abolish them altogether. Provisions of this kind, which have been called "municipal socialism," were generally overseen at the local level.[102]

Infrastructural distribution in the nineteenth century maintained social differences of rank, as well as perpetuating religious and ethnic divisions. The more diverse the population in a given area, the poorer the water supply, sewage systems, transportation connections, and provisions for communication tended to be.[103] Wealthier parts were almost always connected much earlier, and each new institution created social and economic disparities. This pattern recurred again and again, from city to city and state to state. The social aspects of infrastructures came out most fully in projects to redesign urban space; in the nineteenth century, long-standing arrangements in Paris, London, and Vienna struck a limit and had to be reshaped from within. Population growth, the rise of the proletarian masses, disease, social grievances, and mounting social tensions had created the need for change.

Often, a catastrophic event precipitated action. In response to an epidemic of yellow fever, Philadelphia became the first US city to provide public water, a system constructed between 1779 and 1801.[104] In Europe, Vienna pioneered wastewater disposal after cholera struck in the 1830s.[105] Hamburg introduced the public supply of water after a major fire in the 1840s. In Paris, the 1848 uprisings—especially the barricades erected during street fighting—prompted authorities to reconceive urban spaces. Ten years later, the London summer was so hot that the foul stench from the Thames, where the water level had dropped significantly, reached the windows of Parliament (which had to be covered with damp cloth to combat odors). The event has gone down in history as the Great Stink. Like other occurrences, it accelerated the emergence of the modern metropolis in the proper sense.[106]

New experts on public issues who made a name for themselves included William Lindley, who laid the foundations for sanitary provisions in Hamburg and in Warsaw. Another luminary was Edwin Chadwick, who wrote the *Report on the Sanitary Conditions of the Laboring Population of Great Britain* in 1842, a fundamental document for technical solutions to social ills. James Bazalgette oversaw many measures adopted in London after the summer of 1858. Joseph Hobrecht and Carl Ludwig Friedrich von Hinckeldey played key roles in redesigning Berlin. Above them all towered Georges Eugène Haussmann, who became famous for reforming sanitation and patterns of circulation in Paris beginning in 1853. Most pioneers were male, and many of them physicians; in this context, Florence Nightingale, a nurse, warrants mention for introducing exemplary hygienic practices for British soldiers during the Crimean War. When death rates in military hospitals fell significantly, similar measures were adopted throughout Europe.

Most of these personalities enjoyed the support of city governments and broad swaths of the population—even if it materialized in full only after facilities had been completed. Financial worries attended projects, but pluralistic and democratic societies afforded considerable leeway for making adjustments. The extraordinary city planner Robert Moses—the Haussmann of New York—deserves particular notice. Moses and his French counterpart experienced similar fortunes: after years of successful enterprise that brought lasting change, both men wound up out of step with the times. Moreover, they made many enemies, although the latter could at best force them out of office after the fact, with more or less contrived allegations of corruption and mismanagement.

Moses will receive fuller attention later on. Haussmann, for his part, is legendary for making Paris the "capital of the nineteenth century," in the words of Walter Benjamin. Paris stood in competition with London, the largest urban conglomeration and greatest seat of power in the world, but a city less influential in terms of style. Haussmann expanded the grand boulevards, partly to forge corridors through the winding core of the city and to prevent barricades from being erected in the event of unrest. Broad streets and ample parks would encourage the leisurely strolls of respectable, bourgeois citizens.[107]

Haussmann connected train stations outside the center with inner-city roadways. Together with the engineer Eugène Belgrand, he reformed the Paris water supply and updated the sewage system. Harry Lime and other fictitious characters were not the only ones who had adventures underground. In the 1860s, the photographer Félix Nadar descended into the Parisian catacombs; with long exposure times, his pictures captured the eerie, fascinating atmosphere and brought to light the awesome size of a subterranean universe. Haussmann also saw to the construction of central market halls and a slaughterhouse in La Villette, at the edge of the city center. Henceforth, inexpensive meat was available even to workers—along with coal, the principal source of fuel in the industrial age.[108]

At the same time, elements of the population that could not afford rising rents were driven to the periphery and had to use public transportation for daily commutes. The Paris metro arrived rather late, beginning operations in 1900, but once it was up and running, circulation increased by leaps and bounds. London, Budapest, and Glasgow had paved the way; for self-evident reasons, electric motors drove the breakthrough.[109] Today, the

1.2 A journey by boat into the subterranean city is caught on film (from Félix Nadar's photo series *Égouts de Paris* [1860]).

station at Châtelet–Les Halles is the largest hub of mass transit in Europe, with eight hundred thousand passengers passing through every day.

This rhythm of life spread rapidly to other cities; in less time than is generally assumed, its influence extended to rural areas. Quotidian routines changed fundamentally: nighttime lighting and the availability of power and heat meant that the metropolis never slept. At the same time, ordinary activities came to follow a regimented pace.[110] Distances traveled to places of employment and workdays became longer. Middle-class standards of order and hygiene prevailed, shifting vital functions from the street into the home.[111] Public toilets gave pedestrians the opportunity to relieve themselves discreetly.[112] Spittoons were installed to promote cleanliness, and owners of animals were made responsible for disposing of their feces. Since this time, an index of civility is whether one draws attention to oneself through unseemly conduct.[113] Urban environments presuppose an inconspicuous demeanor in the public eye. Even talking about creaturely needs has become taboo for proper citizens. Anything untoward signifies belonging to the lower classes or represents an act of provocation.[114]

Paris, London, Vienna, Berlin, New York, Moscow, and elsewhere became home to plumbing and supply systems that might be centrally planned but still crisscrossed each other in labyrinthine fashion. When the passenger elevator was introduced in the mid-nineteenth century, buildings began to grow higher and higher.[115] As a rule, the layout of underground pipelines was poorly documented and rarely subject to central oversight. To this day, digging beneath the surface of the city—into its unconscious, so to speak—can deliver unwelcome surprises.[116]

This same diffuseness made the subterranean spaces of the metropolis a refuge for those who no longer had a place above. Viennese *Kanalstrotter*—about whom journalist Max Winter reported at the beginning of the twentieth century—lived off leftovers that the urban organism had not used up.[117] In 1956, the Polish director Andrzej Wajda (like Carol Reed for *The Third Man*) made the underground the site of cinematic drama, showing the struggle of the Polish Home Army in World War II. Starting in 1943, heroic resistance fighters used the hidden tunnels and chambers of Warsaw's sewer system to fight the German occupiers. Here, on the screen as in real life, many of them met their doom.

2

LIFELINES OF COMMUNITY: THE EARLY TWENTIETH CENTURY

LA FÉE ÉLECTRICITÉ

The early workers' movement did not take a unified stand on machinery, even though there were many incidents of looms and steam engines being destroyed for fear of competition. Disagreement prevailed about the path toward socialism and a better life in general. But progressive intellectuals hailed electricity. On February 27, 1883, Friedrich Engels wrote to Eduard Bernstein, the leading theorist of the SPD (Social Democratic Party):

> The steam engine taught us to transform heat into mechanical motion, but the exploitation of electricity has opened up the way to transforming all forms of energy—heat, mechanical motion, electricity, magnetism, light—one into the other and back again, and to their industrial exploitation. . . . And [the] latest discovery, namely that electric currents of very high voltage can, with a comparatively small loss of energy, be conveyed by simple telegraph wire over hitherto undreamed of distances and be harnessed at the place of destination—the thing is still in embryo—this discovery frees industry for good from virtually all local limitations, makes possible the harnessing of even the most remote hydraulic power and, though it may benefit the towns at the outset, will in the end inevitably prove the most powerful of levers in eliminating the antithesis between town and country. Again, it is obvious that the productive forces will thereby acquire a range such that they will, with increasing rapidity, outstrip the control of the bourgeoisie.[1]

The second Industrial Revolution had just begun. In the forty years or so before World War I erupted, the technical foundations of the twentieth century were being laid. By the same token, political conflicts of the coming era loomed large. Innovation in the spheres of communication and transport collapsed distances starting in the mid-nineteenth century, in particular between the United States and western Europe.

Whereas the first Industrial Revolution had boosted the productivity of companies, the second Industrial Revolution made a more lasting impact on the population in general. The systematic generation and use of electricity, the development and commercialization of combustion engines, the advent of high-performance materials and synthetic substances, and the rise of modern communications and information networks changed the fabric of everyday life.[2]

Within the space of a few years, inventions proliferated: all manner of electrical appliances, motor-driven vehicles, skyscrapers built on steel frames, dynamite, aluminum, modern printing presses, typewriters, telephones, and devices for recording sound and still and moving images. Other innovations included ballpoint pens, cornflakes, and drinking straws. The twentieth century would boast gas turbines, computers, and nuclear technology,[3] but for the most part it just improved on things already invented, optimizing efficiency and making products cheaper and more broadly available without changing basic designs.

Engels and Bernstein agreed that nonprivileged strata of the population should benefit from the new amenities, but they differed on the political course to steer. Like Marx, Engels believed that forces of production—which electricity clearly heightened—promised the revolutionary transformation of society as a whole. Bernstein held that long-standing conditions and traditions should be taken into account; accordingly, social reform offered the best path forward. The various ramifications of this disagreement on the Left would come to be known as the revisionism controversy. Vladimir Ilyich Lenin implemented the revolutionary approach. As early as 1920, he announced the thoroughgoing electrification of the Soviet Union as a vital step toward communism. Countless large-scale projects for generating energy followed, most of which were geared toward increasing industrial production. However, the planned economy left many sectors of society without a ready supply for personal use—or any real prospect of enjoying the improvements the government had promised (a point to which we will return).

Bernstein and the chairman of the SPD, August Bebel, did not doubt the potential of electricity for effecting social transformation. Indeed, Bebel positively raved about the possibilities offered by modern technology in his 1879 book *Die Frau und der Sozialismus* (*Woman and Socialism*, for which Bernstein wrote the introduction). But in contrast to revolutionaries, Social Democrats held that electrical power would allow abundance to replace scarcity without overthrowing the standing order altogether. As the book's title indicates, Bebel was especially attentive to the role of women in society. He thought that mechanical and electrical appliances in the home and the benefits afforded by industrial kitchens would free women from domestic routines and even allow them to pursue gainful employment.[4]

The goal was achieved, but it took a lot longer than Bebel (or the early women's movement) anticipated. The mechanization of the domestic sphere did not produce only positive results. In the first place, it deprived many servants of their livelihood, and they resisted the introduction of new technology as vigorously as their counterparts in the factories. Moreover—and counter to what many politicians and visionaries assumed—potential users often refused to let electricity into their homes at all. For one, a "better" energy source had been available for some time, first in England and then on the continent. Gas already performed almost all the functions of electricity. The early nineteenth century, for the first time in history, witnessed the illumination of public spaces; gaslight changed how darkness and light were perceived—and on an emotional register, as well.[5] For heating, gas was more than adequate; to this day many people swear by the advantages of cooking with it. Indeed, governments and providers are still trying to put an end to gas as a means of lighting. With some thirty-seven thousand streetlamps, Berlin has the world's largest remaining supply of gas fixtures in operation, and the Gaslichtinitiative Berlin has been working since the 1990s to prevent the city's complete electrification. The arguments are the same as those previously mustered, in the nineteenth century, to preserve oil lighting.[6] It is supposed to be too glaring, too expensive, and too dangerous and also harmful to the environment.[7]

Because gas represented a fire hazard, many people resisted its domestic use; until odorants were added, it was difficult to detect leaks. Lines were more costly to install than water and sewage systems, and numerous accidents occurred. And although gas facilities enhanced the overall quality of life, few wanted to live in their immediate vicinity, which could prove dangerous. In Europe, British companies were the first to supply gas to towns. In the late nineteenth century, municipalities often became providers.[8]

Electrical utilities remained largely in private hands in Europe and the United States, and it was not always easy convincing consumers that offerings were clean and modern compared with the gas distributed by public authorities (which, in the interim, had gained broader acceptance). Industry had its hands full at first. One of its pioneers, Thomas Alva Edison, took a decisive step by illuminating buildings dramatically and making sure that the designs of his devices did not strike the public as too unfamiliar. Edison was a marketing genius who understood the importance of fitting his inventions to standing social practices.[9] Eventually, the use of electricity was established in many spheres, especially when the advantages of compact motors had been made plain. Public transportation led the way, followed by agriculture.[10] By the turn of the century, electric streetcars and subways represented the paragon of metropolitan modernity; politicians believed it a "disgrace" for a city not to have at least one or the other.[11] Commuters adopted the same view. Fares for horse-drawn travel were high; before long, the service came to evoke luxury and private profit interests.[12] Public transportation quickly became a defining feature of life. Streetcars, elevated trains, and subways granted easy access to suburbs and the surrounding countryside and enabled people to reside well beyond the limits of city walls. The social and cultural impact can hardly be exaggerated.[13]

Private households did not immediately share the excitement, and energy providers had to develop and popularize many uses. The modern world became electrified only through currents of discourse and a lengthy back-and-forth between suppliers and consumers.[14] As we have noted, women played a key role in the domestication of electricity; their experience helped provide developers, engineers, and salespeople with ideas about what could be done with electricity in the first place, and they ultimately decided whether a given product or service was desirable or not. Overcoming reservations that homemakers entertained was key.[15] Nor is it by chance that symbols of electrical power had a feminine charge. *La*

Fée Électricité represented magic at work.[16] Indeed, a woman created the most enduring myth about modern wonders. Mary Wollstonecraft Shelley's 1818 novel *Frankenstein* tells how the eponymous hero brings to life an artificial human being by electrical means—only to be undone by what he has wrought. The young author had been inspired by the experiments of the Italian scientists Luigi Galvani and Alessandro Volta.

Electricity went into industrial production almost as soon as it was discovered. Its powers were often equated with the nervous system—hence its image as an elixir of life. This view reinforced analogies between infrastructures and the pathways of veins and arteries; now they included nervous tissue, which stood in for the subtle, spiritual forces animating the organism as a whole.[17] As power lines multiplied and more and more devices were connected to them, the idea only became more self-evident, ensuring the symbolic and practical triumph of the newfound source of energy.[18] Important buildings and sites were bathed in light; private concerns and citizens followed suit to bask in the aura of prestige. Exhibits at world fairs heightened the symbolism, and accidents diminished the appeal of gas. In 1881, at least 384 people died when the Vienna Ringtheater burned down.[19] This was not the first catastrophe caused by gas lighting. (Ironically, one of the largest theater fires ever was sparked by an electric spotlight that short-circuited; the 1903 tragedy at the Iroquois Theater in Chicago claimed over six hundred lives.[20])

Industry, public institutions, and private households ultimately went electric, of course. That said, the process was not complete for decades—even in central Europe, where setting up an infrastructure for the whole population had been a goal since the turn of the century. The United States had taken the lead. After a 1935 visit to New York, the Austrian novelist Stefan Zweig marveled,

> Everything here happens by electricity, that quiet, majestic, and invisible force. It drives the trains, it washes the laundry, it cooks the food, it makes heat and makes cold. It carries one up artificial mountains and back down again. It gives the gift of words, sound, and spectacles. It is unthinkable what would become of New York if this force faltered for even a moment: a missing beat would paralyze every limb, and everything would stand still, grow empty and dark, for it is everywhere—the electricity of this city. If a biologist compared human bodies, I believe he would find more electricity in the nerves and muscles of New York than in the inhabitants of our own world.[21]

Zweig's words took on a prophetic air whenever the US metropolis experienced cardiac arrhythmia. The most famous example occurred on July 13, 1977, when twin lightning strikes downed the power grid. New York was experiencing a heat wave, and routines were already in disarray. Streetlights, traffic signals, and air-conditioning systems failed; elevators, trains, and subways came to a halt; hospitals, government agencies, and airports implemented emergency plans. Looting spread and houses went up in flames during the twenty-five-hour outage. At the same time, blackout parties held by the likes of Andy Warhol and Woody Allen—with flashlights and candles providing a magical ambience—unleashed a tide of creativity and culture. Sheer accident set domino effects into motion that shook the lives of millions.[22]

Twentieth-century urban life was defined by most services becoming electrically operated. Not being connected to the power grid counted as backward—like living without running water. Electrification set the standard for infrastructural provisions because so much depended on it, and providers, commercial enterprises, and advertisers were constantly devising new fields of application. Consumption expanded steadily after the turn of the century. The goal was to create a comprehensive market without costly detours or service deserts.[23] Accordingly, broad social consensus was required that electricity was not only sensible and desirable but, more still, beneficial. On the local level, this meant contending with religious groups—for instance, Hutterite and Amish communities—that considered innovations mere fashions or even violations of divine providence; still today, there are those who resist taking work that God has assigned out of human hands. Objections were also voiced because of the material disadvantages that changes would bring. Setting up technical networks on a massive scale disrupted the landscape, property boundaries, and economic and political structures. Each kilometer of cable and every power pole marked the site of potential conflict.

Societies where relative freedom prevailed witnessed recurring patterns of coercive negotiation. Energy providers would try to avoid trouble by settling things in advance—say, by offering compensation to people living near power lines. To a certain extent, they also acknowledged calls to protect the landscape and environment. When objections persisted, another round of negotiations would usually be held. If the desired results were not obtained, experts would be brought in to hold forth in incomprehensible jargon about overriding needs. Should this step fail, companies would threaten legal action or warn resistant parties that they would be denied service.[24] In socialist states such as the German Democratic Republic, the process was usually more straightforward: land the government wanted was summarily declared to be national property and expropriated; if necessary, refractory elements were charged with speculation and economic crime.

Claims that electrification was universally beneficial often provided the pretext for mandatory connection and use. When the provider belonged to the public sector, administrative measures made it particularly difficult to resist. But once people were hooked up, few, if any, wanted to go back. As one German engineer put matters in 1911, "At first, no one wants a connection. But when the thing is built, everyone gets one. People wouldn't give up running water if you paid them. That's the way it always goes. It will be the same for electricity."[25] The trick was to connect as many users as possible; popular demand would follow. Pricing represented another crucial factor. Rates needed to be low enough to entice reluctant consumers. In turn, when a standard had been established, charges could be raised. However, social and political considerations could influence prices. Inexpensive fares, for example, could accelerate urban development into surrounding areas. Over time, greater population density gave rise to larger power grids with a broad spectrum of users. Because storing electricity is complex and costly, it needed to keep flowing so providers would not suffer financial loss. Different rates were developed in keeping with new scales of operation—say, lower prices at night or outside peak consumption periods. One effect was the extension of working hours; like mechanization and, later, automation, the new regime was at best an ambivalent proposition in the eyes of organized labor.

The decisive factor for the success of electricity in everyday life—and therefore its ascent to infrastructural status—was advertising focused on the domestic sphere. Other energy sources (gas, petroleum, and methylated spirits) had to improve to compete. In 1892, gas lighting managed to regain a small lead over electrical devices because the newly invented incandescent mantle made lamps much brighter while reducing the amount of fuel consumed. (In response, advocates of electricity pointed out that gas lighting used up oxygen in enclosed spaces and left gray deposits on their contents; what is more, they claimed, food prepared with gas had an unappetizing odor and taste.[26]) Another uptick occurred when Johann Vaillant introduced his bath stove in 1894, which became a necessary appointment in better homes both for ostentation and because it was now easy to shower or bathe with heated water. The new business model was to sell appliances at reduced prices, offer payment in installments, or rent them out for a nominal fee. By the turn of the century, thousands of gas ovens were being leased in London. Such campaigns proved even more effective where there was no competition. In 1905, for example, John D. Rockefeller began distributing millions of petroleum lamps in China to prime consumers for ongoing supply by his company, Standard Oil. This ploy was the object of criticism in Alice Tisdale Hobart's novel *Oil for the Lamps of China* (1933, adapted for the screen in 1935), which sat alongside titles about the shady dealings of railroad magnates and the deplorable sanitary conditions in Chicago slaughterhouses.[27] The so-called Rockefeller principle—selling a product at a low price to reap greater profits later—has remained a successful business strategy.

By the early twentieth century, the groundwork for infrastructure-dependent consumerism in the United States and Europe was largely in place. Besides lighting, heating, and indoor plumbing, amenities would come to include vacuum cleaners, refrigerators, and blenders and smaller items such as belt massagers, flat irons, and spin dryers. Promotional campaigns aimed particularly at women extolled hygiene, convenience, and free time to engage in more meaningful activities.[28] In the process, sales and marketing expanded and became more and more professionalized; eventually, it would be a matter of course to have customer service centers, regular advertising in print and other media, and company representatives eager to pay home visits, explain bills, offer advice, and generally sing the praises of products.[29]

Expanding infrastructures also had an impact on residential construction. Twentieth-century housing developments flourished in large part because newly built homes met modern standards. In practical terms, this meant that architects and other planners had worked out how much space was needed per occupant, optimal conditions for cooking and cleaning, and manageable distances from places of work and recreation.

Women's insistence on the latest conveniences became a veritable cliché in advertising.[30]

The history of the telephone exemplifies the rise of electrical infrastructures. As early as 1889, observers declared that "no modern invention has become so established as a means of exchange, or gained such a firm foothold in the civilized world, in such a brief period of time."[31] At first, the telephone was used mainly by the wealthy and people intent on demonstrating social status. Businesses considered it inferior to the telegraph because it left no written traces; its general utility was acknowledged only after a trial period. The phonograph could also transmit sounds; the Urania in Berlin boasted the *theatrophone*, which brought in live operas and classical music. Beginning in 1897, Café Kranzler (also in Berlin) shared an up-to-the-minute selection of pastries with customers via telephone.[32]

To a great extent, the telephone was employed to ends contrary to those anticipated. The sense of proximity that it inspired held particular appeal,

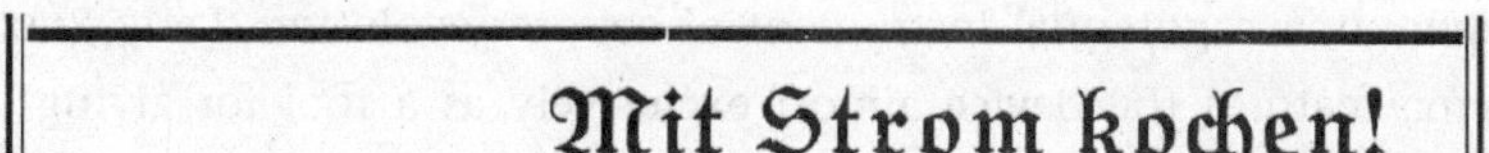

2.1 A 1935 advertisement for domestic electricity promised to relieve women of the hassle of repetitive chores. "It's convenient, pleasant, and healthy. / Dishes prepared this way turn out better and are more nutritious. / The cost of purchasing and operating equipment now is cheap. / Just pay small monthly installments, which you'll barely notice, then this perfect electro-kitchen belongs to you."

and the device became a medium for both erotic activity and spiritual counseling.[33] "As close as if you were there" *(So nah, als wär' man da)* went an early advertising slogan. Nickel-in-the-slot devices at small businesses became social hubs—the forerunners of telephone booths and, later, Internet cafés. Women took advantage of the telephone for social purposes, because it allowed them to pursue domestic tasks at the same time. (In contrast, men were the principle beneficiaries of infrastructural development in the realm of transportation.) Ultimately, people without a telephone at home found themselves excluded from common types of sociability.[34]

Companies underestimated the telephone's social functions; it took quite a while before suitable business models were developed.[35] Political groups were sometimes more attentive. Thus, in late nineteenth-century France, fear arose that the telephone, which allowed free communication for everyone, would undermine standing hierarchies and fuel overly democratic tendencies.[36] In turn, the Soviet government suspected the telephone as a potential instrument of counterrevolution. Early Russian cinema features the device almost exclusively as a tool for giving and receiving orders, not as a means for bringing people together and promoting sociability.[37] For decades, socialist countries delayed the expansion of networks and took elaborate steps, in both human and technical terms, to control the exchange of information. Owning a telephone was a privilege, but someone could always be listening in.

PERIL AND IMPROVISATION

Karl Veidt, recipient of the Iron Cross First and Second Class, served as a chaplain of German troops in northern France between 1914 and 1917. His memoirs describe their military advance in the Argonne:

> Constructing houses of stone and concrete represented another step in the history of humankind that we experienced in condensed and accelerated form. New roads were built. Sawmills with steam and electricity came into operation, as well as cartwrights' and carpenters' workshops. Wells were drilled and water pipes laid in even the most advanced positions. German efficiency and inventiveness found ways and means where seemingly insurmountable difficulties presented themselves. . . . In the end, if permission was granted, one could travel from the front lines to Frankfurt or Berlin. Didn't we go from the age of

the cavemen to the age of steam and electricity in those two years . . . ? Thus did we become one with the land, passing from prehistory to advanced culture.[38]

Apart from documenting German arrogance, these words show just how widespread and entrenched the colonialist view had become. Even a clergyman had adopted the ideology of technology and progress, which had assumed the dimensions of a civil religion by the turn of the twentieth century.

The Berlin Conference on Africa in 1884–1885 played a key role in this development. At the invitation of Otto von Bismarck, the great powers of Europe decided that they had the right to seize land that was, at present, "worthless" and claim it as their own. In concrete terms, possession meant infrastructural development so territory could be put to productive use. The quasi-legal rationale was that the spread of European civilization would benefit the world. History follows the course of improvement, and so-called primitive communities are destined to yield to more advanced ones. Needless to say, it is always foreigners who need to be helped along.

In this spirit, reporting on German East Africa in 1903, the business journalist Paul Neubar declared,

Untamed vegetation without equal devours all works of culture in no time at all, if man does not impose his will on nature through unremitting toil. The railroad is the outward expression of such domination, signifying once and for all that civilized humanity has taken charge and wishes to assert itself. . . . Much more than administration, troops, hospitals, and district offices, the railroad heralds the mother country's resolve to provide her people here with the modern resources and facilities that ensure work will be joyous and success certain.[39]

Statements of the kind were legion. Their authors presupposed that differences between parts of the globe corresponded to unequal levels of development on a universal timeline; standing institutions and technologies proved relative inferiority or superiority. As if to confirm their status, Europeans drilled deep into the earth, mined its resources, covered the surface with new roads and buildings, and relocated populations so conditions abroad would match up with the infrastructure (and, supposedly, law) at home.

Twentieth-century conflicts were to show how "advanced" civilization rested on shaky moral and material foundations. During the First and Second World Wars, the same great powers that had agreed on a civilizing mission were busy killing native populations in Europe and accusing each other of barbarism. Such rhetoric could not have contrasted more sharply with all the talk of shared enterprise but recently celebrated. A standard refrain in accounts made after fighting had ceased was that "The [First World] War started by stopping connections (*Der Krieg begann damit, dass die Verbindungen aufhörten*)."[40] The scope of the world wars reached beyond theaters of combat and into the civilian sphere. In networked societies, conflicts are difficult to contain and tend to become all-encompassing. Ever since, the first step in armed confrontations and power-grabs has been to conquer or destroy centers of communication, transportation, and energy.

The history of infrastructures is neither straightforward nor peaceful. Indeed, military engineering has made incomparable contributions to their development. After all, mass armies require systems of supply and support different from the provisions made for roaming mercenaries of the pre-modern era. Nor is it an accident that, as we have noted, the term *infrastructure* first appeared in the context of a NATO conference. As "the practical art of moving armies and keeping them supplied,"[41] military logistics is a matter of providing for one's own needs, attacking the enemy's resources, and rendering them inoperable (or using them for oneself).

Modern warfare involves spying out points of vulnerability in infrastructures and paralyzing them through focused attacks. By the same token, one's own networks need to be safeguarded and losses mitigated so operations remain as functional as possible. As Pastor Veidt observed, it takes "efficiency and inventiveness" for makeshift arrangements to approach the stability of permanent conditions. Military campaigns are planned in advance, and they seldom play out as anticipated. The infamous Schlieffen Plan of 1905, which counted on the rapid capture of Belgium and France by mobilizing German troops by rail, failed because of unexpected levels of resistance during the initial phase; in turn, the flow of supplies was severely disrupted after tracks in Belgium were destroyed.[42] When a quick and decisive victory—by Christmas 1914, at the latest—failed to materialize, the German army had to dig in and keep fighting for four years. Trenches spread from the English Channel to the Swiss border, and narrow-gauge railroads supplied combatants from lines that were still operational in the hinterlands. As a rule, things worked relatively smoothly up to seven miles behind the front, but beyond this point they fell within striking range of enemy artillery. (Because of the threats faced by horse-drawn vehicles, trucks, buses, and light rail, troops from the colonies were often employed.) The challenge was to maintain chains of command and supply lines from headquarters to the front, and to preserve contact between soldiers and people back at home.[43]

Before World War I, German high command had switched communications to telephones; some six million kilometers of wires and cables were laid between 1914 and 1918. These networks turned out to be more fragile than anticipated, and they were prone to interference. Telegraphy proved more reliable; light signals were also employed when circumstances dictated. At times, messages were even transmitted by dogs or carrier pigeons.[44] The most thankless tasks during the "storms of steel" were performed by scouts, reconnaissance patrols, and field repair units. As Ernst Jünger observed with characteristic matter-of-factness, "While most others strained to leave a shelled zone, the breakdown squad had to enter it calmly and professionally. Day and night, they went into still-warm shell holes to tie together the ends of two severed wires; their job was as dangerous as it was unglamorous."[45] Such operations could be carried out only by intrepid and reliable individuals for whom wartime conditions had become a kind of second nature.

Instincts and habits that guided civilian life had to change from top to bottom also among noncombatants. Military commanders subjected the flow of information to strict control. News and images from the front were released in a form edited to propagandistic purposes. The use of infrastructures was restricted because resources were diverted to meet military needs.[46] Because everyday routines sat firm, measures proved successful only in part. A constant source of concern was how misuse of transportation, communication, and supply facilities worked "counter to the war effort"—to say nothing of sabotage, strikes, and spying. The 1942 declaration by Albert Ganzenmüller (head of the Deutsche Reichsbahn) also applies to World War I:

> The order of the day is for every railroad man to be aware of his duty to do his part in achieving the final goal! All must know that negligence or forgetting of duty on domestic railways harms our soldiers on the front; indeed, every

unfilled car, each supply train that comes late prolongs the war and therefore benefits our enemies.[47]

Women worked where there were not enough men to operate facilities.[48] Prisoners of war were engaged, often without regard for international law, on all sides of the conflict. Assigning them strategically sensitive projects was unadvisable, but they were suited to construction, maintenance, and repair tasks. Prisoners and conscripts also worked in planting and agriculture.[49] From the second half of the nineteenth century forward, humanitarian efforts had aimed to prevent military conflicts altogether or, failing that, to contain them by law so civilians would suffer as little as possible. In fact, however, twentieth-century warfare collapsed the boundaries between spheres of life. In efforts to preserve relative normality on the home front, forced labor was used on an unprecedented scale.

After the formal end of conflict, the management of infrastructures remained fraught—and sometimes offered unseemly benefits for parties instrumental in promoting warfare. Multiple scandalous cases have shown how specialists retained positions of influence in spite of the serious offenses imputed to them. Because expertise is valuable in peacetime too, many ex-Nazis had no difficulty returning to civilian life. Matters of supply, waste disposal, communication, and transportation are important for maintaining order, regardless of regime.[50] Societies with a high level of infrastructural development and administration show a great deal of continuity across systems of government. In political terms, a kind of "antichaos reflex"[51] works to alleviate hardship and prevent the resurgence of radicalism. Especially from the perspective of the professional classes, inadequate provisions for the population at large threaten disaster, and modes of operation geared to military logic are easily recoded for a civilian setting. Just as importantly, domestic and international trade is supposed to spring to life again as quickly as possible.

In Germany, Technische Nothilfe (since 1950, Technisches Hilfswerk) aimed rapidly to return damaged infrastructures to a functional state. The organization was founded in 1919 by the civil engineer and former military officer Otto Lummitzsch to carry out emergency work in realms classified as vital (gas, water, electricity, railways, postal services, and agriculture) in the event of strikes or unrest during the transition from imperial to republican government. Measures regularly provoked controversy during the politically tense Weimar years. Communists, for example, considered the organization a gang of strikebreakers intent on stabbing workers in the back and thwarting legitimate interests.[52]

Damage to civilian infrastructures, especially by bombing, opened new opportunities for architects and urban planners. Relishing the prospect of redesigning cities, they wanted to break up whatever structures had survived. Many municipalities bid farewell to their medieval centers and embraced change.[53] The car-friendly cities that emerged often employed the same experts who, before 1945, had drawn up plans for German settlements that Nazi leaders anticipated building in Eastern Europe.[54] But reconstruction revealed that material infrastructures prove enduring even across deep historical fault lines. In 1951, Ernst Randzio published a study of "underground construction space" in Berlin. The author, who had devoted years to researching transport and circulation in German colonies, found that many urban planners after the war were unhappy that systems of drainage, water supply, gas, electricity, heating, telecommunications, and rails remained halfway intact, which interfered with their own designs.[55] Their counterparts in Japan enjoyed a wider berth, because infrastructures here had for the most part been laid aboveground and needed to be replaced in full.

Thoroughgoing modernization was not forthcoming in West Germany. For example, more than two-thirds of Giessen, a medium-sized university town, was destroyed during the war. In his memoirs, Albert Osswald, who was mayor in the 1950s, recalls that municipal authorities decided to rebuild old streets "in order not to lose valuable provisions already in place, such as sewers, pipes, telephone lines, and electrical cables."[56] With disapprobation, he notes that the decision gave the town only a temporary advantage relative to newly designed settlements. Osswald also finds fault with the university—and with professors' wives, in particular—for wanting to preserve familiar charm at the expense of improvement. He went on to become minister president of Hessen and continued to advocate modernization but met with resistance. In the 1970s, public objection to new buildings grew louder and louder; more was being done to disfigure the landscape now, it was claimed, than had occurred during the war.[57]

In the balance, the majority of the population desired a quick return to services and conditions that had existed beforehand. Successfully improvised measures adopted during the war or its aftermath could be celebrated for having overcome challenges, but normality and a smooth daily routine were paramount concerns. Ad hoc buildings (camps and barracks, for instance) and everything that had outlived its purpose should yield to arrangements that would not recall recent travails.[58] Some of the structures were too massive to be disassembled and stand to the present day. Others were adapted to civilian purposes (like the field rails from World War I that the British had sold to countries elsewhere on the globe, where they continued to be used for decades).[59] Military trucks found ready use for long-distance transport; competition between road and rail is ongoing.[60]

All modern warfare has begun by severing connections between combatants; in turn, equipment continues to be used even after military operations have formally ceased. In Germany, during the revolutionary upheaval in November 1918, radio transmitters that had formerly belonged to imperial propaganda services were found in the possession of insurgents and promptly confiscated, lest they be employed for left-wing agitation. During the Weimar Republic, the government decreed that radio broadcasts were to be nonpolitical. Amateurs consistently violated the policy, laying hold of any frequency available; politicians could do little but look on warily.[61] Makeshift arrangements are difficult to control. Although infrastructural development implies standardization, there are always people who come up with anarchic expedients to escape control.

TIME TRAVEL BY CAR

Orlim Vargas, a native of Ecuador, is probably the last elevator operator in Europe. He works at Les Trois Rois in Basel, one of Switzerland's preeminent grand hotels, where he started as a dishwasher before switching, largely by happenstance, to a position that is unique on the continent. His presence may inspire uncomfortable self-consciousness in guests who are unaware that true luxury means not even pushing a button. Vargas's professional duties include intuitively grasping guests' sense of awkwardness and putting them at ease. Les Trois Rois has long hosted notables, from Herbert von Karajan to Elizabeth II and Bob Dylan. The famous photograph of

Theodor Herzl on the balcony, casting his visionary gaze over the city, was also taken here. A few years ago, the building was restored to match conditions at its opening in 1844. The hotel's other offerings recall an era before the standardized services of the hospitality industry that prevail almost everywhere else.[62]

Switzerland likely has the longest unbroken tradition of grand hotels. These anachronistic institutions bring together many contradictions: movement across international borders and rootedness, public ostentation and private retreat, dynamic spontaneity and an aura of permanence. *Complexio oppositorum* is evidently a recipe for success. Luxury hotels around the world compete to offer discerning guests the highest level of amenities. The Savoy in London, the Plaza in New York, the Ritz in Paris, the Adlon in Berlin, Raffles in Singapore, and the Taj Mahal in Mumbai have achieved mythical status. They all promise the utopia of an ideal existence for the relatively brief duration of a stay.[63] The only drawback seems to be the bill.

Few sites are as closely interwoven with the history of infrastructures as high-end hotels. The rule applies both to interior appointments and to their location in urban space. The product of increased mobility in the age of steam power and railways, they were founded for tourists belonging to the upper classes of Europe and the United States. Grand hotels, which evolved from railroad hotels, sought to embody local color while offering international travelers a predictable experience. Scrupulously trained personnel and the latest technology ensure a maximum of relief, comfort, and security. Before such amenities were broadly established elsewhere, these institutions boasted central heating and hot water, facilities for health and relaxation, electric lighting, telephones, elevators, and the latest in entertainment. Thoroughly up-to-date appointments ensured that the highest expectations would be met as a matter of course. This standard was set not only by well-heeled guests, who competed with each other to demand the best, but also by guides such as *Baedeker*, *Murray's*, and *Michelin*, which have evaluated accommodations and services since the nineteenth century.[64]

Grand hotels offered exclusivity on a level comparable only to luxury liners or Pullman sleeping cars.[65] Yet their doors were open to anyone with enough money, whether members of the old aristocracy or the newly rich. Behind imposing facades, which could be stately or exotic, a host of services stood at the ready; the practical operations that sustained them were carried out as inconspicuously as possible to heighten the apparent magic. "Indexes of urbanity," they were magnets for all who aspired to grandeur themselves.[66] Socializing in this setting presupposed real—or convincingly staged—stature. Staff were not alone in being expected to observe discipline. Inasmuch as exchanges took place "in intimate anonymity and cultivated foreignness,"[67] grand hotels represented the crowning achievement of infrastructural development.

But luxurious accommodations were just one part of the new world opened by infrastructures. Tourism has flourished since the nineteenth century thanks to expanding lines of circulation and communication. With forerunners in religious pilgrimages, the Grand Tour undertaken by young members of the upper crust to learn the ways of the world, and travel by ascendant classes for educational purposes, the first phase of tourism dates to the period between 1835 and 1880. In turn,

> when the railroad made it possible to transport tourists en masse, petty state barriers were torn down and the beginnings of a tourist infrastructure started to emerge (hotels and boarding houses, the first travel agencies, larger seaside resorts and health spas). With the founding of summer getaways, tourism started to include the middle classes. This second phase in the growth of modern tourism lasted roughly from 1880 to 1930.[68]

Neither the construction of railroads nor roadbuilding, which proceeded along parallel lines, simply followed the imperatives of industrialization. Other factors and forces brought masses of people, goods, and ideas into circulation. However, a key impetus for connecting undeveloped regions to modernized areas—whether by rail, highway, boat, or ultimately, airplane—came from travel on personal initiative or organized collectively. Tourists would not accept the excuse of bad weather or seasonal difficulties as an impediment to reaching destinations.[69] Spatial planning and landscape design both exploited and promoted infrastructures. From at least the twentieth century, manufacture and trade ceased to set the course of development. Recreation areas became just as important, and terrain was made tourist friendly so that as many people as possible could travel and experience nature in a pristine state[70]—especially in mountainous regions, but also along coasts, at spas, and on islands. International tourism on Mallorca, for example, began when the Grand Hotel in Palma opened in 1903.

Tourism had a pattern. In the discovery phase, enterprising parties found a location likely to prove attractive. There followed a development period during which transportation and supply lines were put in place. Once these measures had been taken, further expansion proceeded at a pace dependent on the number of visitors; without clever marketing, tourist destinations risked stagnation or decline.[71] Such "infrastructures of happiness" were cultivated in ways as complex, emotionally charged, and fraught as maintaining a luxury hotel. At every turn, they were to respond to visitors' "search for dream worlds" that were also supposed to be authentic experiences.[72] In other words, tourism conditioned travelers to see their surroundings selectively, and even along fictional lines, while fulfilling basic human needs (e.g., diversion and curiosity).[73] As Kurt Tucholsky observed in his essay "The Art of Traveling" (1929), tourists want to "let themselves go" and try out roles different from those they perform in everyday life.[74]

One of the first entrepreneurs to offer an all-inclusive experience was Thomas Cook. Beginning in 1863, he proposed Switzerland as a destination for British tourists. Ever since, this country has thrived on international travel and offered every imaginable convenience: not just accommodations and restaurants for a wide range of budgets (from mountain huts to grand hotels) but also railways, roads, cable cars, and ski lifts. At every turn, the goal is to provide access to beautiful locations—heterotopias that seem to be a world away yet can be reached without great effort. Shortened travel times and relative comfort along the way were basic prerequisites for expanding the *industrie des étrangers* (an expression coined in the nineteenth century).[75] Construction projects like the Gotthard Tunnel, which opened in 1882, soon revealed tensions, however. The image of modern and industrial Switzerland had to harmonize with the traditional picture of an agrarian society at one with its surroundings. "Wherever nature still exists, there are railroads, too" Robert Walser observed drily.[76] And tourists expected both.[77]

At this early heyday of Swiss tourism, the Alps were transformed from an unwelcoming obstacle to free movement into a romantic landscape boasting spectacular panoramas and athletic challenges. Just as hotels competed with each other, regions tried to offer more than rival parts of the country. To the dismay of many local residents and conservationists, more and more remote areas underwent development to attract tourists. This meant more railroad tracks, electric cables, hiking trails, lookout towers, huts, signposts, and guardrails—plus the restaurants and facilities for entertainment, culture, and relaxation. Travel agencies, tourist associations, railroad companies, and after the turn of the twentieth century, automobile clubs propelled change; developments brought major infrastructural projects at almost every level.[78] All the while, everything possible was done to preserve the illusion of the Swiss as a free, pastoral people.[79]

Tourism thrives on a certain paradox. On the one hand, everything is supposed to have as much local color as possible; on the other hand,

2.2 The "Ragaz-Wartenstein cable car" was as comfortable as can be on the way to the beauties of nature. It operated between 1892 and 1964 in the Swiss canton of Saint Gallen.

travelers expect the comforts of home. (Backpackers and the adventurously minded form a group of their own; they usually bring along specialized devices to compensate for the absence of infrastructures out in the wild.) From the beginning, questions were raised whether tourist infrastructures served general interests or the luxurious expectations of a few. As a rule, only when facilities came to depend on the revenue generated by travel were both demands met. Those who remained opposed to all changes could appeal to the conservation of nature as a last resort.[80] That said, the standards of premium hotels with the latest and most technically advanced appointments are not absolutely necessary for tourism. Often enough, the opposite is the case: nostalgia-laden locomotives, winding roads that seem to lead back into time immemorial, old-fashioned ferries, and so on offer charms to travelers glad to leave behind everyday routines. The modern world, where life is fitted to schedules, engenders feelings of loss that can be remedied by visions of past—and supposedly better—days: a "chronotopia" that transforms temporal distance into immanent, physical space where reality has a different texture.[81]

Many travelers prize somewhat outdated features of the places they visit. Rich symbolism inheres in the double-decker buses of London, New York City's yellow cabs, the horse-drawn carriages of Vienna, Hector Guimard's Art Nouveau entrances to Parisian metro stations, the distinctive decorations of Moscow subways, the cable car up the mountain in Rio de Janeiro, and even traffic signs alerting drivers to the presence of kangaroos in Australia. The trinkets and souvenirs that tourists buy evoke an exotic mood back at home. Hundreds of thousands of tokens from the New York subway have traveled the globe, never having been used for their intended purpose, and many of the Royal Mail's red mailboxes have vanished in the same way.[82] A characteristic tension between foreignness and familiarity comes out in the very design of infrastructures.

With a Victorian sense of entitlement, the first British visitors to Switzerland insisted that the people they were gracing with their presence and money should provide the level of comfort to which they were accustomed.[83] Taking a dim view of such attitudes, the historian Heinrich von Treitschke scoffed that the English equated civilization with a ready bar of soap.[84] In fact, the new conditions required adjustment from travelers and locals, guests and hosts, in equal measure. Infrastructures that were developed to meet tourists' expectations changed those of year-round inhabitants. A particularly striking example is the construction of roads in regions that now were thought of as a series of landscapes for visitors to contemplate. Aesthetics proved to be just as important as accessibility.

The rise of the automobile represents a turning point in the history of mass mobility and mass tourism. People had been getting around by bicycle, motorcycle, and bus for decades. The bicycle, in particular, had inaugurated circulation on a broad scale. It was the automobile, however, that was destined to become the symbol of the new era—the center around which an all-encompassing web of economic exchange, circulation, sociability, and leisure would develop.[85] The institutions that emerged in tandem with its ascent form a multitiered infrastructure where contradictory interests prevail—paved roadways, gas and service stations, and roadhouses and countless other, decentralized places of work and leisure.

In contrast to shipping and railways, which were coded in more practical and economic ways, motorcars had built-in ties to sports, recreation, style, and personal autonomy. The automobile brought a new kind of movement: simply hitting the road.[86] In the early days, this privilege was reserved for the wealthy; driving was something for tourism or ostentation before it became a feature of everyday, practical life. Cars have never lost their cachet, however. Routine matters of transport can be accomplished by other means. Cars promise more fun and freedom. Advertisers latched onto this feature from the very start, recognizing that automobiles were a kind of screen onto which fantasies of modern individualism could be projected. The open road doesn't have fixed itineraries, timetables, or conductors. Motorists are captains of their own destiny and don't need to wait for connections or deal with being crammed together with complete strangers.[87] Enthusiasts went so far as to claim that trains were "prisons" and practically "communist."[88] (This rhetoric would provide the model for promoting other innovations in terms of personal comfort and choice.[89])

The automobile thrived on people's desire for distinction. It took money to buy a car, and operating one required specialized knowledge. Athleticism, competition, and conspicuous consumption set the tone for the first phase of automotive history; motorist associations and touring organizations promoted vehicles accordingly.[90] The promise of conquering space proved especially powerful. Indeed, the automobile collapsed the distance between city and country, making it easier and more attractive to live away from the urban hustle and bustle. A mobile population changed the infrastructure of traffic and transportation. Effects included a new system of roadways with motels, drive-in restaurants, and ubiquitous gas stations; by the 1930s, it was the norm for houses to have a garage in the United States.[91]

Resourceful builders, with cars in mind, created parkways, routes intended for noncommercial traffic that led through particularly beautiful regions. Examples include the Blue Ridge Parkway in the Appalachian Mountains. The design was inspired by US landscape painting—which makes infrastructure a medium for the politics of nation building.[92] Even though, in the long term, the parkway concept did not prove successful,[93] features were taken up when planning roads between urban areas and highways that would offer panoramas and vantage points to travelers. Considerations along these lines still shape the routing of many highways in the United States and Europe.[94] Another point of attention was that long, straight roads encourage speeding, tunnel vision, and fatigue; accordingly, since 1939, on-ramps have been built on the model of a *clothoid*, the arc that drivers intuitively follow when turning the steering wheel.[95]

The automobile's built-in connection to leisure and luxury prompted people to consider the invention an unnecessary annoyance; after all, other modes of transport existed. Complaints involved dust, noise, smell, and accidents. Early protest measures could be harsh: stretching steel cables across roads, setting up barricades, scattering nails, and throwing stones at vehicles; innocent parties, the reasoning went, would not be affected.[96] Drivers' often blatant disregard for pedestrians, horses, and old-fashioned forms of conveyance led to road fees in some areas. Motor vehicles were banned wholesale in the Swiss canton of Graubünden until 1925.[97]

Motorists at the helm of advanced technology were inclined to view others as backward, sluggish, and generally stuck in the past. The latter group had no chance of winning the battle. In the 1920s, at the latest, people began to grow accustomed to the new order, especially as fascination with motorized vehicles spread down the social ladder. In the incipient "dromocracy," as the philosopher Paul Virilio has called it, the fast set the rules for the slow. Accidents, now regarded as inevitable, came to symbolize the risks attending the conquest of nature—part of the "heroic paradigm of modernity."[98] Of all modern innovations, automotive transport has likely had the deepest impact on social life, habits, and patterns of consumption. Some people believe they were born with "gasoline in their blood."[99]

For the last hundred years, work, travel, and tourism have formed an indissoluble bond, giving rise to vast numbers of construction projects. Examples include the Grossglockner High Alpine Road, built between 1930 and 1935, which remains one of Austria's top destinations for sightseers to this day.[100] Germany's autobahn system came into being in response to calls from organizations such as the Verein zur Vorbereitung der Kraftwagenbahn Hamburg-Frankfurt-Basel (HAFRABA eV, founded in 1926) for drivers to be able to travel without having to stop at intersections; the autostrade of northern Italy provided the model.[101] The idea was welcomed by Adolf Hitler, who hated the impersonality of railways and once declared that the best hours of his life were spent traveling by car.[102] German highways were designed to blend in with the existing terrain to preserve a harmonious landscape.[103] This initiative encountered resistance, too—at the very least because of its sponsor. Before the first section of the autobahn between Frankfurt and Darmstadt opened on May 19, 1935, activists belonging to the International Socialist League (Internationaler Sozialistischer Kampfbund) drew "Hitler = War" on the road; the prophetic pronouncement was quickly painted over before the führer and press corps arrived.[104] (In fact, equating Hitler's road and warfare was too limited; military experts had long entertained reservations about extensive road networks because they could facilitate foreign invasion.[105] Even worse, they might provide enemy aircraft with a convenient map for targets.[106])

Building highways and long-distance routes was not a German innovation. Moreover, the impulse to cross borders, noted previously regarding waterways, railroads, telegraphy, and electricity, was at work here, too. Almost as a matter of course, the writer and philosopher of technology Eugen Diesel observed in 1930, states found it necessary to "adapt their organization to supranational conditions" with "patent bureaus, scheduling conferences, [and] myriad positions attached to foreign offices for handling shipping, technology, the press, and so on";[107] the United States was making the strongest push for standardization. But despite de facto internationalism, Diesel continued, war threatened to break out at any moment across the globe.[108] Military strategists were absolutely right about national security: convergent infrastructures and standards had created new dependencies that could be beneficial in peace but devastating in the event of conflict. The transformation of a civilian infrastructure into a military one would occur soon enough—and yet again—during World War II.

On the whole, however, an optimistic vision prevailed during the interwar period: infrastructures would bring Europeans closer together. Travel and tourism, many expected, would allow people of different nationalities to meet in person and experience what they shared. The spiritual heirs of nineteenth-century Saint-Simonianism were especially enthusiastic about the prospect of a unified Europe. In their eyes, political fragmentation and isolated nation-states had been the reason for past crises; the free circulation of people, goods, and ideas was an ideal to strive for, in the name of peace.[109]

The late 1920s and early 1930s witnessed plans to link regions and states with a network of roads (as was already partly the case in North America). Francis Delaisi and André Citroën in France, Jan Romein in the Netherlands, Piero Puricelli in Italy, Richard Coudenhove-Kalergi in Austria, and Kurt Kaftan in Germany competed to devise the best model, firm in the belief that technological progress offered the best chance for political and cultural rapprochement. Albert Thomas, director of the International Labor Organization in Geneva, even proposed a system of public works spanning the continent in the hope of establishing an infrastructural framework corresponding to the French foreign minister Aristide Briand's project for a politically united Europe.[110] Joint projects for electricity supply, broadcasting, and railroads were also considered.[111] Other voices called for Europeans to come together to develop Africa, an undertaking that would both spread "civilization" and create jobs.[112] In 1913, the writer Bernhard Kellermann published a novel featuring a massive tunnel for transportation, postal service, energy, and water between the United States and Europe; as the book declared, "Work is the religion of our time!"[113]

Such schemes of technocratic internationalism might seem to lie worlds apart from realpolitik, but political organization lagging far behind connections in trade and infrastructures led to the founding of the League of

Nations in 1920. Countless conferences and innumerable committees in Geneva were devoted to programs that would finally be implemented—after 1945—in the interim, Germany attempted to unite Europe by force.[114]

Switzerland—the center of so much commerce, travel, and planning—stood at the sidelines as the new European order emerged. National independence became its brand. Travel here still offers a form of time travel. High prices have saved this picture-perfect destination from being flattened by the standardized and motorized mass tourism that set in after World War II. Its idyllic charms are out of reach for most people.

ADMISSION AND EXCLUSION

Travel is an education—although the lesson isn't always pleasant. In 1893, Mohandas Karamchand Gandhi, who would later be honored with the title Mahatma, or "great soul," bought a first-class train ticket in the South African city of Durban to travel to Johannesburg. The first half of the journey went smoothly, but in Pietermaritzburg, personnel informed him he had to leave the carriage: Indians were not allowed in first class. The young lawyer never forgot the experience, which played a key role in his dawning political consciousness.

The incident connects South African apartheid and Gandhi's campaign, later in life, against the railroad system in India, one of the essential instruments of British colonialism.[115] Moreover, it offers a parallel to Rosa Parks's 1955 refusal, in the United States, to give up a seat assigned to whites on an Alabama bus. When Parks sought equal access to infrastructural services, police took her away on charges of disturbing the peace. Subsequently, actions organized by Martin Luther King Jr. and others—in particular, the boycott of public transportation in Montgomery—proved decisive for the US civil rights movement.[116]

Not just automobiles but also public transportation services are caught up in matters of social distinction. After all, different classes of travel correspond to different levels of privilege. Most often, price is the criterion. When colonial or racial factors enter the equation, imbalances become more flagrant. Time and again, Gandhi came back to the railways, which symbolized asymmetries at every level of Indian society;

degrading conditions in this sphere harmed the health and morale of the populace as a whole. But just two months after India gained independence in 1947, Gandhi was no longer finding fault with the railroad system itself. Instead, he voiced concern about the way his fellow citizens were treating it:

> The passengers should consider the railways as their own property. They should keep the trains clean. They should not spit and smoke in the trains and should not pull the chain without real need. And not a single passenger should travel without ticket. Then I would be able to say that we have attained true independence.[117]

This appeal to public spirit likely did not help, for Indian society succumbed to greater discord than ever. Before long, Gandhi was assassinated. His ashes traveled by train from Delhi to his ashram at Allahabad—third class. The symbolism was so great that third class was called Gandhi class for a while.

Still today, the kinds of difficulties experienced by Gandhi and Parks occur regularly, whenever arrangements that are supposed to be communal fuel social exclusion. A sense of belonging is predicated on the idea that there are others who do not belong. In the course of the nineteenth century, inhabitants of nation-states nominally came to consider themselves equals. In fact, however, the dialectic of admission and exclusion spread as tension flared between high-minded (or blasé) cosmopolitanism and aggressive nationalism, affecting the most basic aspects of everyday life. Infrastructural development stood at issue, too: the goal of social integration was at odds with that of transcending borders.

Technical networks came to be considered parts of a supposed national organism. Imagery put into broad circulation in the nineteenth century culminated, in the twentieth, in ubiquitous metaphors describing transportation, communications, supply lines, and waste removal in terms of bodily processes, especially the movement of blood and the signals transmitted by nerves. The administration of modern welfare states was couched in biological language (as had been the case for municipal activities, earlier in history). To remain healthy, countries need specialist care—just as patients need doctors. Not everyone was convinced. Thus, in 1937, the German engineer Eduard A. Pfeiffer wrote that the general state of infrastructures resembles "a body affected by internal and external growths, hypertrophic parts, deficiencies, and the like (a little hypochondria may also be in evidence) but still working efficiently, on the whole."[118] Pfeiffer preferred to think of communities as operational units. He and other technocratically minded observers thought it more sensible to speak of a well-oiled social machine than to belabor organic analogies.[119] But whatever the language employed, everyone agreed that infrastructures warranted concern because they could convey undesirable and dangerous elements. It was claimed, for instance, that air traffic had caused epidemics including plague, yellow fever, cholera, smallpox, and typhus to spread with particular rapidity.[120] More rigorous monitoring should be performed at borders, and infrastructural personnel needed closer supervision.

Between the world wars, many countries hailed infrastructures as instruments for forging a vibrant community; in some cases, this meant practices of segregation, as well. Undertakings along these lines were not necessarily carried out by the same persons or agencies, and motivations varied. It is striking, however, how much practices overlapped. In the United States, the private sector (especially railroads) expanded until the end of the nineteenth century. In turn, a new understanding of services and infrastructures emerged. The precedent had been set in 1876, when, in a landmark ruling, the Supreme Court granted a Chicago grain shipper the status of operating a "business affected with a public interest."[121] Transportation facilities, in particular, increasingly came to be items of general concern and to be regulated as such.[122] From the 1890s on, *public utilities* offered a catchphrase to opponents of private ownership—exemplarily, streetcars, electric and water utilities, and high-speed communication facilities.[123] The novelist Jack London, who had the reputation for being a dangerous radical (or even a bomb-throwing anarchist), struck a popular chord and rehabilitated his image by calling for services to be in public hands. Even the Republican mayor of San Francisco (London's hometown) agreed that municipal ownership was a long-standing tradition in the United States.[124]

As a legal concept in the United States, *public utilities* included expectations that people of all classes would receive vital services at a fair and reasonable price.[125] Likewise, in France, *service public* gained currency in the last third of the nineteenth century; the minister responsible, Charles de Freycinet, also sought to guarantee all citizens access to railway travel.[126] The administrative lawyer Otto Mayer introduced the term *öffentliche Anstalten* (public institutions) to German legal language, but it failed to gain wide usage.[127] The term *öffentliche Wohlfahrtspflege* (public welfare care) was preferred until Ernst Forsthoff introduced *Daseinsvorsorge* (provision for existence) at the end of the 1930s—a concept that renewed older notions of *Polizei* (meaning both "police" and "policy")—that is, both providing for basic well-being and ensuring security.[128] Forsthoff distinguished between the *controlled* (*beherrscht*) sphere of individual existence, which he thought was growing smaller and smaller, and the *effective* sphere, which was constantly expanding thanks to developments in communication, transport, supply, and disposal. Governmental administration of general welfare supported the goals of a total national republic (*Volksstaat*) in his professional estimation; because the state took care of people's needs, it could demand complete submission on their part.[129]

This nearly premodern compact of security and obedience, which presupposed a largely homogeneous population, represents a defining trait of fascist regimes during the first half of the twentieth century. However, governments operating by other political codes also claimed power over infrastructures, which the Vienna-born journalist Anton Zischka called the "lifelines of community" in an unpublished manuscript from 1944.[130] Because many services and facilities were still lacking, the Soviet Union embraced them as part of the socialist project to transform society from the ground up. Competition raged even as rivals and distant relatives in Europe and North America exchanged technical know-how.[131]

In the late 1920s, it seemed that investing in public work initiatives would provide the means to escape the massive unemployment caused by the global economic crisis. Building and maintaining infrastructures, which would take many hands to carry out, promised to promote popular solidarity along organic or socialist lines.[132] Newly elected president Franklin D. Roosevelt implemented his New Deal in 1933, and the federal government invested in public works projects on an unprecedented scale. A 1940 report on the initiative by the economist John Kenneth Galbraith (which Roosevelt commissioned) proudly declared, "In the nature of the case, public works expenditure means building for the future, and, to some extent, we have come to gauge the far-sightedness or idealism of States by the extent to which they make these permanent and visible contributions to future welfare."[133] The Public Works Administration had fulfilled its task in almost ideal manner by "combining idle men and idle resources for useful public construction."[134] Model projects of the New Deal were the taming of the Tennessee River and the construction of the Hoover and Grand Coulee Dams.[135] Not only did they renew the idea of internal improvements undertaken in the 1830s; they also represented the realization of modern ideals: creating jobs for people irrespective of their origins and background, domesticating nature, and promoting development in agrarian regions seen as behind the times. These large-scale undertakings generating environmentally friendly electricity provided additional arguments in their favor. (Hollywood gave such projects narrative form and brought out the human drama in films such as Elia Kazan's *Wild River* [1960], most of which revolved around conflicts between settled ways and modern enterprise.)

Italy and Germany undertook similar large-scale projects. To a lesser extent, France and England did so too, but these countries faced the twofold task of developing territories in holdings abroad while setting up basic infrastructures at home; resources had proved scant since World War I.[136] Much was left in private hands, but efforts were made to bring infrastructures under centralized, administrative control. In 1939, Britain became the first country to nationalize the electricity supply in full.[137]

For all the similarities, concrete goals and how projects were ideologically justified had important differences. The 1940 Galbraith report affirmed that investments in infrastructure should serve only peaceful goals. This was undoubtedly true at the time, but changes were not long in coming. In 1943, Oak Ridge became the site of a research facility because of the abundant electricity readily obtained from the Tennessee River. Before long, the National Laboratory was conducting pioneering work to develop the first atomic bomb. From the outset, National Socialist Germany had aimed to facilitate military expansion, and infrastructural projects were initiated—in particular, the founding of the Reich Labor Service and highway construction. The state demanded that private citizens and energy providers be connected to the same networks. In addition to an obvious symbolic charge, this measure had legal implications because infrastructures now served to boost political compliance. In 1937, Jakob Nagel, state secretary in the Reichspost Ministry, summed up the vision: "The Deutsche Reichspost is not a commercial enterprise. Its work is not governed by considerations of profitability. It serves the people as a whole! Its purpose is to promote the cultural tasks and needs of the state and the national community, not to make money."[138]

This goal of the common good dressed up in populist rhetoric promoted social integration by consolidating and intensifying processes of exclusion.[139] Anti-Semitism and racism had been mounting since the end of the nineteenth century. Some restaurants, hotels, and clubs boasted of being Jew-free. Since 1895, for example, the Kölner Hof in Frankfurt's railroad district had advertised to "Aryan" guests with a sign at the entrance that read "Jüdischer Besuch verboten" (Jewish visitors not allowed).[140] Various bathing facilities on the North Sea and Baltic coasts also made a point of keeping away Jews and people of color.[141] As with South African petty apartheid, the aim was to prevent people from meeting and interacting with members of other nationalities. Parallels include Jim Crow laws in the United States. The open disenfranchisement and persecution of Jews in Germany began in 1933.[142] In almost no time at all, the lengthy process of emancipation was reversed. Measures differed from policies of separate but equal in that they never stopped drawing lines of division: German Jews were first denied the right to vote, then step by step deprived of everything else—including freedom of movement and the right to work or own property.

The exclusion of foreign elements found its most conspicuous expression in prohibitions to use parks, benches, bathing facilities, public transportation, restrooms, and hospitals. Affected parties were stripped of the safety net of health insurance. Undesirable groups were also not allowed to own radios; not coincidentally, these devices were called *Volksempfänger* (popular receivers—that is, machines intended "for the people").[143] Territories under German occupation implemented similar policies. In Paris, for instance, persecution began in 1940 with a directive that Jewish metro riders occupy only the last car of the train (contemptuously known as the synagogue).[144] As access to networks supplying basic services became increasingly precarious, it proved harder and harder to earn a living. Meanwhile, in the Netherlands, even a clogged toilet could pose an existential threat to people in hiding, like Anne Frank and her family.[145]

By 1935, all Jewish employees of the Deutsche Reichsbahn had been dismissed. From this point on, special trains to the border of the Reich and seaports accelerated Jewish emigration. Measures were enacted to allow Jews to use public transportation only with a special police permit. Such practices made official policy what had been cruelly joked about at the turn of the century, when anti-Semitic seals for letters were sold and fake fourth-class train tickets were marked "Jerusalem. One way."[146] Nazi propaganda celebrated discriminatory measures by justifying them. Thus, Veit Harlan's *Jud Süss* (1940) showed the titular villain imposing customs duties on the roads and bridges of the principality of Württemberg. The message was clear: Jews had long abused institutions to enrich themselves, to the detriment of ordinary people. As World War II unfolded, the Reich Security Head Office Referat IV B4—where, among others, Adolf Eichmann and transport officer Franz Novak were employed—organized the transfer of European Jews to concentration and extermination camps. The model was taken from the system developed for conveying livestock to slaughterhouses. (Tellingly, the practice went against the logic of the war effort by tying up resources vital for operations.) Travel to Auschwitz and other camps really was a one-way ticket, sometimes with a brief, exploitative detour—for instance, the I. G. Farben factories, where, in a cruel irony, artificial rubber and gasoline were produced to keep automotive traffic running. Victims of the regime were made to compensate for the sluggishness of the rails in spheres unrelated to deportation and extermination. The image of the track leading to the gates of Auschwitz has become an icon of twentieth-century horror.

The efficiency of the Deutsche Reichsbahn provides an extreme example of the misuse of infrastructure systems; a purely functional orientation readily serves inhuman purposes.[147] Large-scale, impersonal, and anonymous organizations make possible the crimes of "desk perpetrators."[148] Despite this misuse, other developments during the Third Reich command notice: the beginnings of mass tourism and the regime providing citizens with automobiles (*Volkswagen*), radios (the aforementioned *Volksempfänger*), and other standardized goods. In particular, the autobahn system represents a mixed infrastructural achievement. On the one hand, the construction of highways can be understood to have provided a functional matrix for community on the national scale. On the other hand, these roadways stood at odds with the military logic that otherwise prevailed, because they were not really suitable for wartime use. Indeed, the tourism that flourished in Nazi Germany was meant to create new socio-spatial relationships and thus, ultimately, to dissolve local and family ties.[149] Paradoxically, autobahns became a specifically German institution symbolizing the freedom of postwar society because they are open to all and, at least in principle, impose no limits on movement.

3

THE MEASURE OF MODERNITY: THE LATE TWENTIETH CENTURY

LOST FEELINGS

In 1987, a senior member of the Kaufmann family in Germany's Ruhr valley reflected on purchasing a refrigerator thirty years earlier:

> When I can buy something as nice as that, it makes me happy, even today. You'd always walk past the refrigerator, brush over it with your hand and open it up. It was shiny. There was butter inside, and sausage, still as fresh as the day before yesterday. You know, I think there are so many happy feelings when times were hard that are—lost, I'd say. . . . Nowadays, everything is just there, automatically. The television, refrigerator, washing machine, even the car—it's all just there.

In the middle of parental musings, the son interjected, "At least you don't pet the refrigerator anymore."[1] Scenes like this, with domestic appointments providing reference points for family life, were legion in the decades following World War II. Collective memory and personal memory are not the same. Whereas the former tends to be dominated by political events, the latter follows different rhythms. Individuals are more likely to remember moments in which the shape of everyday existence changed than what historians set down as decisive for peoples, nations, and humankind.

Devices and events that change perspectives and broaden horizons prove especially memorable: a car, a trip by boat or plane, or the first refrigerator, radio, or computer in the home—even if, with hindsight, the contraption was ridiculously heavy, bulky, and expensive. Looking back at earlier routines, habits, and domestic items inspires thrills of nostalgia. Objects make the flow of time emotionally tangible, particularly as we find ourselves surrounded by more and more gadgets that seem to have shorter and shorter lifespans. Material things offer reflections of history, and some of them can become external carriers of memory. The same rule applies to music, sounds, and sensations associated with formative periods of life.

Infrastructures themselves display this quality only to a limited extent. Who remembers when a certain road was paved or a new power line was installed—unless the work involved proved exceptionally inconvenient? Consumerism has thrived since 1945, with new and better products always coming onto the market. A fundamental change involving infrastructures, energy, and material resources took place after the war, and it was more fundamental than is commonly acknowledged. Environmental historians have spoken of a 1950s syndrome and even a new epoch.[2]

The nineteenth century witnessed the rise of heavy industry and smokestack-ridden landscapes—coal, iron, steel, and railroads forming Coketown clusters, a description that tips its hat to Charles Dickens's *Hard Times* (1854). Smoke and ash had discolored buildings made from red bricks so much, the novelist observed, that they bore black patterns "like the painted face of a savage."[3] Along similar lines, Detroit has been called Motown since at least the 1940s. This title might be applied to other cities, too. The new urban landscape emerged through forms of energy other than coal; oil, natural gas, and nuclear power fueled development. Assembly lines, electricity, automotive and air traffic, and chemicals and fertilizers set the tone.[4] Instead of stone and iron, plastic and concrete define modern times.[5] Much of what European societies hoped for between the wars—stability, predictability, recognition of individual achievement, and commensurate remuneration—has been made possible by Motown clusters. In this framework, governments invested money from steadily mounting taxes in infrastructural projects meant to be accessible to all. In fact, however, the opportunity to enjoy their benefits was a matter of purchasing power; thus, policies promoted and sought to distribute opportunities as evenly as possible, both socially and spatially. Planning and maintenance shifted to the public sector, and households concentrated on consumption. Accordingly, buying the first refrigerator, car, or television came to be highly emotional events attesting directly to improved quality of life.[6] Longing for creature comforts made people overlook the uniqueness of the moment and how rapidly horizons of information and modes of experience were expanding.[7]

Eastern European countries had the *kommunalka*, or shared living arrangements, where appliances or the first televisions were installed; elsewhere, community centers of villages had similar spaces.[8] Little by little, arrangements were adapted to the single-family household. The trend toward individualization was the same in western Europe but more pronounced: people stopped relying on public transportation as soon as they could afford a moped, motorcycle, or car. As the growth-oriented and self-compounding dynamism of coketown clusters transformed into motown clusters, everyday life and routines underwent further change. Motown clusters also followed the principle of mass production, accelerating the rise of a consumer society governed by circular logistics systems constantly renewed with minor adjustments.

"I'm always in a race with the junk yard!" Arthur Miller had his doomed salesman declare as early as 1949: "I just finished paying for the car and it's on its last legs. The refrigerator consumes belts like a goddam maniac. They time these things. They time them so when you finally paid for them, they're used up."[9] This mode of limited functioning, or planned obsolescence, has long been criticized for ecological reasons.[10] Consumer goods that could last a lifetime—or at least as long as they can be repaired—are instead made so they will wear out physically and seem to be out of date well before then. Products are behind the times as soon as new models come onto the market. In 1954, the designer Brooks Stevens defined the business calculation as a matter of rousing "the consumer's desire to own something a little newer, a little better, a little sooner than is necessary."[11] This strategy entailed a specific relationship to garbage, to which we will return later.

In the 1950s, business sought to stimulate, control, and streamline the needs and wishes of buyers. Policy makers, however, embraced the idea of making resources broadly available to promote equitable social relations. Both factors combined to make a high-energy society.[12] Government, suppliers, manufacturers, and consumer-citizens formed a "coalition of wasters."[13] After all, it seemed there was enough to go around. Petroleum, which was brought to Europe via ships and pipelines, provides a case in point. It was sold at a price that undercut not only wood, coal, gas, and kerosene but even nuclear energy (which was just starting to be produced); flooding the market led to increased demand.[14] Consumerism met with a particularly favorable constellation of factors, too. War-weary people were focused on personal advancement, and the growing supply of consumer goods and resources like petroleum helped them realize their goals. High levels of consumption served as a means for capitalist societies to mark their superiority to the rival system in the east.

Before long, the refrigerator formed the centerpiece of the household and domestic life.[15] Once it was possible to keep a stock of food at hand, eating habits changed. Family members could help themselves as they wished, and shared meals no longer determined the rhythm of the day.[16] And because the refrigerator could store large quantities of food, it made sense to transport large purchases with the car. Thanks to the constantly expanding logistics of consumer goods—epitomized, since midcentury, by newly invented shipping containers—all kinds of food became available year round.[17] Until this point, the cycles of domestic agriculture or what grew in private gardens had determined what appeared on the table. In isolation, household appliances simplified, accelerated, and streamlined domestic activities. In the aggregate, they forged new frames of reference, networks, and challenges. Not only did such amenities presuppose global supply chains of commodities; they also relied on water, sewage, electricity, transport, and communication systems, all of which contributed to further growth of the market.[18] The demands of ever-increasing consumption rose to the point that they continued even after environmental concerns came to dominate public debate in the 1970s. Now, old devices were replaced by designs that were supposed to be more efficient—which entailed new cycles of commerce. As a rule, the amount of energy required for operation was taken into consideration, but not the energy required for manufacture.

Sometime between the late 1940s and the early 1970s, depending on the country and political system, modern consumerist mentality formed.[19] Popular outlooks were based on the postwar boom that occurred in Western (and, for the most part, Eastern) economies, state-run social welfare initiatives, and supply-side economics. Indeed, the notion of infrastructure might be described as a philosophy or set of basic assumptions.

The word slowly made its way into common parlance, expressing the dominant theory that societies would modernize as a matter of course under the right conditions.[20] Hereby, *infrastructure* became synonymous with all that was lacking in supposedly backward regions.[21] Nor did this assessment concern just overdue colonial efforts or international development plans (an effort often dated to the inaugural speech delivered by the US president Harry S. Truman in 1949). Planning and infrastructure were deemed the way to elevate and standardize life throughout the nation-state.

Ideological rivalry fueled the process. Spurred by the Marshall Plan and the unified zone of economic exchange that began to emerge in the mid-1950s, western Europe witnessed the rise of social welfare policies; public ceremonies (usually performed by politicians) opened roads, sewage treatment plants, city halls, sports centers, cultural institutions, and recreational facilities. Such events signaled that there was plenty to distribute. Competition for urban refinement and cosmopolitan distinction thrived as municipalities looked sidelong at neighbors who might

3.1 Politicians dreamed of opening ceremonies such as this one for a bridge and new stretch of highway (Neuenburg, Baden-Wurttemberg, 1959). Photograph by Willy Pragher. *Source*: Staatsarchiv Freiburg.

boast even better organization. The common good, the legitimation of state authority, economic advancement, and provision for the needs and wants of life took material form in infrastructure and associated policies. These priorities complemented the consumerist mindset almost perfectly.

Science, business, and politics eagerly adopted the language of infrastructure and the supposedly neutral standpoint it represented.[22] Given the consensus that infrastructures organized space and society, projects were developed to provide for population groups as equitably as possible, especially those considered to be creatively and entrepreneurially engaged. The goal, the German federal minister of the interior Hans-Dietrich Genscher declared in 1970, was "to make life more pleasant for citizens, to increase their freedom, and to improve chances vis-à-vis competitors in other communities or regions."[23]

Along these lines, a village in the Westerwald region that participated in a nationwide contest in 1971—*Unser Dorf soll schöner werden* (Let's make our village more beautiful)—proudly announced,

> Central water supply and full sewerage have been in place for a long time now, and a sewage treatment plant with five other communities is in the planning stage; dust-free garbage collection has been taken care of. Our well-managed kindergarten is to become a focal point for children from neighboring municipalities as well. In addition to the sports field, a new gymnasium is under construction, which, with adjoining multipurpose rooms, will also serve as a village community center. The renovation of the outdoor swimming pool is in the works.[24]

In those years, when tax revenues were abundant, even small communities could join the race for a higher living standard. Most politicians in the postwar decades shared the belief that quality of life depended on generous provisions for infrastructural development. High-ranking authorities—for instance, the European Economic Community, or the European Union after 1994—supported initiatives, granting rural areas in particular seemingly inexhaustible funding. In this way, agriculture became largely mechanized, changing country life to an unprecedented extent.[25]

Military facilities also played a role in connecting remote areas to infrastructures.[26] Decades earlier, railroads and tramways had turned habitations outside city limits into suburbs.[27] In turn, automobiles accelerated and compounded processes of urbanization and suburbanization.[28] Mailboxes at the side of roads, drive-in movie theaters and restaurants, and shopping centers at the edge of town are so many icons of the age of individual mobility.[29] Without cease, postwar reconstruction inaugurated projects for expanding and modernizing roads, public transport, water and sewage systems, telephone lines, radio and television towers, and even seaports and airports. Planners and politicians focused on creating lasting and reliable conditions for life and work. If the state ensured a robust infrastructure, thriving communities would follow. Ernst Forsthoff, who had coined the term *Daseinsvorsorge* under different political auspices, welcomed the stabilizing effect of governmental administration. Eventually, however, he feared that it would entail an unbridled sense of entitlement.[30] He had a point: expectations developed during the 1950s have lasted to this very day.

In the United States, such ideas were criticized relatively soon. Because the economy here had taken off in the 1940s, the first signs of wear and tear appeared earlier than elsewhere, in both material and intellectual terms. The machinery set in place to fuel urban and rural modernization encountered resistance—as did the plans of Robert Moses, who, on the model of Georges Eugène Haussmann's transformation of Paris in the nineteenth century, had overseen the redesign of public spaces in New York since the 1930s. Protests greeted his efforts to reshape the densely populated borough of Bronx by introducing freeways in the 1950s. High-modern infrastructural planning and the motown cluster—both of which Moses tried to implement directly—had hit a limit.

For the next generation, the same public official who had justified infrastructural expansion in the name of providing a better future represented the embodiment of technocratic soullessness.[31] With 150,000 apartments, 600 playgrounds, countless parks, zoos, and swimming pools to his credit, in addition to streets, parkways, tunnels, bridges, power plants, world exhibitions, and the United Nations headquarters—Moses, in his seventies, was blamed for destroying neighborhoods, communities that had evolved over generations, and historic sites and causing ecological harm. In the 1960s, the architectural critic Jane Jacobs mounted an initiative against plans for building a road through Greenwich Village.[32] Moses had heard the arguments before, but in his eyes they were outweighed by the advantages of public facilities and services. Considered by many as the "man who killed New York," he was accused of being undemocratic and even racist.

In fact, his vision fell in line with a long-dominant code. As the sociologist Bernward Joerges puts it, "A reformist, liberal-paternalist elite on the supply side" would attend to "upwardly mobile middle classes on the demand side"—a group whose ambitions were epitomized by "new automobiles, the materialization of the American dream of mass production, mass consumption, and mass culture."[33]

After his ouster, Moses published his memoirs, *Public Works: A Dangerous Trade*; the title expresses regret that his worldview had fallen behind the times. Less prominent city council members and planners had similar experiences.[34] From the 1960s and 1970s on, the high-modern ideal of infrastructure has proved increasingly tenuous. "America in ruins" has become a popular political refrain. Even as budgets shrink, calls are made to invest in roads, bridges, and other public resources.[35]

Demands to improve the economic basis of social life have defined infrastructure from the start. Thus, a series of articles published by *Der Spiegel* in 1961 struck a polemical tone by calling Germany an underdeveloped country" and noting that the federal government had spent some 5 billion German marks on foreign development. Meanwhile, authors contended, the approximately 1,000 marks that each West German was paying per year to promote civilization" among "colored tribes in deserts and jungles" were needed at home. "The infrastructure of the Federal Republic" had plenty of "dangerous shortcomings"; equipment and conditions in hospitals and schools needed improvement, as did the "varicose vein network of federal German traffic routes."[36] The series concluded by calling for far-reaching reforms of the financial and tax systems. Racism notwithstanding, the view entered the mainstream; the government, whether local or federal, is regularly asked to do more (i.e., spend more money) to maintain and expand infrastructures domestically before funding projects elsewhere.

Equitable, middle-class societies developed—at least to some extent—in western Europe as a result of policies of *Daseinsvorsorge*. But even if services and markets stood open to all, the potential for social conflict persisted, now fueled by the individual quest for distinction. Moreover, a busy and mobile society could place excessive strain on networks and undermine the advantages they offered.[37] Two years after the article series on infrastructure, *Der Spiegel* featured a cover story on traffic on German roadways. From one day to the next, friction was mounting between inexpensive Goggomobils, which almost anyone could afford, and the stately vehicles produced by Mercedes-Benz; worse still, some motorists were addicted to speed, whereas Sunday drivers rolled along at a leisurely pace. "The aristocratic German autobahn," which had been "built for 200-kilometer-per-hour racers," was being compromised by traffic jams that made everyone equal, and equally miserable.[38] Remedial measures proposed by planners included desynchronizing work schedules, parceling out commuter flows over the day, and vacation periods that differed from one federal state to the next. New roadways were built with so-called crawl lanes (*Kriechspuren*) for slower traffic and bypasses to circumvent regular sites of congestion. Despite these efforts, the environmental activist Horst Stern's dictum was regularly confirmed: "Sow streets and reap jams" *(Wer Straßen sät, wird Verkehr ernten)*.[39]

A similar dialectic governed consumer goods. Simply buying commodities did not always make life easier or more convenient. Indeed, increasingly complicated technology and associated logistics could lead to more trouble than it was worth. The 2009 novel by the French engineer Alain Monnier, *Givrée*,[40] draws on everyday experiences to tell of the pitfalls of technology and modern communication. Marie, an ordinary Frenchwoman, has purchased a new refrigerator. The thermostat turns out to be defective, so she wanders through a jungle of hotlines, only to get lost in the inferno of customer service. At long last, she receives the message that the spare part needs to be shipped from Indonesia—which in turn unleashes a cascade of side effects and chain reactions. Assorted witnesses to Marie's travails step in to help, but misinformation only spreads. In the end, no fewer than seventeen replacement refrigerators wind up landing in her kitchen.

CONTROLLED COLLECTIVES

Stafford Beer liked to quote an observation made by the British prime minister Harold Macmillan, to the effect that "estimating a country's economic performance with statistics is a bit like trying to catch a train with last year's timetable." A business economist by training, Beer was one of the key figures of operations research—a field inaugurated decades earlier, in the 1930s, when the British military sought to optimize decisions by collecting a maximum amount of information and feeding it into mainframe computers. After World War II, the private sector embraced mechanized information processing, and in the 1960s the practice was pursued in a political context.[41] Beer was invited to Chile in the early 1970s to assist Salvador Allende in building a socialist society. Along with the Chilean engineer Fernando Flores, he received the commission of setting up a system of cybernetic synergy—Cybersyn—to steer the economy.

Cybernetics seemed the way of the future in the 1960s, when the race between political systems was still undecided. Given that domestic and international connections would only expand, and in light of increasingly advanced interfaces between human and machine, it was the order of the day to develop mechanisms to control both as efficiently as possible and automate important processes of labor, communication, and exchange.[42] For Allende, Cybersyn promised to coordinate Chile's economy, which had been largely nationalized. On November 12, 1971, he met with Beer and Flores, who proposed linking the country's most important companies by telegraph. Incoming data would be processed by an IBM 360/50 computer. As soon as a key index—productivity, energy consumption, or manpower—deviated from the norm, headquarters in Santiago would be notified so appropriate measures could be taken. On March 21, 1972, the government computer began processing up to two thousand telex messages a day from four hundred plants across the country. In October, transport operators went on strike, and Chile's logistical system threatened to collapse. Thanks to data gathered, officials managed to mitigate negative consequences for the population. In Britain, the enthusiastic headline of the *Observer* read "Chile Run by a Computer." The next year, in September, the computing center was even housed in the presidential palace, La Moneda. But just three days after the move to the palace, the military staged a coup, and the experiment in high-tech populist rule came to an end.[43] Ever since, Allende has been a mythical figure among socialists for having returned the US-dominated transport and energy sector to the people, only to be undone by US interference.

For a spell in the 1960s and early 1970s, socialist statecraft seemed to represent the fast lane of global history. The Eastern Bloc seemed firmly established, and sympathy for socialism was growing. Many former colonies were apparently following the course charted by the Soviet Union. The Sputnik satellite symbolized the triumph of science and technology, auguring well for the future of the Second World. Meanwhile, the West was looking for ways to move beyond Cold War confrontation to cooperation. In 1966, President Lyndon B. Johnson proposed stronger connections between West and East. Ideological differences notwithstanding, the two sides should work together on global problems. Meteorological institutes, earthquake research facilities, international air companies, and news agencies had long been pushing for worldwide computer networks.[44] Finally, in July 1977, it seemed the time had come. The International Institute for Applied Systems Analysis in Laxenburg, near Vienna, conducted the first data exchange among Austria, Poland, the Soviet Union, and the United States. The experiment lasted for all of three weeks. The communication lines of the Eastern Bloc, which made use of the telephone network already in place, simply were insufficient.[45]

From today's perspective, the outcome is hardly surprising. Although Lenin had called for communist lands to be thoroughly electrified, the discrepancy between ideal and reality remained enormous. For want of adequate intelligence, the West had long focused on the impressive achievements of Eastern Bloc countries in the realm of industrial production, which was animated by so-called barrel ideology, or mass production. A clear view was further distorted by Soviet propaganda, which celebrated the pioneering role of socialism in world history.

Infrastructure and socialism should have fused seamlessly.[46] After all, the point was to equalize access to supply, disposal, transport, and communication, which all belonged by right to the people. In practice, however, the Soviet Union had huge gaps between theory and practice from the outset. A vanguard claimed to lead the process of revolutionary transformation, which meant communication went from the top down. Soviet leadership would remain suspicious about information circulating horizontally until Gorbachev's belated glasnost campaign. Under such conditions, social and technological innovations could not spread freely or be implemented effectively. In contrast, the Soviet Union and other socialist states concentrated on heavy industry. This policy led to rapid success along the lines of the first and, in part, second Industrial Revolutions. However, progress toward a socioeconomic order based on knowledge and information was sluggish, and steps toward a service economy were even more halting. Political leaders preferred large-scale infrastructural planning and showcase projects; eventually, the latter represented weak points in the system because they had not been designed to be self-regulating and still were governed by central planning.[47] Overly strict control of science and technology created disadvantages vis-à-vis the open and competitive societies of the West.

The unequal development of telephone networks (which prompted cooperative efforts with the West to fail in 1977) made this fact painfully clear. Under state-socialist regimes, telephones were intended to serve the chain of command more than to facilitate open communication.[48] Over time, the area covered by networks shrank, and the quality of transmission decreased. In the West, the medium had come to serve free exchange and spontaneous socialization. For populations in the East, in contrast, telecommunications made it more obvious than ever that socialism was not speeding ahead. Until the 1980s, the rate of expansion was miserably low, increasing somewhere between 5 and 20 percent a year, depending on the country and region. International calls were almost impossible without a long wait and operator assistance. Those who had the good fortune to own one of the few telephones available were regularly importuned by acquaintances. In a further drawback, secret services monitored lines. In dissident circles, it went without saying that using the phone was risky.[49] Given the dearth of connections, it frequently proved impossible to make an appointment by phone; people commonly dropped by for visits unannounced. Many apartments had pens and pads of paper on the door; if the occupants were not at home, one left a written message.[50] The general quality of transmission was poor, because copper was lacking and the iron substituted for it did not hold up to the effects of weather. Repairmen always had plenty to do.

Even if one managed to get a line at home, obtaining personal numbers was no small matter. Information was available at kiosks, but many didn't bother to seek it out because transmitting messages in person, by mail, or through trusted contacts seemed safer and more reliable.[51] In 1989, thousands of communities in the German Democratic Republic (GDR) still had no public telephones. Where they did exist, they were often vandalized. A good quarter of public telephones were out of service at any given time. Needless to say, the same conditions did not affect the army, police, security apparatus, party, military forces, or state-run companies, which had special networks of their own.[52] One suspects that authorities delayed the expansion of dynamic infrastructures like the telephone system to maintain control. The same holds for roadways, which deteriorated in quality the farther east one went. Streets in the Soviet Union, which often remained unpaved and therefore vulnerable to bad weather, were notorious for their potholes; travelers spoke of driving "on square wheels."[53] State socialism never fully knew automotive circulation and all that it implies—mobility, communication, and broadened horizons.

Projects were also undertaken with an eye to little but public prestige. Thus, despite having to make a significant detour, the principal highway in Bulgaria passed near the small town of Pravets, where the head of state Todor Zhivkov had been born, to mark a site of national pride.[54] Albania built numerous roads where there was no significant traffic at all.[55] Yugoslavia went to great lengths to create its Brotherhood and Unity Highway. This *autoput*, an important route for guest workers traveling between Munich and Istanbul, came to be known as a death trap in the 1970s. *Der Spiegel* surmised that the different kinds of vehicles caused all the accidents: generally, Western cars were superior to Eastern ones, which prompted their drivers to become petty tyrants on the "nightmare road of the continent."[56] As far away as Switzerland they came to be known as Balkan speeders.[57]

Vehicle production in socialist countries initially lagged behind roadway construction. Although the narrow range of different types of vehicles offered limited opportunities for distinction, the car became one of the most sought-after luxury goods and a primary incentive for achievement. Over time, the situation reversed itself; as more and more people owned cars, roadbuilding did not proceed apace.[58] Automobiles remained largely the prerogative of men, especially older ones. Military service provided the chance to obtain a driver's license, but waiting times for a new vehicle were notoriously long.

Central committees in Eastern Europe recognized that mass motorization entailed centrifugal social movement, evident in everyday life thanks to many smaller developments. Soviet truck drivers, for example, liked to turn professional mobility into personal freedom: their fondness for driving around all day to no particular end quickly became a matter of popular lore.[59] Even though the GDR in the late 1950s had plans for a dense highway network, actual measures for improving—and regulating—social and economic activity experienced delays again and again. The state prioritized the progressive expansion of inner-German border fortifications, then followed up with an ambitious housing construction program under General Secretary Erich Honecker.[60] Personal mobility remained limited; in historian Axel Dossmann's phrase, people got around "with the handbrake on." And to the extent that East Germans went on vacation, organs of governmental control sought to prevent them from making contact with people from abroad. Parking lots and rest stops on motorways posed particular challenges for the police and secret services. All the same, party leaders managed to get the Federal Republic—that is, West Germany—to pay for maintaining routes (in addition to providing many things in short supply).

In Eastern European countries, people continued to rely on railroads as their primary means of transport. The Soviet Union even witnessed the renewal of large-scale projects first conceived in the nineteenth century. In 1974, for example, Leonid Brezhnev pushed forward with constructing the Baikal-Amur line, which ran parallel to the Trans-Siberian Railway. That said, efforts to mobilize Soviet youth in a manner comparable to the 1920s and 1930s, when there had been great excitement for massive undertakings, did not pan out.[61] Some of the most peculiar episodes of infrastructural development occurred in Germany. By the time the Berlin Wall was erected on August 13, 1961, numerous measures had already been taken to interrupt travel; the border between the two halves of the country ran for 1,378 kilometers—the exact length of the highways the GDR had inherited from the National Socialist regime. Everything that passed between West and East—whether human beings, consumer goods, or information—was subject to strict control. Since the Berlin blockade (June 1948 to May 1949), the divided city's infrastructure had also been the object of gamesmanship.[62] The legendary airlift during the blockade showed that disruptions of circulation and communication could be remedied with enough political resolve.

The expansion of the border within Germany produced an elaborate infrastructure. What came to be considered one of the best-secured frontiers of the twentieth century—and a monument to the inhumanity of rival political orders—was, of course, never as impenetrable as either side claimed.[63] For decades trade proceeded, with movement back and forth that never attracted much notice.[64] Eastern and Western Germans exchanged everything, including garbage and blood reserves.[65] Despite the barrier, people traveled and even fled, and spying flourished. In retrospect, the inner-German border is an example of consistent and brutal but ultimately ineffectual efforts to hold the world apart. Both states were reasonably modernized, so it proved all but impossible to block flows from one side to the other. Willy Brandt's Neue Ostpolitik, or New Eastern Policy, sought to open the border progressively by measures including the Pass and Transit Agreements, encouraging intra-German trade, and generally easing human relations. Communication between West and East was also promoted by means of radio and television. The GDR undermined such efforts by jamming transmitters, introducing new antenna systems for residential buildings, and making it a crime to listen to enemy stations. But lasting success proved impossible. From the early 1970s at the latest, Western television was more or less tolerated.

The 1975 Conference on Security and Cooperation in Europe, held in Helsinki, affirmed the international right to exchange information. For many citizens of the Eastern Bloc, the agreement represented an opportunity to break away from opinions dictated by the state apparatus. Western media started reporting more about Eastern European countries, where news circulated in dissident circles. In a parallel development, secret services assumed almost grotesque dimensions on both sides. For instance, in 1989, a single department of the GDR Ministry for State Security had the means to listen in on four thousand lines simultaneously; meanwhile, in the Federal Republic of Germany, almost a hundred thousand connections were subject to targeted surveillance (*Zielkontrolle*).[66]

Western intelligence services also eavesdropped on neighbors in the East. In the name of preventing the spread of communism, the mail and telephones of the West German population were monitored, although the practice relaxed in the 1970s as privacy and data security came to be acknowledged as rights.[67] Countries in the Eastern Bloc increased surveillance in direct proportion to the population's attention to Western movement, communications, and information. It didn't help that countries like Hungary and the GDR intended to close gaps relative to the West by focusing on production (for instance, the housing program spearheaded by Erich Honecker, mentioned earlier). The rigid planned economy and efforts to export goods to obtain foreign currency blocked thoroughgoing modernization. In many areas, development remained stuck at 1960 levels, and the lack of resources for maintenance became a constant source of complaint. Bottlenecks occurred especially in the infrastructure of infrastructure, that is, basic services and facilities indispensable for other operations. The sluggish expansion of transportation networks, service and repair stations, and industrial supply posed numerous problems.[68] As late as 1986, after four decades of state socialism, experts in the GDR boasted that 90 percent of citizens had access to a central supply of drinking water.[69]

Although the Eastern Bloc possessed abundant supplies of oil and gas, fuel consumption was excessive, partly because energy prices were subsidized. Temperatures were largely regulated by opening and closing windows, not by thermostat. Energy was practically the only area in which a surfeit existed, and overloaded networks, power outages, and a general culture of neglect were the norm. Shortcomings were particularly evident in areas without an ideological dimension; cemeteries, for example, were prone to fall into disrepair.[70] In the late-Soviet period, signs declaring "Ne rabotaet" (Out of order) were a feature of everyday life.[71] The jarring contrast between political slogans displayed on posters and signs and dilapidated infrastructural facilities came to represent the decay afflicting late socialism. From the late 1970s on, popular discontent found expression—albeit on a modest scale—in relation to environmental concerns. Here, in particular, state socialism was clearly failing to provide for the public good, notwithstanding ritual proclamations about representing the vanguard of progress. Finally, the nuclear accident at Chernobyl in 1986 and official efforts to reassure the populace signaled, once and for all, that safety measures had long been neglected. By the end of the decade, protests were occurring throughout the Eastern Bloc, most of which were sparked by infrastructural shortcoming or failure. Citizens demanded oversight independent of Soviet control and the right of local residents to have a say.

In 1989, development in the GDR stood 40 percent lower than that of West Germany.[72] Particularly when it came to microelectronics, the country had fallen far behind. Following reunification, infrastructural development exploded. Other Eastern European countries had to wait until they joined the European Union to experience a comparable boom. In the 1990s, the classic infrastructural ideal of the postwar decades, which had long been called into question in the West, experienced a renaissance in the East. Politicians such as Hans-Dietrich Genscher and Helmut Kohl, who promoted this turn, belonged to a generation that had grown up during the heyday of infrastructural expansion.[73] The lands of the former GDR, where technology installed in the early twentieth century was still in operation, received the most modern telecommunications network in Europe. New highways and roads were built as part of the German Unification Transportation Project (Verkehrsprojekt Deutsche Einheit). Like other initiatives, they proved to be outsized; planners failed to anticipate the scale of deindustrialization and migration that occurred. There were even proposals that eastern Germany be used for experiments diametrically opposed to the goals of state socialism—that is, for decongested, downsized, and decelerated living spaces exempt from pressure for constant growth.[74]

Erich Honecker, the former chairman of the State Council in the GDR, spent the last years of his life in Chile. As noted previously, during Salvador Allende's brief tenure as president, the country had acquired the reputation of being a socialist bridgehead at the heart of the ideological enemy's sphere of influence. A harsh turn occurred in 1973; now, Chile stood at the forefront of neoliberalism. (Stafford Beer went on to lead a relatively modest life between Wales and Toronto, presiding over the World Organization of Systems and Cybernetics; his advice was sought whenever left-wing governments in countries such as Mexico, Uruguay, and Venezuela tried—mostly unsuccessfully—to manage their societies more efficiently.)

POWER RESERVES

In 1962, the Iranian author Jalāl Āl-Ahmad published a book about a disease he called *Occidentosis*. The inciting incident for the work was the imminent White Revolution, whereby Reza Shah Pahlavi sought to westernize his country. The shah took his cue from the model offered by Mustafa Kemal Atatürk's modernization of Turkey, beginning in the 1920s. Āl-Ahmad was convinced that a kind of Euromania had seized Iranian elites and poisoned them. In his eyes, Western mass culture and technology represented a Trojan horse that would unleash dire consequences: people would be uprooted, feel isolated, have mistrust, and become cynical, and the masses would experience wholesale proletarianization. Last but not least, he warned, changes would entail the emancipation of women, to the detriment of family life and traditions that had stood the test of time.[75]

By Āl-Ahmad's estimation, 90 percent of the Iranian population disapproved of the westernization project. As an antidote, he called for boycotting westernized schools and radio stations and enjoined his countrymen to refuse to pay taxes.[76] It is not difficult to see the spirit of the 1979 Iranian Revolution in this and the makings of Islamic fundamentalism at work. Āl-Ahmad took issue with what he considered underhanded offers to join the global economy and cosmopolitan society. (Proponents of westernization thought that Iran would quickly replicate the material development of Europe and the United States by connecting to their networks of exchange.) Infrastructure stood front and center—a boon for some, but a bane for others.

Europeans had justified nineteenth-century colonialism as a civilizing mission of religious inspiration. Even if it had that origin, it didn't take long for the hunger for the resources in overseas El Dorados to commandeer supposedly noble efforts. Imperialist expansion sought to maximize the value of distant lands by implementing measures to ensure growth, productivity, and progress in keeping with European priorities. The ready alibi was that territories abroad would make something of themselves. Refractory elements were regarded as unfortunates at the sidelines of advancement—when they were not consigned to extinction as members of an unproductive race. In other words, colonialism attempted to transfer European principles of achievement, the functional division of labor, and trade to as many areas of the world as possible. Infrastructure would tie everything—and everyone—together, ultimately to universal benefit, but certainly for the good of colonial powers.

Infrastructures—chiefly, medicine, weaponry, and transport—made forays into unknown parts of Latin America, Africa, Asia, and Oceania calculable risks.[77] Subsequent expansion relied on economic development, especially by means of ports, routes down waterways, railroads, and telegraphy.[78] In many accounts from a European perspective, railroads were directly equated with civilization. Investment on a massive scale sought to ensure that colonies would prove profitable. It is striking how often private entrepreneurs invoked infrastructural projects when requesting monies from colonial governments or administrators back in the "mother country."[79] Two factors undermined the legitimacy of colonialism. For one, the way projects were carried out contradicted the humanitarian rhetoric and claims of promoting freedom, equality, and fraternity. For another, promises made to colonized peoples that they would be granted more autonomy in exchange for participating in World Wars I and II were not honored.

When, finally, Europeans were forced to withdraw, they left behind systems of communication, transportation, supply, disposal, and administration—the dowry of the civilizing mission. In the 1950s, they initiated programs of development aid, especially with a view to ensuring basic nutrition and health care. The impetus was still the "struggle between progress and darkness, rationality and superstition, science and religion."[80] Infrastructures have remained the focus of development efforts to this day, because they are the area for which governmental institutions bear responsibility; in contrast, manufacture and trade are supposed to be matters of private initiative.

Poverty in the Third and Fourth Worlds continues to be defined in terms of the absence of minimum levels of water, food, shelter, and education. Not just humanitarian agencies but also international organizations have the goal of remedying the situation. A 1994 report, "Infrastructure for Development," by the World Bank described the overall state of affairs in gloomy terms but affirmed that infrastructure would offer the best way out; project management should be prudent, cost effective, and focused on users, not based on prestige or scale.[81] Still today, the key words in the organization's yearly updates are *equipment, construction, plant, port, road, irrigation, steel, highway, railway*, and *dam*.[82] All the same, some 2.6 billion people worldwide continue to cook exclusively with wood or dung; almost as many lack adequate sanitation; 1.3 billion have no electricity; and 0.8 billion remain without a reliable supply of drinking water.[83]

To be sure, open colonial ambitions to control spaces and resources have vanished from the language of international politics. But decolonization has raised the question, again and again, of whether the foundations for connecting non-European lands to the networks of the world economy amount to continued dependency. Infrastructures in particular seem to be a kind of storehouse of power. Not only are they long lived; they are predominantly geared toward export. In consequence, they perpetuate asymmetries between poor and rich countries, well after political autonomy has been achieved. Indeed, one factor in decolonization may well have been the realization that countries under the thumb of the Global North did not need to be formally ruled from outside for profits to be made. Overt domination is always a liability. In this regard, infrastructures represent anything but a neutral proposition. They are linked to calendars and schedules, legal and administrative structures, standards of information, supply, and trade and a mentality that prioritizes productivity.

At the turn of the twentieth century, the Bhadralok Debate among Bengali intellectuals addressed the appropriate stance to take vis-à-vis European economy, science, and technology. Articles that appeared in the *Dawn*, a journal published in Calcutta, laid out three paradigmatic positions.[84] The first expressed respect, if not awe, for European enlightenment and analytic science. Proponents called for their countrymen to overcome backwardness by adopting the Western model of success as fully as possible. Along these lines, policies of reform and modernization have indeed been undertaken across the globe, often following the example of the early Soviet Union, Kemalist Turkey, or the Iranian regime under the shah. The second position, overlapping in part with the first, advocated employing Western technology while preserving traditional values and local cultural practices. This middle road has been charted particularly well by the Japanese since the Meiji Restoration. The third position agreed with the diagnosis made by Jalāl Āl-Ahmad decades later: propositions from the West should be rejected outright because they create an alien and morally suspect world. This position is often associated with emphasis on spiritual knowledge of nature and medicine—a comprehensive worldview superior to the compartmentalized thinking of the West.

The ambivalent connection between young nations and other parts of the globe has given rise to internal tensions. In some countries a fundamentalist mood has prevailed, and others have adopted more open and accommodating approaches. Nor are Muslim societies alone in attempting, on the basis of religion, to resist the pull of materialism and its promises.

Among Christians, there are Anabaptist sects such as the Amish, Hutterites, and Mennonites; likewise, Orthodox Jews strive for a life beyond excessively worldly comforts. These communities shy away from innovations that bring people into contact with profane enticements (for instance, motorized vehicles and private telephones).[85] In Europe and the United States, life in isolated, rural conditions has come to represent an alternative to the mainstream. In the eyes of those who value things as they are now—and seem always to have been—modern hyperconnectedness brings nothing but alienation and anonymity, which threaten self-reliance.[86]

But such austerity was not really an option for countries that gained independence after 1945. Elites in young nation-states, in particular, wanted to make the most of the infrastructures, communications systems, and production sites left behind by colonial rulers—to nationalize them and employ them to their own ends.[87] Exploiting natural resources was especially important, being a ticket to integration into the world economy. Industrialized countries now depended on their raw materials: oil, metals, minerals, and certain foodstuffs. Yet adjusting to the rhythms and patterns of global trade proved difficult. Usually, governments lacking political and economic experience could not manage without assistance from foreign investors, entrepreneurs, aid organizations, or consultants. In turn, these groups vied jealously for advantage and influence.

The resulting constellations often failed to meet criteria of efficiency or economic viability because momentary interests and political configurations prevailed. Such a dilemma is not unusual, but discrepancies between the Global North and the Global South often proved too stark for functioning infrastructural systems to emerge. Societies in Latin America, Africa, Asia, and Oceania had not evolved in the same way as those in Europe and North America; here, operative notions of infrastructure met with radically different conditions and conceptions of social organization.

Dam projects in Africa exemplify the forces pushing and pulling in opposite directions. Some had been planned in the late-colonial period. When implemented after decolonization, they symbolized national independence or pointed toward the welfare-state model imported from Europe. Economic conditions immediately following independence suggested that standards of living would continue to rise, and populations expected great things from their governments.[88] For all that, promises and prospects of rapid modernization and an equal share in the anticipated surplus rarely materialized.[89]

Examples include the Volta River Project in the Gold Coast, a former British dependency that has been called Ghana since 1957. The Scottish explorer Mungo Park proposed the project in the late eighteenth century, and the colonial administration in London drew up plans in the early twentieth. The Akosombo Dam, which would hold back the largest artificial lake in the world, was constructed between 1961 and 1965 under the first postcolonial administration. Kwame Nkrumah, Ghana's first prime minister and later president, saw himself as an African Lenin, invoking the latter's plan to electrify the whole Soviet Union.[90] Nkrumah was more right than he might have wished, because he became an early representative of the many leaders who would turn into despots.[91] A comprehensive infrastructure complex emerged for the Volta Reservoir: deep-sea ports and electricity grids, railroad lines, roads, bridges, and telegraph lines. News agencies and radio and television stations also moved in, which, per the ruler's dictates, were to report uncritically on government activities. Nkrumah sought rapid industrialization along Soviet lines. However, his large-scale projects—including a nuclear power initiative—and long-term development plans put the Ghanaian state in a difficult position. Instead of becoming economically independent, this country rich in resources amassed financial obligations, which only increased when world market prices plummeted for cocoa, one of the main exports. In 1966, Nkrumah was ousted in a military coup that took place as he was visiting North Vietnam and China.

Events in many African countries followed a similar course—when preexisting domestic trouble did not foil industrialization and infrastructural development before they started in the first place. Dam projects, which could not be realized without international aid, in particular were caught between political fronts. Examples include the Aswan Dam in Egypt and, later, the Cahora Bassa Dam in Mozambique.[92] The Kariba Dam in the Central African Federation is another case. Before construction was actually underway, win-win rhetoric set the tone: foreign investors, local populations, the newly independent state, and nature itself would only benefit, it seemed. In one fell swoop, the "primitive" conditions of Africa would be overcome, and a multiethnic community would live on the shores of the new reservoir.[93] But building operations on the Zambezi River showed that very little of what was hoped for would come to pass. All parties acted perfectly rationally in terms of their own interests, but conflicts and work at cross-purposes undermined the coherence of the overall project. No sooner had the dam been completed—against all odds—than the Central African Federation broke up, becoming Zimbabwe, Zambia, and Malawi, because of disputes over the use of energy and other resources it would generate.

Western planners' expectations that infrastructures would align developing nations with the priorities and cultural horizons of their own countries were rarely fulfilled. The Trojan horse harbored many surprises. When offers were taken up, local populations adapted materials to their own ends, integrating them into everyday activities in unexpected ways. Thus, power cables came to serve as clotheslines, and traditional donkey carts were fitted with axles and wheels from motor vehicles.[94] It's not a given that technology is neutral and will work the same everywhere. Everything that goes into communication, transport, supply, and disposal amounts to a social and cultural artifact. Elements combine into infrastructures only when supply and demand run on parallel tracks and users incorporate them into routine practices.[95] For this to occur, the new offerings cannot offer technological benefits alone. To prove viable, infrastructures need to consolidate the relationships that define a given community.[96]

Needless to say, major infrastructural differences exist among the countries of the Global South. Traffic is chaotic, the water supply can be unreliable, and problems with communication or electricity sometimes take up a significant portion of everyday life. Official claims about modern services do not necessarily mean they are kept in working order.[97] A host of images depicting the Third World—largely from a Western perspective—are supposed to capture the discrepancy between tradition and modernity: chimneys smoking in the jungle, dark-skinned people in European uniforms, telegraph cables extending over vast expanses of untouched desert, and so on.[98] One hears fairly often that modernization swept over traditional societies too rapidly, leading people who previously were integrated into communal life to become uprooted.[99] On the assumption that further development is required, international experts tend to appeal to education. Seemingly neutral offers of basic facilities and services imply hidden, even unconscious justifications for intervention.[100] Under colonial rule, but afterward too, infrastructures brought asymmetrical power relations.

It is never entirely predictable what effects new technologies will produce.[101] In Zanzibar, for instance, the introduction of electric light, cooking equipment, and television changed family routines. Anthropologists noted that women cooked less and became more independent. Men, in contrast, spent more time in the home. Simultaneously, however, domestic quarrels occurred more frequently; people did not sleep as long, and intimate relations seem to have waned. Older people felt more secure because of electric light, and girls started attending school at a younger age. Finally, visits by tourists increased—a development that locals viewed in negative terms until the material benefits were plain.[102]

Time and again, experts have analyzed why certain public institutions failed to work or gain acceptance. Among other things, they found that many migrants from the countryside to the city were unable to change their habits and routines quickly and needed time to recognize the advantages of technical infrastructures. Administrative incompetence at the local or national level—often the result of poor pay or nepotism—contributed to widespread skepticism about anonymous services, and personal networks continued to be preferred. Indeed, a gray area around infrastructures often makes it difficult to discern merits and disadvantages, which leads to lax attitudes about paying for services. If strict measures are not taken, not only private citizens but even municipalities and national agencies may disregard bills. Moreover, when facilities are not reliably maintained, local populations will simply help themselves to electricity—or boycott public utilities altogether.[103] Already by 1955, observers remarked that technology reliant on large amounts of energy requires adjustment on the part of the populace, in addition to training professional staff.[104] Evidence from supposedly developed, industrialized societies abounds that rules are neglected as soon as vital supplies are at stake.

Since the nineteenth century, realpolitik has prevailed where the flow of resources is concerned, especially petroleum.[105] Overt and undercover operations in countries that have abundant raw materials or are important for military strategy are legion. Besides the overthrow of Salvador Allende, the most fateful intervention was perhaps the ouster of the

Iranian prime minister Mohammad Mossadegh in the 1950s, when British and US intelligence services conspired to reverse the nationalization of the Anglo-Iranian Oil Company. Pure power politics decided who owned natural resources, and estrangement between the Global North and South mounted. (Incidentally, Jalāl Āl-Ahmad was a great admirer of Mossadegh.)

Surprisingly, countries rich in raw materials rarely join forces to exert pressure on industrialized ones. One instance—the most instructive example for the Global North—occurred in 1973, when the Organization of Petroleum Exporting Countries (OPEC) enacted an embargo that caused the first oil crisis. For whatever reason, comparable measures were not taken again, but the sight of empty highways—symbols of modernity that had suddenly become useless—was burned into the popular memory of the West.[106] Ever since, industrialized countries have proceeded more cautiously, seeking to moderate and streamline energy use and obtain it from more varied sources (including nuclear facilities).

In the twenty-first century, rivalry for resources is not the same as it was in the nineteenth and twentieth centuries. China, for instance, was a hotly contested prize during the heyday of colonialism and the focus of imperialist railway projects from Europe and North America.[107] Today, as the world's second-largest economy, China is in a position to build an export infrastructure of its own—along the old Silk Road, among other places. Although occasionally enlisting thugs domestically to drive recalcitrant populations from valuable land, the Chinese government prefers smile diplomacy when pursuing global projects of building railroads, ports, and pipelines.[108]

China is making headway in Latin America—for example, by constructing the 5,300-kilometer-long Ferrovia Transoceânica between the Atlantic coast in Brazil and the Pacific coast in Peru[109]—and setting up operations in erstwhile heartlands of colonialism. In recent years, Chinese investors have poured more money into Great Britain than any European country. The China Investment Corporation has been responsible for massive projects including the sewage system in the county of Kent, electricity supply lines in large parts of England and Wales, the expansion of Heathrow Airport, and the high-speed rail line between Manchester and London. "While industrialized Western nations remain stuck in protracted democratic processes," a 2014 report concluded, "Chinese investors open a broad view of the future."[110] Are infrastructures once again becoming reservoirs of power for imperial ambitions?

RETURN OF THE REPRESSED

The Zabbaleen of Egypt have achieved a certain prominence thanks to travel journals written by Westerners and photo sharing on the Internet. These people, some sixty thousand in number and belonging, for the most part, to the Christian Coptic community, collect around three thousand tons of trash on the streets of Cairo every day. This amounts to less than a third of the waste generated by one of the largest megacities in the world. With raffia baskets and donkey carts, men and boys roam the streets in the shadows of the night in search of recyclable material. The material gathered is sifted through, separated, and processed (if necessary), usually by women and girls. *Zabbaleen* means "dung heap," having obvious derogatory connotations. But Western observers voice admiration and occasionally show signs of a guilty conscience; it seems the Zabbaleen remind them of something that occupies a space below the ordinary threshold of consciousness. *Garbage* is a relative term—indeed, a social construct—naming what no longer can be used for political, economic, or moral reasons. Its presence evokes disorder and danger.[111]

The Zabbaleen have counterparts in almost all the megacities of Africa, Asia, and Latin America. They deal with waste in a way that once was widespread in the United States and Europe but has almost disappeared today. Nineteenth- and twentieth-century waste management systems developed in parallel to general mechanization. Now, garbage is supposed to be invisible, and its odors neutralized or masked. That the Zabbaleen rummage around in trash and even let their children play in it inspires chills of horror for many observers. But it also commands respect, inasmuch as these people take care of what others cannot even face.[112]

Cairo's waste collectors reportedly manage to recycle 85 percent of sorted material, some of which is sent to China for further processing. The rate compares more than favorably with European countries' best recovery rate of 10 percent. The Zabbaleen may be viewed as ecological heroes, then.[113] In premodern and modern times, in Europe and elsewhere, everyone has always recognized that matter and energy belong to a cycle whereby one thing can be transformed into another. This awareness has often been pushed aside, however, in favor of dumping refuse elsewhere or stockpiling and forgetting it. The same societies that boast of scientific and technological know-how have revealed a remarkable blind spot.

Even in "throwaway societies," the reuse of devalued materials never vanished entirely. In times of scarce resources—especially war or natural disaster—the practice has been revived.[114] But on the whole, dumping, burying, or burning waste is the norm. It seems easier, cheaper, and most importantly, quicker to banish it from sight. Accordingly, disposal of what has outlived its perceived usefulness represents an enduring problem for industrial societies.[115]

"Waste," an observer noted decades ago, "is so ubiquitous that it will soon become the central metaphor for culture."[116] Together, its material, metaphorical, and even more, sociopsychological aspects constitute the flipside of modern society.[117] Biological analogies have long illustrated as much; it is common to speak of households and cities performing functions of elimination, as if they were actual organisms. In the nineteenth century, in keeping with medical opinion and standards of cleanliness, efforts ensured this would occur as inconspicuously as possible.[118] Affective and behavioral norms changed as sanitary conditions became the standard for measuring foreign environments and people. Moreover, any delay in the removal of garbage (say, in the event of a strike) or reminder of unseemliness (for example, gas leaks or proximity to a sewage plant) now counted as a grave problem. As a representative treatise on urban health published in 1897 remarked, "Preventing the accumulation of waste products in the vicinity of human beings is not only a main requirement of health care but also of culture in general inasmuch as it has advanced beyond a certain minimum level and no longer tolerates offensive or unpleasant sensory perception, especially when longer in duration."[119]

Emotionally charged images regularly override rational concerns. Urban trash—vermin, disease, homelessness, mountains of refuse, polluted water, flooding, and damaged pipes—continually comes to the fore while also being banished from consciousness, or repressed.[120] Cleanliness has come to represent a value in itself, and not just in relation to garbage. Its importance is evident on a microlevel—for instance, every time we enter a bathroom, a "domestic bulge in the flow of water systems" that amounts to a psychically and culturally shaped space of transition. Here, water finds almost ceremonial use in the bath or shower, washing away the tensions and conflicts of day-to-day life in the modern world.[121] Similar psychological processes are at work in the dialectic of buying and discarding, eating and throwing away, and giving and receiving gifts (that is, activities that revolve around abundance, social distinction, and ritualized affirmation). Water is laden with cultural significance, from the wastefulness of watering lawns and flushing toilets to the cult of personal hygiene and expensive bottles of mineral water.[122]

Until well into the twentieth century, liquid and alluvial materials seeped into the ground or flowed into the sea via pipes, canals, streams, and rivers. This is still the case in many parts of the world, and every effort is made to keep runoff moving and free of toxins. Infrastructural development straightened rivers and built concrete banks. Some waterways have even been reserved for wastewater; for example the Emscher, a tributary of the Rhine, became an "artificial intestinal outlet of the Ruhr region" in the nineteenth century.[123] Sewage farms were created around large cities for wastewater processing. Ultimately, the oceans were deemed fit to treat sewage on an enormous scale, and other types of waste were dumped in them, as well. Huge islands of plastic eddying in the Pacific now pose a challenge as great as nuclear waste and the toxic materials that have accumulated in many great bodies of water.

The easy business of banishing coarser and more solid materials from view finds symbolic representation in garbage chutes, colloquially known as trash eaters (*Müllschlucker*) in German. One simply throws household refuse into a hole, and it vanishes down a shaft.[124]

Collecting trash in public areas, a task often performed by minorities or so-called guest workers, is a low-status activity. People who do so are often viewed as the dregs of society. In the nineteenth century, inmates of workhouses took care of waste disposal and sewage farms. By the logic of the day, such hard labor would "clean" them in a figurative, moral sense.[125] The landfills that were introduced for discarded materials and garbage could never be far enough away from urban centers, even though logistics demanded that they be located reasonably near. Refuse was also frequently burned at special plants; at the time, no one thought of generating electricity from it.

Before the advent of manufactured abundance, reusing materials was the norm in Europe. Leftovers were fed to animals, broken furniture and tools were fixed, organic refuse became fertilizer, and scrap metals were melted down and reprocessed. None of these practices ever disappeared entirely, but over time they were relegated to positions of minor importance.[126] When an early form of waste sorting was introduced in Charlottenburg near Berlin in 1907—with separate collection points for kitchen refuse, rags, broken crockery, wood, metal, ashes, and sweepings—the practice proved too costly; ten years later, it was discontinued.[127] During World Wars I and II, collecting raw materials essential for military operations worked reasonably well, but it was only possible because of massive propaganda campaigns ("I gave gold for iron," went one German slogan) and concerted effort by agencies such as the War Raw Materials Department (*Kriegsrohstoffabteilung*) started by Walther Rathenau in August 1914.

As a rule, recycling rates decline steeply in times of peace and economic growth. Over the course of the second half of the twentieth century, ragpicking and related trades disappeared from everyday life. In a parallel development, however, junkyards for industrial, household, and automotive products that had fallen into disuse proliferated. Beginning in the 1970s, the limited quantity of natural resources became apparent to many Europeans, which led to innovative efforts to use materials more efficiently or replace them with new ones. Just as significantly, awareness grew that too much is produced that cannot be recycled or used again. The outstanding symbol for this state of affairs is noncompostable plastics—millions upon millions of shopping bags. But less visible problems also exist: contaminated water, soil, and the "heavens above." Simply building taller smokestacks won't do the trick.[128]

For decades, waste disposal infrastructures aimed primarily at getting objectionable substances out of the sight and environment of the parties who had produced them. At best, items discarded by Western consumers might find a second (or third) life somewhere else. Thus, cars considered out of date were shipped to Africa.[129] In what is colloquially known in Germany as garbage tourism (*Mülltourismus*), refuse traveled great distances domestically and internationally.[130] All the same, pollutants continued to flow into shared water, soil, and air. Most people accepted, if grudgingly, that this was simply the price to be paid for the conveniences of modern life.[131]

The habit of shifting the undesirable consequences of infrastructure elsewhere has been the object of criticism in western Europe and the United States since the 1970s. International information exchange in the fields of environmental protection, meteorology, and climate have made it abundantly clear that many ecological problems are global in scope. In the short term, or at the local or even national level, turning a blind eye might be possible, but it is plain that troubles will not go away. Phenomena such as melting glaciers and polar caps, expanding deserts, dying forests, the growing hole in the ozone layer, and warming oceans have, for some time now, been acknowledged as signs that changes are necessary. As more and more countries pursue infrastructuralization and industrialization, the urgency only heightens.

A gap exists between actual conditions and the promises of technological and scientific progress. Priorities have shifted as industrialized societies geared toward consumption confront limited natural resources. Cultural critiques have multiplied, with bountiful self-recrimination over environmental matters. But so have constructive efforts that renew ecological practices of the preindustrial age.[132] In this context, debates about nuclear power are instructive. In the 1950s and 1960s, nuclear energy was seen as freeing the world from dependency on fossil fuels and reducing harmful emissions. The future seemed bright. But it soon became apparent that medium-term aspects represented a liability. The technology was highly complex, with operating risks that required specialized personnel; the possibility of human error was very real. Moreover, questions about processing and disposing nuclear waste hadn't been solved. In keeping with high-modern tradition, dealing with problems was deferred until further advances had been made. Reactor accidents in Harrisburg (1979), Chernobyl (1986), and Fukushima (2011)—plus dangers posed by highly improvisatory practices of storing or dumping waste—dispelled the illusion that finding answers could wait.

The history of industrialized societies has made it plain how difficult it is to break out of the "coketown" and "motown" configurations discussed earlier. These infrastructures depend on compounding industry and spreading worldwide. In the high-modern heyday of growth, efforts to protect nature and heritage against encroachment (by power lines, roads, and dams, for instance) were dismissed as obstacles to progress, or even as relics of a premodern sensibility. Only since a measure of self-reflection has come to prevail have reservations gained traction.[133] Tellingly, former waste disposal facilities have adopted environmental practices and gone green by reusing infrastructural waste such as sewage, sludge, and industrial by-products.[134] Yet problems persist, especially when it comes to dangerous material. Exported by wealthier nations, it has accumulated at places such as the Agbogbloshie electronic waste dump in Accra, which has gained the unfortunate distinction of being a "toxic city."

Today, planning must address the question of whether new infrastructures will prove socially acceptable, sustainable, and environmentally sound. Because projects are supposed to serve the common good, they have prompted citizens to examine fundamental issues critically.[135] Since the 1970s, protests against nuclear power have included an element of resistance against centralized power structures and the loss of democratic representation. New highways and airports raise questions about nature, climate, and noise and, behind it all, highlight that life is permanently accelerating. Incineration facilities draw attention to the ubiquity of chemical products and risks of contamination in a society of planned obsolescence. The construction of dams raises concerns about conservation and protecting water from pollutants. Large-scale building projects, commercial development, overland cable lines, landfills, and power plants invariably give rise to objections that the landscape and natural world are being disfigured.[136]

Politicians, planners, and engineers have regularly voiced frustration about the importance assigned to nontechnical factors. But in a pluralistic democracy, decisions are made through dialogue that often confounds sweeping, unilateral designs. The allocation of public funds has also been subject to greater scrutiny in the wake of prestige projects undertaken from the 1950s to the 1970s, when so-called *Hallenbadeffekte* (literally, indoor-swimming-pool-effects)—that is, (overly) generous infrastructural provisions—were evident even in tiny villages. In industrialized countries today, undertakings must fall in line with demographic fluctuations.[137] Otherwise, buildings and services will fail to be used to capacity and will need to be adapted to serve other purposes—or even dismantled altogether. Finally, public budgets grant less economic leeway to projects.

Ever since the idea of a postconsumer society gained currency, many urbanites have sought to be more ecologically responsible and involved. *Waste reduction, lower emissions*, and *recycling* are the watchwords.[138] And as these practices have come to be more widely observed, waste itself has become a commodity—as biomass, for instance. Unsurprisingly, then, the Zabbaleen in Egypt are now facing difficulties. The city of Cairo, which wants to exercise greater control over municipal life, has contracted Spanish and Italian firms to set up a solid waste collection system.[139]

II

NODES OF INFRASTRUCTURAL DEBATE

4

PUBLIC, PRIVATE, OR PARTNERSHIP? HOW INFRASTRUCTURE IS ORGANIZED

PUBLIC HANDS, INVISIBLE HANDS

People are always looking for a free ride, literally or figuratively. I remember a story I read in the seventh grade about one clever scheme in England. People short of money would send letters without paying postage; the Royal Mail expected recipients to cover the costs. Enterprising individuals marked envelopes in a code they had worked out with addressees beforehand. The secret writing in plain view conveyed essential information: whether the sender was well, if plans were proceeding, when the sender might be expected to come back home, and so on. When the postman came, the recipient would take a close look at the envelope, feign disapproval about the sender, and decline to accept it. A message had been transmitted at no expense.

The story must have been set before May 1, 1840, when the United Kingdom introduced postage stamps to prevent this kind of trick. The measure, initiated by Sir Rowland Hill, is one of the earliest systems of prepayment on record. The first stamp was the One Penny Black; it bore the likeness of a young Queen Victoria in profile. One year later, it was replaced by a version in red. Because cancellation marks didn't stand out clearly enough against the black ink, some people had managed to reuse stamps.[1] No one anticipated the generations of collectors to come. Postage was simply a matter of financing public services and preventing misuse of the mails. By the end of the nineteenth century, the Blue Mauritius, which entered circulation seven years after the One Penny Black, was one of the most legendary—and valuable—stamps of all time.

To this day, the issues attending the irregular use of public services are debated, sometimes with ideological fervor. History offers no ready answers to questions. After all, infrastructures have been set up and operated under varied circumstances. Why do some people always flout the rules and refuse to pay for what is in everyone's interest? Why do scofflaws even meet with tacit popular support—like the Spontis in 1970s Germany who called for free public transportation ("Zero fare, zero fare, or we'll bend the rails!" *[Nulltarif, Nulltarif, sonst biegen wir die Schienen]*, the slogan went).

Time and again, social scientists and practically minded planners have examined turnstile jumping, abuses of the honor system, and similar misconduct. Ultimately, behavior comes down to moral conventions and attendant circumstances. Thus, in an agricultural context, the historical development of common land use has received attention. How do individuals behave when property belongs to a whole village? Can shared use of forests or pastures work at all?[2] The questions bear on the idea of a "second nature," or culture, governing human life in society and provisions for general welfare.[3] Their urgency is constantly renewed in the realms of communications, transportation, supply, and disposal. What is the proper balance between costs and revenues if a service is an infrastructure—that is, not a luxury but an arrangement that should be available to all?

The example of the stamp illustrates that systems of payment constitute worlds of their own: admission slips, tickets for trains and buses, coins, tokens, phone and smart cards, and more recently, payment apps all serve the same purpose, even though they function differently in concrete terms. They can thrive for years and even decades, only to be replaced abruptly; more than a few such relics have become charged with symbolism and nostalgia.[4]

In fact, simple-seeming schemes of payment are highly complex. On the one hand, they represent individualized accounting that occurs on a case-by-case basis; on the other hand, they are the same for all customers and represent a standardized scheme. As a rule, only experts are fully aware of this system's drawbacks. Revenues are subject to economic fluctuation, and services for the broader public often pay off in the long term only because they are subsidized. Accordingly, planners try to develop schemes of payment for regular use well in advance—for instance, subscriptions and monthly or annual passes. Such arrangements have the added value of making personal stakes in general services more visible—evidence that infrastructure is of vital importance for the routines of everyday life.[5] An alternative payment model is the fixed-term contract. Especially for water and electricity, metering devices record consumption, and charges are levied according to fixed or flexible tariffs. Complex mechanisms quantify how much is used, calculate costs, and collect money for providers. Consumers, for their part, must understand the billing process and be willing to pay. In the event that they do not acknowledge fees or fail to meet financial obligations, services may be cut off.[6] Finally, there is the model exemplified by parking meters. This variant requires managing time and attention. As with coin-operated meters used for water, electricity, and gas, responsibility rests with users. Such devices were introduced in England to avoid charging a flat fee and to allow flexibility—a measure intended to make services such as gas available to less prosperous segments of society.[7] Since then, they have been a means of biopolitical control throughout much of the world.[8]

In 2000, the worst-case infrastructural scenario became reality in Bolivia, culminating in the *guerra del agua*, or water war, in Cochabamba. As a precondition for further aid, the World Bank had demanded that local water supplies be privatized. Aguas de Tunari, the company that took charge, raised prices so drastically that they ate up more than a quarter of many households' income. A general strike ensued, with deadly violence. Eventually, privatization was reversed. The 2010 film *Even the Rain* dramatizes events. Critics argue that deregulation threatens basic human rights on an even broader scale in the future.[9]

Municipalities, provinces, states, and government in general have an alternative to levying charges on an individual level because they can collect taxes without accounting for every aspect of their use. In 1883, the cultural historian Wilhelm Heinrich Riehl described the practice's origins:

> The great reform of road construction in the eighteenth century coincided more or less exactly with sweeping administrative reforms. . . . The highway was the first visible, tangible object to make it clear to ordinary people that in the

modern state the individual must first give up personal advantage in order to receive it again, with interest, from the collective.[10]

Besides ensuring security through the military and police, the central task of government remains the construction and expansion of infrastructures. One of the strongest arguments for taxes is that they provide for the common good. The overall trend of state institutions has been to shift from demonstrations of power and authority toward providing services that benefit everyday life and derive legitimacy from their functional value. Governments face the problem, however, of winning public acceptance for expenditures, because budgets are enormous and no one can keep track of how much goes where. Citizens demand a say in how funds are allocated.

Every year, the German Taxpayers' Association (Bund der Steuerzahler) draws up a balance of public monies being used inefficiently or wasted outright. A constant complaint is that taxes introduced for a specific purpose continue to be levied and employed for other projects. Thus, a tax on sparkling wine introduced in 1902 to finance the construction of the German fleet and the Kiel Canal remains in force to this day; it brings the federal government hundreds of millions of euros annually. Another example is the money citizens have been asked to pay since 1991 to show solidarity and pay for reconstructing the eastern part of the country. Positive effects of infrastructural development are extolled, but few people understand what is being financed—for instance, whether the vehicle tax goes to build roads or something else entirely.

As early as 1863, in light of increasing provisions for social welfare, the economist Adolph Wagner posited the law of increasing state activity. Critics challenge the implicit paternalism and inefficiency of structures that emerge when government is assigned the task of seeing to the common good.[11] Since the 1970s, disagreement has prevailed in western Europe about the role of the state. As in discussions of internal improvements in the United States, participants have fundamentally opposed positions. Tension surrounds the tangled relationships between individual and community, private and public welfare, economy and society, and much more. Disagreement often centers on taxes—which, as a matter of numbers, would at least seem to belong to the realm of objective fact. Advocates invoke the history of infrastructural development to buttress arguments, but sometimes to surprising effect. Thus, in 1974, the neo-liberal economist and Nobel laureate Ronald Coase cited the example of lighthouses. Contrary to widespread assumption, he observed, lighthouses were not originally intended to serve the public interest; in England, they were built on private initiative and financed by harbor dues.[12]

Debates are thoroughly shaped by ideology. Conceptions of human nature and society are so different that neutrality is impossible. Even if cultural attitudes about free rides, tax dodging, stealing electricity or cable, train hopping, and so on are subject to change, the moral postulate persists that citizens should contribute to communal life.[13] Is it ethically and economically permissible to use somebody else's bus or subway ticket if it's still valid? Some people play hide-and-seek with inspectors and officials to resist state power or big business and sympathize with rule breakers. Others deem this activity a crime against everyone that merits punishment.[14] Vandalism and wanton destruction admit relatively straightforward judgment (although one may interpret the causes very differently).[15] Carelessness—an effect of the little attention infrastructures are usually granted—may also play a role. Nonpayment of fares costs some 250 million euros a year in Germany. Monitoring and checking measures cost another 100 million. An estimated 3 to 3.5 percent of all passengers routinely travel without a ticket; steady increases in penalties seem to make little difference.[16]

Since the 1970s, the practice of free riding has bolstered arguments for privatization throughout the Western world—and, more recently, elsewhere too. Deregulation promises to remove many services from a state apparatus that is supposed to have become bloated. Beginning in the nineteenth century, government took on more and more tasks on behalf of citizens, from municipal and regional utilities to operations on a national scale such as railway and postal systems. The institutions were staffed by reliable civil servants with a distinct ethos and corporate identity; they also represented an enormous investment. The Deutsche Reichsbahn, for instance, was "the most valuable asset of the Reich"; following World War I, it was collateral for reparation payments to the Allies.[17]

These organizations emerged in the high-modern era of infrastructural development—between about 1880 and 1960—in keeping with the overall tendency of nation-states to devise and implement large-scale

4.1 A warning against vandalism, from *Postmerkbuch für den Schulunterricht*, was published by the Reich Postal Ministry, 1938. "Anyone damaging telegraph or telephone equipment is liable to prosecution. (According to Article 317 of the Penal Code, a prison term from one month to 8 years.)"

planning. With the exception of Eastern Europe, however, governmental steering of economic activities was never uncontroversial. Eventually, management shifted to the private sector for many reasons. For one, optimism about the future began waning in the 1970s. Birth rates stagnated—or even declined—which necessarily affected infrastructures. The pool of taxes levied in the decades following World War II ran out; finally, the expense of maintaining services (which had constantly expanded during this period) swallowed up funds—as did the high costs of staffing and pensions.

Infrastructures had enabled government to extend to more and more areas of life; in the eyes of wary citizens, this development threatened to create a world administrated from top to bottom. A critical view of governmental planning, mounting bureaucracy, and undemocratic rule by professional experts seemed both timely and necessary. Economists who would come to be known as neoliberal initiated a publicity campaign taking aim at ossified and overstretched institutions of the welfare state. In the words of Herbert Giersch, government should promote "personal responsibility" instead of an "early retirement mentality." The invisible, self-regulating hand of the market, the argument went, would produce better results than public administration; business reacts more quickly and flexibly to technological innovation and prevents the "service deserts" that plague state-administered spheres. Economic action by the government should be in a subsidiary capacity at most, or in areas where the market is clearly faltering. Prosperity can be maintained only through international competition, and regulations impair entrepreneurial freedom.[18]

Opponents of this position argued that the very word *infrastructure* means that institutions belong in public hands—the state.[19] Especially in Latin America, left-wing governments nationalized services in the name of reducing dependence on North American and European companies. As events in Chile show, these measures did not necessarily proceed smoothly. However, they represented powerful declarations of sovereignty. In turn, large-scale projects promised to bring prestige. Although renewed reliance on foreign investment roused distrust, such sentiments could be used to political ends.

Infrastructure projects were fraught propositions, then, and they needed to pay off quickly. Once they were up and running, they could be operated by governments at relatively low risk. As amortization proceeded, profits usually rose. For the most part, however, money flowed to lenders. Deals with international financiers that had seemed reasonable more and more resembled exploitation (at any rate, it was easy to present them as such).[20] Examples abound of development leading to dependence on foreign banks; indeed, the business of financing transport networks built and operated on state initiative is one of the main reasons that a global capital market emerged in the first place.[21]

The political consequences are easy to see. Transport and utility facilities were often nationalized in a manner resembling a coup, which could provoke international crisis. The textbook example occurred in 1956, when British, French, and Israeli forces undertook punitive action against Egypt over the Suez Canal. Elsewhere, in many African and Asian countries, expropriations proceeded incrementally. First, new price regulations and contracts were introduced, and then limits were imposed on work permits for foreign managers (or the opening of facilities was delayed) until investors had bought their way out. As the *Harvard Business Review* noted in 1995, an abiding feature of infrastructural investment is that higher profits also bring higher risks.[22]

This is not the place to determine whether the public or private sector tends to do a better job; the aim here is to show how different forms of organization are reflected in the everyday lives of users. In many areas where private providers have stepped into what was a public monopoly, greater choice has resulted. But because prices and offers vary with the market, consumers bear the added burden of looking closely at options and devoting time and energy to decisions. The endless offers to consumers from electricity and mobile-phone providers exemplify the problem. Navigating a veritable jungle of fees is trying, and customer support frequently falls short. Outsourcing services to private enterprise only compounds difficulties. In a word, making everything a matter of consumer choice does not necessarily represent an improvement. Contrary to advertised goals, business often prioritizes operational efficiency over user satisfaction.

Privatization and deregulation in the 1980s and 1990s yielded mixed results. Successful efforts contrast sharply with failed ones.[23] The privatization of British Rail is generally considered to have been a disaster. In the early 1990s, under Prime Minister John Major, the government broke up the public railway system into about a hundred separate companies. Now, buying a ticket valid for areas served by different enterprises all but amounted to a kind of IQ test. Prices rose rapidly, and delays and missed connections multiplied. Moreover, travel became more dangerous, because regular investments in safety were not forthcoming. In 2005, the London *Times* cited a study that had calculated the total time of trains running late since privatization to add up to some eleven thousand years.[24] And in the end, public subsidies actually increased. In 2001, the Railtrack group went bankrupt; some of its holdings were then reincorporated into Network Rail, which is publicly owned.[25]

It is common for private firms to disregard matters such as equal accessibility and public mindedness. In response, governments have enacted new laws. Since the 1990s, there has been less talk of *Daseinsvorsorge* (see chapter 3) than of basic or universal services, which should remain available under all foreseeable conditions, corporate profit interests notwithstanding.[26] By the same token, governments are supposed to prevent providers from abusing their control of networks to put competitors at a disadvantage.[27] Such measures have come under fire, however. Points of controversy include the "creation of equal living conditions" called for in the German Federal Spatial Planning Act (*Bundesraumordnungsgesetz*) of 1965.[28] This legislation made public infrastructures the means for ordering and integrating social space and population groups—including rural ones—to a degree that critics found unacceptable.

Private providers put economic considerations above political or social objectives, which entails other problems. Once deregulated, services became concentrated in places with a high density of users, where prices are lower, whereas prices in less populous and less networked areas are higher.[29] This is known as splintering urbanism, or development that yields some districts with better and some with worse services and rates. Privileged regions emerge with premium users, most of whom are globally interconnected, and less fortunate regions must content themselves with more primitive arrangements.[30]

Given the scant agreement about who organizes infrastructure best, government, local authorities, and business have been joining forces to form public-private partnerships. In this arrangement, private firms—usually large companies with liquid assets—take on the work of construction; examples include the stretch of highway between Hamburg and Bremen, the A1 Hansa line. In exchange, local or federal government offers advantageous terms for paying off investments; a common variant is for the company to receive tolls for a set period.[31] Similar deals have been struck for schools and related facilities (cafeterias, for example).[32] Experiments with leasing have also been made. Wastewater plants, streetcars, exhibition halls, and sewer systems are sold to investors, then the government pays to use them.[33] Some municipalities have sold off property—for instance, public housing—to fill their coffers. Another proposal is to set up infrastructure funds on the model of hedge funds, which were gold mines before the financial meltdown of 2008.[34] In light of uncertain returns, the idea has gained little traction.[35]

The advantages and disadvantages of privatization probably need to be assessed on a case-by-case basis. At any rate, critical voices have grown louder in recent years. The main complaint from consumers is that they now have to deal with much more than simply getting connected to services. One is far less likely to be visited by a company representative who wants to read the meter than to be contacted by someone selling a new plan for mobile communications, electricity, water, or financial management. Customers are constantly having to make decisions while also attending to things that once were done for them by more or less friendly personnel. Facilities have thinned out to such an extent that it takes forever just to find a mailbox or post office; on top of that, people are asked to use automated services for postage. It seems hard to believe, but printed envelopes will one day become collector's items and change hands for millions of dollars, like Blue and Red Mauritius stamps.

ACTIVE PUBLIC SPIRIT

On July 5, 1942, the Japanese military placed a zero kilometer marker in Nong Pladuk, Siam. The town, two hours west of Bangkok, was to be the starting point for a railway between the Gulf of Siam and the Indian Ocean; the other end would lie 415 kilometers away in Thanbyuzayat, Burma. Building the line would enable the Japanese to avoid the sea route around the Malay Peninsula, which was especially perilous in wartime, and facilitate military attack on British India. Some sixty thousand British, Dutch, American, and Australian prisoners of war performed the work. Further contributions came from almost two hundred thousand forced laborers, including Japanese soldiers who had lost their honor after desertion or other offenses. Track material was brought from conquered territories known, in the aggregate, as the Greater East Asian Sphere of Prosperity—a designation comparable to the contemporaneous Greater Germanic Empire in Europe, half a world away.

Construction proceeded at a ruthless pace, and the line could be traveled by December 1943. Western prisoners called the Thailand-Burma line the Death Railway.[36] Each crosstie was said to have cost a human life; over one hundred thousand Asians and twelve thousand Allied prisoners died in the course of the project. Many of the 688 bridges along the way crossed the Khwae Yai River, one of which provided the setting for *The Bridge on the River Kwai* (1957), a film decorated with multiple Oscars. The ultranationalist Yasukuni Shrine in Tokyo still houses the first locomotive to have made the journey. After the war, the Thai government had to rebuild almost the entire line. The forced laborers had shown little concern for quality; indeed, they had done poor work on purpose so the railway would not operate for long.[37]

Few infrastructures built under duress have received as much attention as the Death Railway, but using forced labor for transportation projects was widespread practice. Detailed figures are unavailable, but according to Congolese lore, for example, the construction of the Matadi-Léopoldville railway in the 1890s came at a terrible cost; "every railroad tie represents the life of an African, every telegraph pole that of a European."[38] The British explorer Henry Morton Stanley summed up the reasoning behind such undertakings: African countries without a railroad are not worth two shillings; *with* one, however, their value defies calculation.[39]

Particularly onerous tasks were often performed by prisoners of war or inmates of concentration camps. Work included draining swamps, digging canals for irrigation or drainage, quarrying stones, and constructing roads.[40] All work was difficult and thankless; the humanitarian provisions

4.2 Forced labor was used for the "common good." Here, prisoners build rails at Buchenwald, spring 1943. United States Holocaust Memorial Museum, courtesy of Gedenkstaette Buchenwald.

of international law, which sought to restrict such practices beginning in the late nineteenth century, went largely unheeded. Forced labor was justified by the general benefit it was supposed to provide; moreover, the claim went, hard work had educational and morally cleansing virtues.

Forced and punitive labor was especially rampant in colonial settings, where people were supposedly averse to doing much at all.[41] The practice drew on the European ideology of productivity. Self-sufficient ways of life in Africa and other parts of the world vexed the self-appointed missionaries of civilization, not least because they stood in the way of creating an international labor pool. Therefore all conceivable methods for ensuring that tasks would be performed "freely" were enlisted. In the process, traditions of serfdom and vassalage were revived that had thrived in the premodern era. Such feudal relics persisted in Europe, too, albeit in weakened form—for instance, in road construction—until technological advances and specialist engineering made them unnecessary.[42] The last echoes today include municipal regulations about keeping public sidewalks adjacent to private property free of leaves and snow.

In alleged or real emergencies, crises, and states of exception, disrupted or nonoperational systems of communication, transport, supply, and disposal require repair by simpler arrangements. In particular, war and reconstruction demand hands-on work until damaged infrastructures can be repaired. Drastic methods have been employed to this end. For instance, in the months immediately following the end of World War II, pedestrians in German cities could be detained at random and forced to clear away rubble so that roadways could be used. Such work traps (*Arbeitsfallen*) proved quite effective. Because families were waiting and there was no way to get in touch with them, detainees worked extra hard so they could get back home.[43] Similar incentives might have motivated participants in Eastern European *subbotniks*, a Russian word referring to services generally performed on Saturdays for community building. Work could include clearing and building roadways, squares, and parks and activities for improving living conditions in general. Lenin's 1919 essay, "A Great Beginning," provided the playbook, glorifying heroes in the rear who volunteered to help the socialist struggle. Often, however, the scheme simply compensated for shortcomings of the planned economy. In the 1950s and 1960s, efforts to maintain public spaces and facilities frequently turned into long-term assignments and part of the economic planning itself.[44]

In the German Democratic Republic, the Nationales Aufbauwerk (NAW) was founded in the early 1950s to mobilize people to perform unpaid labor. The inaugural project was constructing the massive apartment buildings on East Berlin's Stalinallee. Later initiatives included parks, sports facilities, and cultural institutions. Inducements included socialist competition—that is, striving to outdo one another in the name of a shared future. Social engineering fused with individual incentive in efforts that proved more or less voluntary. To get a telephone connection or a new apartment, participation in, say, the *Mach mit* (Join in!) initiative would be helpful. Detractors said the acronym NAW stood for "Fools at Work" *(Narren am Werk)*. In time, the program was replaced by the Volkswirtschaftliche Masseninitiative (literally, People's Mass Economic Initiative), in which labor collectives vied for distinction.[45] Forerunners of organized forms of community service were legion—for instance, campaigns during the interwar period in Bulgaria or the Voluntary Labor Service (Freiwilliger Arbeitsdienst), which started in Germany in 1931. After 1933, the latter turned into the paramilitary Reich Labor Service (Reichsarbeitsdienst). Workers reclaimed wasteland and built settlements, dams, and roads. The objective was to strengthen young people physically and morally by having them work on projects that would benefit the public.[46] Similar mobilization occurred in the framework of state socialism—for example, during the construction of the Baikal-Amur highway in the Soviet Union.[47]

US history offers parallels in projects such as the Civilian Conservation Corps, which combined infrastructural expansion with environmental protection. The New Deal created jobs to put young people to work and counter the ills of the Great Depression. Between 1933 and 1942, nearly 3.5 million people received the honor of being "soldiers of labor."[48] From 1935 on, the Works Progress Administration (later, Work Projects Administration) brought gainful employment to another 8 million people. New Deal initiatives became models for an economic policy, later associated with the British economist John Maynard Keynes, that sought to counter cycles of boom and bust. State investment, which stepped in where private investment was lacking, mitigated economic crisis and, in the process, gave the United States landmarks such as the Golden Gate Bridge in San Francisco. The creation of the country's largest electrical utility in the Tennessee Valley also belongs in this context. Indeed, even unemployed artists received commissions to design public spaces or provide publicity for the New Deal.[49]

Time and again, the success of these projects has been invoked against those who argue that the state should refrain from economic intervention. Calls for Keynesian infrastructure programs are regularly made in times of crisis and widespread unemployment. After all, proponents observe, infrastructures are not wholly public or wholly private in the West: governments can step in, at least to some extent, without undermining free enterprise. (Admittedly, definitions of what belongs to infrastructure can be rather loose.)

Users of infrastructural facilities and services do not necessarily register the differences between public and private management; indeed, it often takes a specialist to do so.[50] Mixed-economy ventures are widespread in public works, and the practice started long before public-private partnership became a catchphrase. Administering economic interests and public services under the auspices of a special-purpose company has generated a broad array of organizational and legal entities with advantages and disadvantages that are the subject of ongoing debate.[51] Of course, it makes a difference who builds a facility, who rents or operates it, and who owns it. When contracts are not clear, projects are prone to charges of corruption (as has been the case with the privately developed A1 highway between Bremen and Hamburg or the Skye Bridge in Scotland).[52]

Other bodies for developing and administering infrastructures include agencies such as the US Bureau of Reclamation and the Dutch Rijkswaterstraat. Both were founded to ensure governmental oversight of hydraulic engineering but gradually came to manage a host of other projects. The charter of the Dutch agency, which began in 1798 as the Bureau of Hydraulic Engineering, declares, "Rijkswaterstaat. Working to make the Netherlands safe, secure, attractive and accessible for all."[53] Another example, from Austria, is the Autobahnen- und Schnellstrassen-Finanzierungs-Aktiengesellschaft. In Germany, a company for financing transport infrastructure was founded in 2003. Since 1998, the Bundesnetzagentur (Federal Networks Agency), which formerly regulated telecommunications and postal services, has been responsible overseeing electricity, gas, and railroads.[54] Such bodies, although financed by taxes and massive in scale, do not oversee all infrastructural activities; usually they are limited to organizing projects of national significance.

Even in times when governments played an active role in the economy, alternative schemes for building, operating, and maintaining infrastructures were pursued. Cooperative models—which have a long and storied tradition—were often revived. Indeed, well cooperatives maintaining sources of water survived until the nineteenth century. Recently, the principle has been rediscovered in a completely different sphere; in 2014, it was proposed that some seventy thousand cooperatives team up to renovate the dilapidated highways and bridges in the heavily traveled area around Heilbronn.[55] Many gymnasiums and indoor swimming facilities built during the heyday of governmental engagement now are run by local sponsoring associations. The *Energiewende*, or turn to renewable energy sources and sustainable practices, has given rise to the idea of citizen dividends (*Bürgerdividende*): residents along the highways of the future are supposed to invest in operations guaranteed to earn interest. Of course, such appeals to self-interest are intended to mitigate potential resistance to new roadways.[56]

What now is known as civic engagement crystallized in the sphere of infrastructure. This phrase and related terminology refer to privately organized and financed projects—for instance, buildings funded by donations and erected by volunteers. Similarly, endowments (*Stiftungen*) and other forms of patronage sponsor social and cultural projects. Such undertakings may or may not belong to the realm of infrastructure in the narrower sense. Many facilities are cofinanced by profits from state-operated casinos—an arrangement that fuses charitable activities and gambling. Despite a somewhat disreputable activity being involved, the idea has spread. Bangladesh, for instance, has witnessed efforts to build latrines on private initiative with funds from public lotteries.[57]

The final variant is to rely on the private sector for building and operating facilities. This approach has predominated in transportation and communications but is also used, to a lesser extent, for utilities and waste disposal. Since the nineteenth century, specialized workforces have enabled a select group of construction and civil engineering firms to thrive by building roads, canals, railways, bridges, and tunnels; laying cable and pipes; and setting up and operating power stations and communications facilities. Commanding detailed knowledge, highly trained personnel, and international experience, these companies have been key players in globalization. Projects beyond a certain scale are almost certain to involve companies such as Hochtief, Bilfinger, URS Corporation, HDR Inc., and Bechtel.[58]

The extent of public and private involvement in infrastructure differs from one country to the next, where economic culture sets the tone and course.[59] But governmental authorities never stand idly by. The state needs to buy land for projects—or expropriate it, when owners are unwilling to sell. France, for example, has had a law on the books since 1833 to allow for this eventuality when constructing railways.[60] Central administrations also collect tolls, levy charges, and allocate tax monies for development.[61] Impact fees were first introduced in Hinsdale, Illinois, in 1947, to lay sewer lines. "Growth should pay its own way" went the all-American slogan.

In 2010, with empty municipal coffers, the mayor of Niederzimmern (Thuringia, Germany) struck on an idea that seemed crazy at first. He decided that town residents should own potholes. His *Teer muss her* (something to the effect of "We need some tar!") campaign proved a resounding success. At fifty euros per pothole, he managed to fix all the town's dilapidated roads. Since then, a pothole monument has been erected and a local musician has chronicled events in song.[62] A free ride, one might conclude, is more likely to come from civic-minded residents than from fare dodgers.

5

PRESTIGE PROJECTS: SYMBOLIC VALUE AND FAILURE

THE VISIBILITY OF THE SELF-EVIDENT

We live in an environment where most of what we see is man-made. Perhaps on your summer vacation you have a chance to hike the Appalachian Trail, but the rest of the year you drive through refinery row on the New Jersey Turnpike. . . .

A few centuries ago—say, on the American western frontier—a quite different view prevailed. Nature was seen as savage, hostile, cruel. Mountains and forests were barriers, not refuges. The lights of civilization were a comforting sight. . . .

One winter afternoon a few years ago I stood by the side of a highway near Gallup, New Mexico, looking on a classic vista of the American West: red sandstone buttes rising from a valley floor. It was the kind of landscape made famous by films and paintings and postcards, not to mention all those Marlboro Man advertisements. But this particular vista had something more. In front of the cliffs, and towering over them, were several cylindrical spires that I recognized as petroleum fractionating columns; off to one side was a grove of gleaming white spherical tanks. The towers and tanks belonged to a plant for processing and storing propane, or liquefied petroleum gas.[1]

This reflection, which the science writer Brian Hayes offers in *Infrastructure: A Guide to the Industrial Landscape* (2014), calls to mind John Gast's painting *American Progress*. Most people consider sites that intrude on pristine environs an eyesore. Taking an unconventional tack, Hayes offers a guidebook for "the world we have made for ourselves"; so far, his book is unique in kind.[2] The opening sequence points to a fairly recent shift in popular perception and moral sensibility. Nature is something inherently good and worthy of preservation, a haven for supposed authenticity. "Second nature"—that is, the world of human artifice—represents an ongoing problem requiring attention and adjustment.

As noted in preceding chapters, the vast majority of infrastructures escape our notice. Usually, we are glad this is the case. When they do catch our eye, the reason is frequently irritation: a continuum that is supposed to keep flowing begins to falter, prices seem inflated, and it appears that resources are being squandered. Infrastructures range from wholly invisible to full-blown spectacles, with every shade in between.[3] They all have signature features, and at times, political calculations lie behind them. Many projects are undertaken to no end other than showing off.

Infrastructures signify connectedness and modernity. Conversely, they can stand for technical standards that have fallen behind the times; consider landline telephones, which still need to be plugged into the wall. Sometimes, tricks are employed to fill the gap between generations of technology. Thus, when the mania for cell phones started in the 1990s, Latin America witnessed a brisk trade in conventional telephones resembling the new devices. By using them in the presence of others, status-conscious people sent a message that was inaudible but perfectly clear.[4] Cell phones provide prime examples of the aesthetic, functional, and social aspects of so-called terminal equipment, where the vagaries of fashion and the market prevail. In the realm of gadgetry, look and design are not the only factors; a trend toward miniaturization is also evident. User-friendly exteriors house an ever-increasing array of functions that, especially with advances in digitization, are becoming less and less material. Electrical and satellite technology belong to the same picture.

As a rule, only the entry and exit points of infrastructure—and sometimes points of transfer—are subject to aesthetic considerations. Railroad stations, airports, metro hubs, and other intermodal sites receive titles charged with historical significance and symbolic value. Like street names, they can occasion vigorous public debate. Generally, actual networks of traffic and transportation are visible only during construction—when they tear apart landscapes or disrupt routines. Nor does getting used to their sight always prove easy. Canals, railways and attendant facilities, roads, power lines, and more recently, wind turbines have all provoked complaint. Objections include the sense that a unified and organic configuration of space is being damaged by unwelcome reminders of how industrial society works. On this score, conservationists and environmentalists agree with pessimistic conservatives who view every utility pole and paved road as a sign that the world is not (or no longer) their own. Henry David Thoreau's *Walden* (1854) is a classic for those who would sooner lend an ear to a "different drummer."

Even people of a more neutral disposition take offense at sensory impressions they find unpleasant and cannot get used to, especially from sewers, water-processing facilities, and landfills. The visible, audible, and olfactory aspects of waste disposal—for instance, street cleaning, garbage collection, and emptying latrines—are performed, as much as possible, during the night or in the early morning. The ringing of church bells and chiming of clocks connote welcome familiarity for some.[5] However, not all alerts are welcome. Wailing sirens are a case in point—especially for people who lived through bombing raids. Tests of emergency systems conducted in the decades following World War II reached 101 decibels, which members of the younger generation did not enjoy, either.[6]

Some sounds belong irrevocably to the past. Gone forever are post horns, which, until the nineteenth century, announced the arrival of mail coaches and gave the sign to ready a change of horses, open city gates, or simply clear the way. The words of the poet Wilhelm Müller (1794–1827) describe the burst of sentiment they could rouse:

The horn is blowing from the street.
But what about it makes you leap,
My heart?
There's no letter coming for you:
Why is it you pound as you do,
My heart?
The mail's coming from the city,
Where lives a girl dear and pretty,
My heart!
Are you racing that way to see
Whether all is as it should be,
My heart?[7]

It seems doubtful whether a similar state of excitement—which requires waiting, after all—could be inspired by email or WhatsApp. At any rate, even when horses and carriages were still the norm, people complained about the noise from busy traffic. To remedy the situation, engineers experimented with a host of different paving surfaces, including wood and rubber. Satisfactory results were not forthcoming until the materials for wheels changed and tires were filled with air.[8]

Loud aircraft became a permanent nuisance, at the very latest, when jet engines were introduced in the 1950s.[9] Well before this point, squealing brakes and automobiles skidding around corners provided a source of

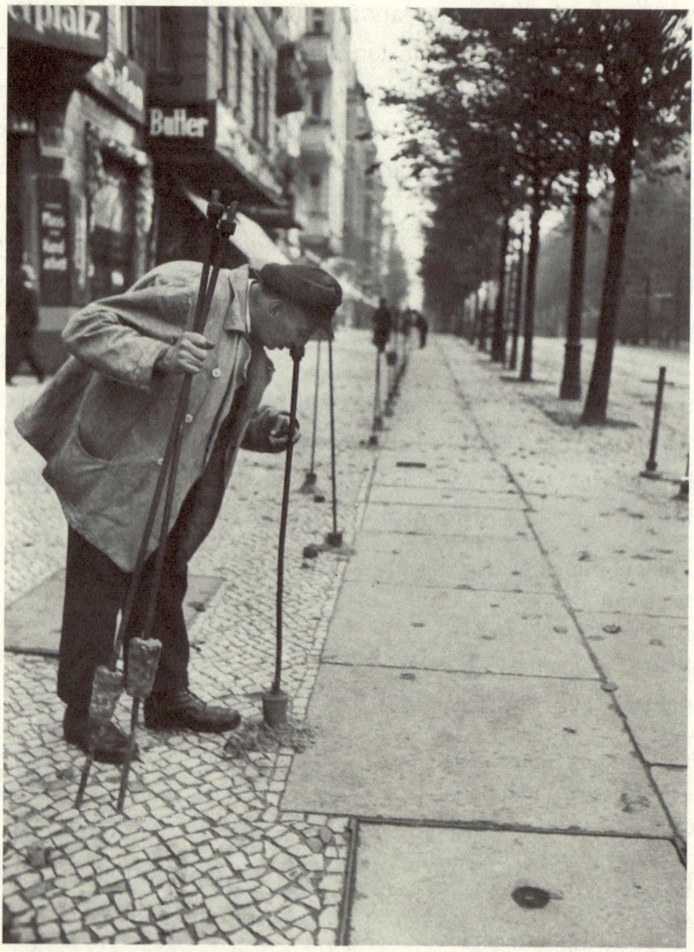

5.1 A Berlin gas detector in 1921 performs his job, identifying leaks in underground gas pipelines as early as possible. Photograph: Friedrich Seidenstücker (1882–1966). © Copyright bpk / Art Resource.

grief—as did streetcars alerting pedestrians to their approach. As Hildegard von Spitzemberg, baroness of Württemberg, lamented on a visit to Berlin in 1898, "The ringing of all these vehicles is ear-splitting." Unworldly provincials, she added, would occasionally fall into each other's arms in relief that they had survived the experience.[10] On behalf of overtaxed nerves everywhere, the philosopher Theodor Lessing undertook a campaign before World War I against the uncouth and noisy behavior of city residents.[11] His fight for "aural hygiene" bore little fruit, however.[12] Since then, cars have become much quieter, but the crush of traffic has increased.[13] Beginning in the 1970s, sound barriers began to be put up here and there—which amounted to impositions of another kind. Conversely, cars came to provide a refuge with sound environments of their own, adjustable at will.[14] High-speed trains now offer a sterile and soundless experience, and windows can no longer be opened.[15] Many passengers cut themselves off from the outside world even more with headphones; even conversations between other people have become annoying.

The visibility and aesthetics of infrastructures vary culturally and historically.[16] During the nineteenth century, it was a point of pride to make them as conspicuous as possible; water and gas tanks, power plants, and other facilities were towering fetishes and temples of modernity that boasted grand designs and elaborate ornamentation. It was practically a commonplace to compare public buildings on a certain scale to medieval structures. This was certainly the case for railroad stations such as Milano Centrale or Grand Central in New York City. Time and again, water towers and dams were hailed as "cathedrals of progress."[17] The Murray Hill water reservoir in New York, which was built in 1842, exemplifies the style of Egyptian Revival.[18] The Abbey Mills Pumping Station in London, constructed after the Great Stink, was lauded as a "proud mixture of Venetian Gothic and a touch of the Byzantine."[19]

But even before World War I, the abstraction of matter-of-fact capitalism was slowly replacing ostentation.[20] In the cities of the interwar and postwar periods, when functional considerations prevailed, residents perceived infrastructural arrangements—for instance, sewage treatment plants; heating, pumping, and electrical stations; prisons; hospitals; and police and fire buildings—as nuisances. Sober facades predominated, when facilities were not pushed out of urban centers altogether.[21] Similar measures were taken to reduce aboveground power lines and cables, which often spread across cities like giant spider webs.

It has been argued that moving infrastructural provisions underground exemplifies the concealment of power diagnosed by Pierre Bourdieu.[22] Sewers, electrical cables, and the like served as indicators of social justice—or its absence—until they were banished from view; now, there is no telling who is connected and who is not.[23] According to the urban historian Lewis Mumford, military considerations and security concerns dictated undergrounding measures.[24] Whatever the reason, efforts have multiplied to conceal the inner workings of cities, particularly in western Europe. In the United States and eastern Europe, it remains common to lay power lines, telephone cables, and to some extent, pipelines above ground; doing so makes construction and repair much cheaper (and demonstrates the abundance of resources). But even when infrastructure networks are materially present, routine usually pushes them below the threshold of consciousness. Their unrecognizability, according to Paul Edwards, follows from the temporal regime they establish for users. Infrastructures are set for the present and encourage people not to think about the past or future. They exist on the mesolevel of historical time, as it were—a register distinct from daily chronology.[25] Similarly, the sociologist Susan Leigh Star has observed that infrastructures are integrated into social structures, cultural patterns, and technologies in a way that escapes subjective perception.[26] Habituation makes them disappear.[27]

In this context, Robert Musil's observation about monuments applies to infrastructures, too. "Something seals them off from attention, which runs off them like water on oilskin, without stopping for even a moment."[28] Infrastructural provisions, especially older ones, often refer to the circumstances of their creation—another quality they share with monuments. Public fountains display an array of commemorative images, and manhole, or utility vault, covers present information about when and by whom they were made.[29] Bridges have plaques that name their builders. Many structures in Germany long sported a swastika, occasioning controversy time and again. The presence of the symbol of National Socialism was especially unwelcome when the two sides of the Hirschberg bridge over the Saale River were joined to herald rapprochement between the Federal Republic and the German Democratic Republic in the 1960s.[30]

Wherever one turns a watchful eye, signs point to infrastructures that ordinarily escape notice. Utility sites are full of small placards in different colors that provide information to construction crews, fire departments, and emergency services, indicating where a pipeline or conduit for gas, sewage, or electricity runs, say, or where the nearest water hydrant can be found. Hikers sometimes stumble on them even in remote locations. (Hayes describes how these markers disenchant the landscapes of the "Wild West.") Some signs scream out, literally or figuratively. Fire alarms, for instance, are impossible to ignore. Before the advent of cellular communications, telephone booths were painted brightly so they would be as visible as possible. Emergency call boxes built on private initiative seek to be equally conspicuous. Examples include the stations that have been set up in Germany since 1969 by the Björn Steiger Foundation, which honors the memory of a boy who died waiting for help after a traffic accident.

Navigating roadways requires visibility and a sense of overview. Clear demarcations of space and an orderly flow of traffic are necessary if travel is not to present excessive risks. To this end, a throughgoing system of conventions has developed over time. The first step was to make sidewalks in order to clear the way for carriages and horses (the nobility's preferred mode of conveyance). In turn, separate lanes minimized confusion by establishing zones where different speeds—and levels of social distinction—prevailed.[31] The nineteenth century witnessed the advent of clear street names and house numbers to provide better orientation.[32] Along the same lines, public spaces were inscribed, so to speak, so that codes of proper conduct would be observed.[33] Today, Germany employs 650 different kinds of signs on the seven hundred thousand kilometers of its roadways—some 25 million markers in all.[34]

Of course, the rules governing traffic should also be clear across borders. Over decades, international efforts have been made to synchronize not just technology but also icons, so that no knowledge is required of the language spoken where they are displayed. The League of Nations and the International Organization for Standardization played an early, and pioneering, role.[35] At present, the ISO oversees no fewer than eighteen thousand technical standards.[36] The "walk right, stand left" rule on escalators to the London Underground has been in place for about the last hundred years; the model of giving parties in a hurry their own lane has spread across the globe. The schematic timetables and line network plans devised by Henry C. Beck in 1933 have also become the international norm.[37] All the same, national and local peculiarities persist—and not just when it comes to which side of the road to drive on. Examples include the different ways travelers comport themselves in traffic and at bus or train stations, although most people intuitively adjust to local convention.

As early as 1840, Edgar Allan Poe's "The Man in the Crowd" made the danger lurking in anonymous urban streets its theme; since then, countless works of literature and film have revolved around the sudden appearance and disappearance of mysterious individuals in crowds. The corollary, for police and authorities, has been to develop the means for identifying travelers and controlling undesirable mobility; license plates, passports, and driver's licenses all serve this purpose. More recently, communications have been monitored through telephone numbers, email and IP addresses, and social media accounts. Systems of travel and exchange require calibration so that the whereabouts of suspicious (or simply missing) parties may be determined. The impersonal cracks in networks of transit and communication have long been a resource for criminals. Thus, in the nineteenth century, the time lag between telegraphed messages and physical movement was exploited to fraudulent ends in business, betting, and stock markets. But new communications technology also made it possible to seize fugitives who might otherwise have escaped. In 1910, for instance, the notorious Dr. Hawley Crippen, who murdered his wife, eluded English police by boarding a ship bound for Canada; when the captain received a telegram, however, the killer was caught at the next port.[38]

Unsurprisingly, then, some of the most pressing problems for infrastructural policy remain tracing, monitoring, and overseeing public and private behavior—while avoiding the abuse of sensitive and personal information.[39]

PRESTIGE AND POETRY

A few years ago, a sourcebook on the XXII Summer Olympic Games (1980) was published in Russia. The volume documents how infrastructures can benefit from exceptional events, but also how they can fail for the same reason. The Soviet Union had applied to the International Olympic Committee to host the international competition in 1956; not until 1974, when a favorable political constellation emerged during a period of détente, did Moscow receive the commission. Now, the Soviets had to invest in facilities and accommodations. Transportation and communication needed improvement, security and energy supplies had to be bolstered, and measures for minimizing environmental impact were taken. The wheels were promptly set into motion. But preparations coincided with economic stagnation, and Soviet leaders recognized, as early as 1976, that meeting expectations would prove impossible. The Games were nearly canceled. Ultimately, they were held after all, because the loss of face would have been too great.[40]

Like many large-scale projects, the undertaking followed a course of its own. Cost-cutting measures and time pressures meant that standards of quality sank and undermined the soundness of facilities. Plans for expanding the highway ring around Moscow and other roadways fell behind schedule. So that public transportation would be available to foreign visitors, authorities limited its use by Soviet citizens during the Games. Further snares resulted from other pragmatic decisions. One residential complex under construction was converted into an Olympic village, and the organizing committee found it necessary to import almost all the television technology for broadcasting events from the United States. Worse still, almost as soon as the slapdash arrangements were more or less done, in 1980, the United States and a sizable number of Western countries announced that they would not be participating because of the Soviet invasion of Afghanistan the previous year. All the same, former residents of the Eastern Bloc remember the Olympic Games as one of the last great events before the collapse of the Soviet Union —a grand, symbolic display of state socialism. Without much competition, the Soviet Union and its allies had many occasions to celebrate victory. Much the same was to occur four years later in Los Angeles—albeit in reverse—for Western nations. The United States would also come to appreciate that massive projects of this kind bring short-term investment but rarely pay off in the long term.[41]

Economic analysts have concluded that neither richer nor poorer countries should expect major events of this kind to fuel growth. Benefits are minimal for wealthier hosts, and the burden on budgets for less well-to-do nations proves excessive.[42] Thus, the 2004 Olympic Games in Athens contributed more to Greece's national debt than to infrastructural development.[43] The 2014 Winter Games in Sochi, Russia, swallowed a full $50 billion. This was the first time they were held in a subtropical city, but a good part of expenses went to building infrastructure, not to creating conditions suitable for athletic activity. Many facilities did not find a use afterward. It would be wrong, however, to regard staging major events as exclusively prestige oriented and always ill advised in financial terms.[44] A vast array of motives and expectations factor into the equation. Whether these have been realized can prove very difficult to determine.

Before the Olympic Games were revived or soccer and other world championships were held, Universal Exhibitions, which showcased industrial achievement, catalyzed intensive development.[45] The first such fair took place in London in 1851. From this point on, exhibitions in cities such as New York, Paris, Vienna, Philadelphia, Chicago, and Milan all featured the latest technological achievements.[46] A common feature was the infrastructural logistics developed in preparation for events. Thus, the first line of the Paris metro went into operation during the Exposition Universelle in 1900; it coincided with the opening of the Gare d'Orsay, Gare des Invalides, and Gare de Lyon, which heralded a new age of travel and transport to and from the capital. More than fifty million visitors could also admire the new Pont Alexandre III, symbolizing the alliance between France and Russia.

Efforts on this scale are undertaken in competition with other countries; demonstrating the latest in what is technologically possible provides political capital. Massive ventures reflect state power "because they address a fundamental problem of politics: Creating commonality out of diversity."[47] Projects win over the population by forging a collective, corporate identity. *Euergetism*—the practice of ruling through festivals, charitable acts, and donations—has persisted from premodern monarchies to the democracies of the present day. Many facilities and networks now in existence represent the aggregate of small- to medium-scale projects that have been synchronized and operate in concert. In retrospect, it may seem that thorough planning underlies the immense technological systems coordinating social life. But as a rule, this is only the case when governments pursue a program of accelerated modernization or the consensus emerges—in a specific constellation of political power—that it is necessary to catch up with advances elsewhere.[48]

During the 1920s, policies in the Soviet Union embraced infrastructure as a means of tying together experts, political leaders, and the general population. To gain acceptance, measures not only had to promise abundance; they had to do so in terms anchored in technological know-how and be based on practices of everyday life admitting ready comprehension. When announcing the great initiatives of his communist government, Joseph Stalin engaged an army of writers—whom he called "psychic engineers"—to give dams and power plants their blessings in appropriate language.[49] During the irrigation of the "hungry" Caucasian steppe, Soviet propaganda constantly deployed images of multiethnic cooperation—for instance, at the huge construction site of the Nevinnomyssk Canal—to promote the sense of a common cause. Organic metaphors abounded. In this way, the transformation of desert into land for growing cotton represented both a renewal and a return to the source. (In the 1970s, when it became clear that the depleted Aral Sea was drying out and disintegrating into smaller bodies of silty water, the same organicist imagery was turned against political leadership; now it evoked disease, infirmity, and decay.[50])

The professionals in charge of designing and carrying out projects wish for optimal conditions; after the fact, they regret only that circumstances were not more favorable. Technocratic ideology sets the tone. The premium on efficiency does much to explain why infrastructural planners have time and again proved susceptible to the lures of authoritarianism.[51] In this light, we can understand, at least in the abstract, Albert Speer's enthusiasm for Adolf Hitler's vision of turning Berlin into Germania, a megacity with ten million inhabitants. The dictator had declared,

> When one enters the Reich Chancellery, one should have the feeling that one is visiting the master of the world. One will arrive there along wide avenues containing the Triumphal Arch, the Pantheon of the Army, the Square of the People—things to take your breath away! It's only thus that we shall succeed in eclipsing our only rival in the world, Rome. . . . In ten thousand years they'll be still standing, just as they are, unless meanwhile the sea has again covered our plains.[52]

The desire to overwhelm others has rarely been voiced so bluntly, but military parades and ceremonial pageantry regularly accompany the unveiling of large-scale projects. The basic reasons are the same in Germany and elsewhere. Such measures have proved their effectiveness over and over—especially when colonial powers have sought to demonstrate the superiority of European culture.[53]

Countless large-scale projects in Latin American, African, and Asian countries show that technocratic ideology is thriving. As ever, it presumes that economic, technical, and social problems can be solved in one fell swoop with massive development and construction. This is precisely the view that has been called into question in western Europe and the United States since the 1970s; here, only incorrigible dogmatists would claim that one best solution exists for every problem. Initiatives have given rise to more and more political and social conflict; this turn need not be understood solely in negative terms, because it is the expression of a free, pluralistic society where issues are confronted directly.

The more that official rhetoric stresses the symbolic value of a project, the more likely it is that the public will voice objections about details, and possibly the very idea of the project. Recent examples from Germany include major projects such as the Stuttgart 21 railway line and the Berlin Brandenburg Airport. Up to the 1970s, large-scale infrastructural undertakings had an iconic value and stood for scientific and technical rationality; since then, they have turned into rallying points for protests. Almost without fail, opponents claim that decision-making has been undemocratic and the parties involved are somehow corrupt; defiant taxpayers sometimes welcome the chance to rebel against authority.[54]

But even in pluralistic societies, governmental and private bodies continue to pursue projects that promise symbolic returns. Such initiatives require coordinating an ever-larger number of actors motivated by different considerations and goals. Financial risk is involved because funds must be spent on schedule. Politicians push for pet projects to coincide with election cycles. Companies sometimes aim for greater profits by delaying construction and pocketing contractually guaranteed monies.[55] Building codes and fire protection ordinances (for instance, those that hindered the new Berlin airport) represent an additional factor. Regulations at the local, national, and international levels can reinforce each other, yielding a labyrinth only specialists can navigate. At times, no one has an overview of the whole.[56]

Attentive planners have concluded that the classic high-modern scheme—integral projects designed for decades of unchanged operation—should be abandoned. Projects open to unforeseen and even unlikely developments prove more successful—especially when a balance is struck in advance between different priorities. Factors include technological innovation, shifting cultural values, social events, and considerations of sustainability. Plans should allow for correction and adaptation.[57] Experience makes it advisable to take the possibility of intersecting processes into account. When parties involved create irreversible conditions or conclude informal agreements in order to advance their own interests, procedures that work in the abstract are often thwarted, undermined, or bypassed altogether. Plan-approval protocols now in place have reduced these risks and offer opportunities for involvement at different levels. Still, a fair share of infrastructure planning winds up being decided by mediators—or even a judge—because members of the public manage to exploit a wide range of options for raising objections; to some extent, this happens because of NIMBY (not in my backyard) attitudes.[58]

Studies offer detailed analysis of why projects run over budget. Added expenses creep in when responsibilities are not clear, the technology employed is incompletely developed, the wrong incentives are provided to contractors, planning is too rosy, or special interests are given priority. A prime case is building regional airports, which often represent "landscape maintenance" to serve political calculations.[59] Past projects that went well can achieve mythical dimensions and curse new undertakings because planners think the same results can be obtained again. Greater accountability is recommended for companies and agencies involved.[60]

That said, there is also precedent for such measures going awry. In the early years of the Soviet Union, for example, even slight deviations in the course of construction could lead architects and engineers to be sacrificed on the altar of planning.[61] Nor are dramatic professional risks unknown in Western countries. In Great Britain, under the Town and Country Planning Act of 1947, planning officers were appointed to supervise construction activity on homes and castles. They made life so difficult for people who wanted to build that on June 20, 1991, an angry homeowner shot one of them dead. By chance, the incident was caught on film—and the shooter became a national celebrity.[62]

6

CRUMBLING BRIDGES: THE LIFE CYCLES OF INFRASTRUCTURES

TEMPI PASSATI

In 1985, an opinion piece by the features editor of a leading German weekly quoted the words of Goethe, when he was serving on the privy council in Weimar, referring to work being performed in the area behind the train station. Attentive readers were puzzled: Goethe could be credited with great foresight, but the first German railroad (between Nuremberg and Fürth) didn't start running until three years after the poet had died, in 1832. There was no way he could have spoken of a train station. The board of the journal seized the opportunity to get rid of the editor, Fritz J. Raddatz, who was a difficult personality and had sometimes played a bit fast and loose with details. The novelist Martin Mosebach had a similar experience a few decades later. A book he published in 2014, which was set in 1991, had characters using cell phones as a matter of course. In fact, digital networks did not go into full operation in Germany until the following year. Worse still, the author presented laptops, email, and the Internet as features of everyday life.[1] Although these incidents demonstrate how busy writers might fail to get the facts straight, infrastructures do have a tendency to scramble the sense of time.

Infrastructures come into being, age, and sometimes vanish. As long as they seem self-evident, that they do not last forever is easily forgotten.

At most, people remark how portals, facades, equipment, and design vary with architectural and industrial fashion; because stylistic changes correspond to temporal stratifications of life and history, they can evoke nostalgia. Another time comes into view through old station buildings, vintage cars, double-decker buses, and even outdated computers or cell phones; some of these relics become valuable commodities. Usually, the same does not apply to networks themselves, unless they find new, modified use—as was the case with England's waterways. Very few people feel sentimental stirrings when they encounter old telephone or utility poles, and the thought of lead pipes in plumbing or rust on bridges provokes anxiety.[2]

Conversations in which older people recall times when things were not so easy can be fraught. New technologies often give rise to disagreement, with members of the younger generation objecting to the tiresome habits to which their elders cling by force of habit. In particular, the field of agriculture long resisted mechanization and electrification; those who stuck to their ways were regarded as behind the times.[3] To this day, when younger people give a new device to parents during the holidays, it is far from certain that the parents will use it.

Naming new technologies after functional predecessors evokes cultural memory; consider *steam horses*, *airports*, *email*, or even *information superhighway*. Design is frequently modeled on forerunners to convey a certain dignity: walls of dams have been made to look like castle ramparts, and railroad cars have been given the appointments of luxurious carriages. Not too long ago, television sets were built into the frames of classic furniture, and video cassettes were housed in covers reminiscent of books. Ring tones and digital icons likewise allude to the analogue world of yore.

When media changes rapidly, nostalgia for vanishing forms of experience builds up quickly. In 1980, the British band the Buggles charted a worldwide hit with their single "Video Killed the Radio Star." The next year, it was the first song broadcast on MTV. By 1984, the rock group Queen's "Radio Ga Ga" was reflecting on the changes music videos had wrought: "We watch the shows, we watch the stars / On videos for hours and hours / We hardly need to use our ears / How music changes through the years." Along the way, the song invokes broadcasting events such as Orson Welles's 1938 radio play *The War of the Worlds* and Winston Churchill's rousing wartime speech "This Was Their Finest Hour." Today, the preformatted content of social media all but ensures that people will interact only with others of the same age and sensibility, unless they modify settings and explore other offerings. By 2015, anyone born in 1980 or earlier already felt a bit like a dying breed: "We still know how to switch channels on the TV set, roll up the window in the car, deal with tangled wires, call information, tape over old music, and so on."[4]

Specific infrastructures shaped the experience of different generations in the nineteenth and twentieth centuries, providing media for self-reflection and self-definition.[5] In particular, advances in transportation and communication extended individual mobility and expanded horizons. The rule applies to steamships, railroads, interstate buses (in the United States, the Greyhound line has been in operation since 1914[6]), bicycles, motorcycles, and automobiles. The dial on early radio receivers opened up a world, even if faraway places like Hilversum or Luxembourg still seemed to exist only on the airwaves.[7] Within a given FM broadcast area, stations helped consolidate collective identity.[8] Cinema and television provided generations of viewers with things to talk about for days on end, and the structure of programming shaped temporal and social routines. Here, too, the trend toward heightened mobility was in evidence. Record players, tape machines, and jukeboxes were stationary, but eventually portable radios, Walkmans, MP3 players, and smartphones combined freedom of movement with the possibility of entertainment.[9]

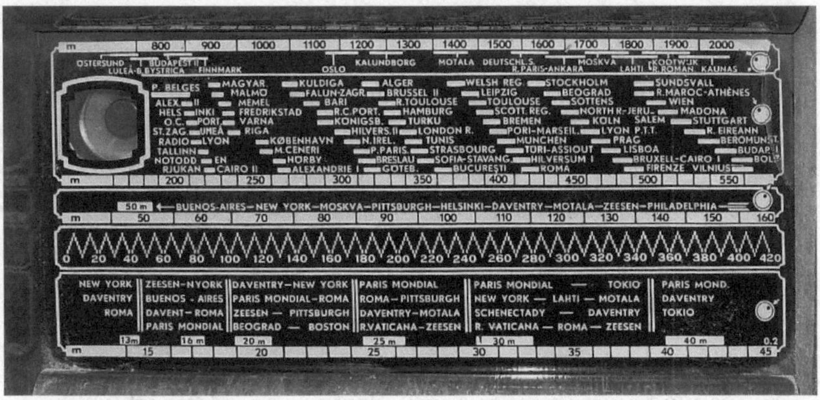

6.1 Radio frequencies of the analogue era included locations worldwide, places both known and unknown, which listeners could select in the twinkling of an eye.

As early as 1913, the Nuremberg philologist and journalist Wolfgang Riepl recognized a rule governing media and their evolution. Since then, the history of communications has substantiated Riepl's law:

> The most simple means, forms, and methods are never fully displaced or decommissioned by even the most perfect and most highly developed forms once they have become established and are found useful; rather, they are preserved, but now compelled to search for other tasks and fields of application.[10]

New media rarely replace old ones in full. In general, they make forerunners change in response to the competition they offer or push them into the background. The dynamic has recurred throughout the history of infrastructures. Railroads posed a challenge to inland canals, trucks took over many services provided by railroads, radio networks performed the tasks of wire telegraphy, email did much of what postal correspondence had done, and so on. Some forms of communication have disappeared almost entirely—for instance, the pantelegraph and optical telegraphy.[11] At the same time, older technology can prove restrictive and make innovations operate at an anachronistic level. Thus, the announcements over intercoms at train stations still occur at frequencies of up to 3,400 hertz—which often means that messages are incomprehensible; this limitation dates back to the days of analogue telephones, which could not handle higher capacities.[12]

Experiments in England from 1853 on provide a striking example of the second life that infrastructures can lead. The system of pneumatic tubes developed here expanded into ramified networks in many European cities. At one point, Berlin had ninety pneumatic post offices, four hundred kilometers of tubing underground and two thousand kilometers on the surface.[13] Around 1870, in London and New York, the idea of transporting passengers by "pneumatic dispatch" was entertained. Efforts were soon abandoned: such conveyance was no faster than systems already in operation, and it required travelers to be willing to ride on "ghost trains."[14] When Berlin became a divided city after World War II, the pneumatic tube system lost its importance, although it continued to be used at some workplaces in East Berlin for many years. Subsequently, many conduit routes came to house telephone and electrical cables.

The long life of infrastructures is evident in continued reliance on existing pathways and routes, especially after wars and during transitional

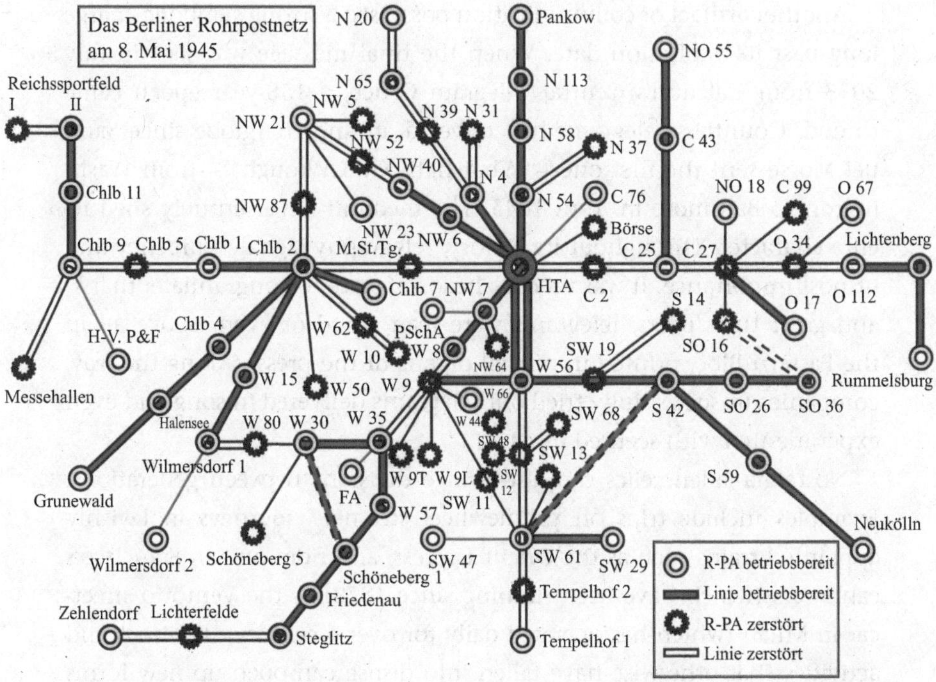

6.2 Before 1945, Berlin and other large cities were home to pneumatic tube networks that carried letters at lightning speed. *Source*: Collection telosgraphein.

periods. Many of today's provisions for traffic, supply, and disposal follow patterns that evolved over centuries—or even millennia. The spaces adapted for such purposes prove much more enduring than the various systems set up in them.[15] A 1988 survey found that "the man-made backbone of our society is probably well established and unlikely to change."[16] Technological innovations are superimposed on old arrangements, so to speak; thus, the Internet piggybacked on the telephone to create an overlay network until fiber-optic cables came into use.[17] That said, the power of technical advances to penetrate the field of existing services tends to be overestimated; in contrast, projections about the length of time that old systems will continue to be used alongside new ones generally fall short.[18] In confirmation of Riepl's law, the German postal system still accepts telephone calls, faxes, or emails that then, via snail mail, are passed along to intended recipients.[19]

Another artifact of communication possessed a strong symbolic charge long past its expiration date. When the final message was sent in July 2013 from Calcutta's Central Telegram Office, a 158-year epoch came to end. Countless telegrams had traveled around the globe since Samuel Morse sent the first one—"What hath God Wrought"—from Washington to Baltimore in April 1845. The medium never entirely shed its elite character. Throughout its history, telegraphy signaled urgency and utmost importance. It was used by heads of state to congratulate, thank, and greet their peers; telegrams were read aloud or even, above all in the Eastern Bloc, printed in official organs of the press. (Along the way, companies unsuccessfully tried out telegrams delivered in song and even experimented with scented missives.)

Nostalgia-laden relics can also bridge divisions between generations. Examples include trips on paddlewheel steamers, journeys in lavishly appointed trains such as the Orient Express, and rides on San Francisco's cable cars (which have been running since 1878) or the Ventotto streetcar in Milan (which has operated daily for over a century). Routines and activities that otherwise have fallen into disuse can open up new forms of experience with an age-old pedigree and have the effect of traveling in time. This is why they appeal to tourists. Route 66—the Mother Road—between Chicago and Santa Monica has been granted a new lease on life and is advertised as the path to modern-day adventure. Travel on the few remaining night trains would have long since vanished had not so many films and novels been set on them.[20] And a ride on a horse-drawn carriage now seems de rigueur on visits to Vienna, if only because the leisurely pace bespeaks the luxury of having time and money to spend.

Although infrastructures of bygone eras offer more than the opportunity to relive the past, this feature is worthy of note. Recall the previously mentioned efforts in Berlin to keep gas lighting on city streets; advocates deem gas a "natural" light and a valuable cultural inheritance that "has a relaxing effect and creates atmosphere."[21] Initiatives elsewhere seek to preserve lighthouses, even though the satellite-based GPS has long performed the function they once provided. For example, the Red Sand lighthouse in the Weser estuary and the lighthouse at Cape Finisterre in Galicia have only gained in local, regional, and national significance.[22] They have become sites of remembrance, historic landmarks, and even the setting for works of literature and film, due to what they evoke offering contrast with the social isolation, technological unreliability, and general disorientation of modern times.[23]

The rudiments of abandoned infrastructures—bridges that seem to stand there for no reason, overgrown stretches of highway or railroad tracks—also serve as markers of transience.[24] Sometimes nature seems to be reclaiming space from human hands. Subway cars that have outlived their usefulness in New York are sunk along the East Coast, where they become artificial reefs that attract marine life, to the delight of fishers and divers.[25] But most remnants of old infrastructures are built over or dedicated to new ends. Old facilities have been transformed into institutions serving social or cultural purposes—slaughterhouses, gasworks, water towers, train stations, and more.

Bunkers have roots anchored especially firmly in the past,[26] and they are too massive to remove. Nor can one be certain they will not be needed again. Nevertheless, in the 1990s, the municipal government of Beijing opened up the approximately five thousand bunkers beneath the city for civilian use by renting out individual units. It is estimated that one million people live there today—for the most part migrant workers from the countryside. Instead of paying the staggering rents for conventional housing, they save money to send home to their families. Hygienic conditions are considered satisfactory; the only misgiving is the possibility of a gas leak or fire breaking out.[27] Switzerland, a small country harboring any number of mysteries, has built nearly three hundred thousand bunkers in recent decades. The government declared that every piece of property over a certain size needed an accompanying shelter. So that these spaces would find use under nonemergency conditions, some have been turned into music salons, quarters for asylum seekers, and library archives. In one, conceptual artists have opened the first zero-star hotel. (Civil defense law specifies that spaces must be able to return to their intended purpose within twenty-four hours.[28])

Unsurprisingly, Albania has the most enigmatic shelters. Anachronistic Stalinism isolated the country completely. The nation's leaders felt threatened from all sides; enemies lurked everywhere, even within. In 1972, Enver Hoxha, the head of state, saw to it that standardized bunkers were built throughout the country. Upon his death in 1985, the target number of

750,000 still had not yet been reached, but Albania's economic resources were exhausted. Since the early 1990s, facilities have been repurposed by an international host of creative types looking to turn them into something original. Some of the bunkers at beaches have become changing stations and others house livestock. Others still have been decorated by artists, transformed into hamburger stands, or, with the addition of doors and windows, made into basic bed-and-breakfast accommodations.[29]

WAITING FOR SERVICE

Poetic capital has been minted from unforeseen and fantastic events in the anonymous networks and impersonal landscapes of city life. The exemplary case involves an albino alligator that, according to a *New York Times* article published in February 1935, had made its home in the sewers.[30] If *white elephant* is synonymous with failed large-scale projects, *white alligator* is shorthand for urban legends. Robert Daley's 1959 book *The World Beneath the City*, popularized the story.[31] In 1980, it was adapted for the screen in *Alligator*; a decade later *Alligator II: The Mutation* appeared. These movies belonged to a spate of Hollywood productions that, for at least two decades running, focused on dangers associated with infrastructures, exploiting the anxiety of people who inhabit artificial modern environments.

Disaster films always unfold in the same way. A representative sampling of familiar types of people, with everyday hopes and conflicts, is introduced; then, catastrophe sweeps away all normality. Invariably, one clear-sighted individual (or sometimes a small group) has tried to warn authorities, only to be brushed aside. Most screen time is devoted to portraying the rage of destruction. The main role falls to a heroic party who possesses the necessary qualities for facing the exceptional circumstances: alertness, empathy, decisiveness, and above all, an unshakable will to persist despite hopeless odds. The basic dramaturgy, which admits infinite variation, was prefigured by *King Kong* (1933), *San Francisco* (1936), and the Japanese *Godzilla* movies. In the 1970s, the genre came into its own when films portrayed realistic, contemporary settings in which everything is already barely holding together. Often, disaster strikes systems of mass transportation—for instance, ships, railroads, or subways. Airplanes have proven especially popular. Things can go wrong because of technical glitches or human error; more recently, terrorist attacks have played a prominent role. Variants include skyscrapers in flames, malfunctioning elevators, collapsing tunnels, bridges and dams, and threats to supplies of water and electricity. Scholars break down the basic plotlines into "The Ship of Fools" or "The City Fails," which contrast with scenarios where a wholly external cause (say, alien invasion) starts the trouble.[32]

Cinematic myths would hardly enjoy so much success if they were purely a matter of fantasy. Real disasters have attended technological advances, and reactions to them reveal both general human fears and concerns specific to different cultures, societies, and geographic locations. The possibility of mishap haunts normal functioning and operations. What once counted as divine *fatum* came to be viewed, especially in the course of the nineteenth century, in terms of secular causes. Now, accidents are more likely to be exceptions that might be averted through better planning and maintenance. Trained cadres of safety engineers, insurance agents, lawyers, and ultimately, judges promise to read the future.[33]

The sinking of the *Titanic* and other transportation disasters became the epochal signature of the late nineteenth and early twentieth centuries.[34] The collapse of a railroad bridge on the Firth of Tay in Scotland on December 28, 1879, provided the German novelist and poet Theodor Fontane the occasion to reflect that whatever is "wrought by human hands" is nothing but "dross" (*Tand*). It has come to be taken for granted that a certain number of deaths will occur on roadways.[35] Indeed, unfortunate incidents—exploding gas or steam boilers, collapsing bridges, and broken dams, for instance—have led to better engineering and know-how.[36] But as the authors of a 1937 assessment of urban services observed, there have been "countless sacrifices of human life"; designs "had to fail before the experience gained" enabled the "thoroughgoing regulation of construction and operations."[37] One of the most understudied aspects of infrastructure is the learning process behind it all. Infrastructures materialize knowledge. Experts and personnel responsible for construction, maintenance, and upkeep occupy a paradoxical position. Their work usually takes place outside the public eye, and they are supposed to interfere as little as possible with routine operations. Looking back on two hundred years of experience, a 1976 report by the American Public Works Association noted that achievements have occurred with little fanfare. For all that, they represent nothing less than the foundations of US civilization.[38]

Wherever parties directly involved are present—for instance, workers on construction sites blocking traffic or lobbyists demanding a different allocation of tax funds—the public often reacts with displeasure. As early as 1880, the journalist Julius Rodenberg complained about ongoing work on Berlin's sewer system, "which has for years been churning up the city now here, now there, turning it into pits and mountains of sand."[39] Urban legend would have it that projects are scheduled for the beginning of the vacation season so that drivers will appreciate that even more investment is needed.[40] The phrase *under construction* has become synonymous with inconvenience and delay—but also a sign of affluence and good things to come.[41] High-modern construction projects were supposed to be undertakings of limited duration that would improve undesirable conditions and facilitate normal life. Accordingly, Rodenberg conceded that work was necessary to "ensure that Berlin's tremendous growth" would benefit from "free circulation and development."[42] More recently, however, unfinished projects have become a permanent state of affairs. The multitude of networks now in existence means that something is always breaking down or needs to be updated. Level-headedness and patience are required of the public. We address strikes and purposeful disruptions in chapter 7; for the moment, our attention falls on unintentional interruptions that draw unwelcome attention to operations.

The problem of aging infrastructural facilities was first addressed in the United States. Here, development had been relatively concentrated and continuous, and the adverse effects of war were not pronounced. The booming economy led to heavier use and wear and tear. Ecological concerns emerged earlier in the United States than elsewhere, as did debates about decaying facilities. Rachel Carson's *Silent Spring*, which appeared in 1962, sounded a wake-up call about the natural world. A later title, *America in Ruins* (1981), did much the same for human-made environments, arguing that infrastructural arrangements were being run down faster than they could be replaced.[43] Ever since, talk about how the United States is falling apart has flourished. In 1982, *Business Week* declared *infrastructure* the new buzzword.[44] Doomsayers can invoke numerous ominous events—for instance, the blackouts in New York that occurred in 1965 and 1977 or the city's power failure in 2003. Such events have spawned Hollywood movies and urban legends (e.g., the supposed rise of the city's birthrate after outages).[45]

Bridges, in particular, feature in discussions of overloading and material fatigue. Suspension bridges collapsed under the weight of military columns in the nineteenth century; engineers recognized too late that synchronized marching generated dangerous vibrations. A particularly grievous collapse occurred in Angers, France, on April 16, 1850, and 226 people died. Even though the bridge had opened just eleven years earlier, its poorly insulated wires were corroded. The investigating commission was headed by Jules Dupuit, who would later be known as the first economist to evaluate the efficiency of public works contracts. Other sites of misadventure include the Quebec Bridge over the Saint Lawrence River, which collapsed during construction in 1907 because deadweight had been miscalculated. The seventy-five victims included thirty-three members of the Mohawk Nation, who were often employed in building because they were said never to suffer from vertigo.[46] The Tacoma Narrows Bridge in Washington State has become the textbook case of flawed planning. In 1940, after just three months of operation, it began to sway perilously when strong winds blew; before long, it fell altogether. The design now provides a cautionary tale to students in introductory physics courses the world over.[47]

The Silver Bridge over the Ohio River, which collapsed in 1967, is more typical of problems. It remained standing for forty years—until greater traffic volume and heavier vehicles proved its ruin. In 1927, the Ford Tin Lizzy had weighed in at just under seven hundred kilograms. In the following decades, the average weight of cars doubled. More still, the big-rig trucks celebrated in contemporary pop culture could be up to twenty-seven thousand kilograms.[48] During the evening rush hour, minor corrosion in a supporting pillar led the structure to give way, and forty-six people plummeted into the river. The bad news triggered a flurry of efforts to assess implications in both the short and the long terms. Just a few months after the disaster, the Silver Bridge reopened to traffic, but President Lyndon B. Johnson took the cautionary measure of ordering 1,800 other bridges that were getting on in years to be inspected.

Whenever unfortunate events occur, professional analyses of the causes contribute to the international knowledge base of engineers. That said, it is not always possible to act on findings right away because of local obstacles; since the late 1960s, shrinking public budgets and existing policies have compounded difficulties. Even now, funding tends to go to ostentatious new projects rather than the upkeep of existing structures. Repairs cause annoying delays; even when they are successful, the benefits remain largely invisible. To get political wheels moving, it often takes dramatic vocabulary that conjures up threatening scenarios and astronomical levels of investment. One consistently effective strategy is to argue against falling behind other communities, regions, or nations.[49] The attrition of consumer goods ranging from light bulbs to automobiles has long received attention (albeit less in terms of improvement than profit). It is extremely rare for infrastructures to be examined, much less planned, with the same considerations in mind.[50] Nor does the fabled mechanism of modern invention seem to apply. Practices and products that have been forgotten are marketed as innovations; business thrives on cosmetic changes masquerading as something new. Since 1997, organizations such as the Component Obsolescence Group have sought to mitigate the effects of shorter and shorter product life cycles by ensuring the availability of parts for repairs in areas such as space travel, military equipment, railways, and power plants.[51]

But infrastructures do have natural lifespans. Early on, railroad ties lasted no longer than five to seven years. At any given point between 1845 and 1870—when the total amount of track in Europe increased tenfold, to eighty thousand kilometers—15 to 20 percent of railways needed repair. The demand for wood shrank timberlands at an unprecedented rate.[52] Today, when best practices are observed, most infrastructural edifices using metal or concrete have a service life of fifty to seventy-five years.[53] These periods can prove considerably shorter, however, because of increasing use rates and loads. For much of their history, no systematic provisions have been made for repairing or replacing the individual components of infrastructures. As a result, roadways, data networks, and railroads have worn out prematurely, as they receive more and more use, and demands for comfort, reliability, and environmental responsibility mount. These circumstances drive up maintenance costs and increase the difficulty of introducing new services.

By nature, infrastructures seem ill equipped for demographic change. Planners and politicians presume that populations and rates of consumption will continue to rise. As one observer noted in 1992, professionals display "a deep aversion to shrinkage or reduction, which evidently evokes associations of decline or regression"; such "perspectives trigger deep unease in a society oriented on growth from top to bottom."[54] After all, not just infrastructures have life cycles; so do the people who use them, in ways that differ from one generation to the next. In Europe, most planners, politicians, and ordinary citizens grew up and were shaped by the extraordinary prosperity characterizing the decades after 1945. Conditions allowed governments to plan and implement infrastructural developments with few hindrances, and the rhetoric used to justify measures went unquestioned. But over time, facilities have aged and come to require more and more elaborate maintenance. Political discourse started to shift from universal coverage toward criteria such as reliability and sustainability. Attention now falls on outstanding repairs, maintenance crises, and avoiding lapses of service and operation—not on creating equal living conditions.

The task is all but Herculean. The American Society of Civil Engineers estimates that a quarter of the roughly six hundred thousand bridges in the United States harbor structural defects. One-third of major roadways are in poor or mediocre condition. More than four thousand dams have serious safety problems. Some $2.2 trillion needs to be invested in infrastructural measures.[55] This assessment may seem another overstatement intended to prompt political action. However, the bitter reality is evident in events such as the 2007 collapse, during rush hour, of an eight-lane bridge in Minneapolis, when thirteen people drowned in the Mississippi River.

A 2013 study conducted by the German Institute for Economic Research (Deutsches Institut für Wirtschaftsforschung) calculated that the annual investment required for maintaining the country's infrastructure stood at seventy-five billion euros. The federal government is responsible for ensuring the integrity of 38,000 highway bridges; states and municipalities are charged with maintaining another 120,000. They are normally inspected for defects every three years and undergo a major review every six years. Such measures require restricted use and intermittent closure; commuters have to find new routes, and new congestion ensues. The inconvenience is especially great seeing that it affects trucks bearing heavy goods, which need to find bridges that can handle the strain.[56]

Arthritic traffic routes reveals unwelcome conditions long in the making—for instance, that the German rail network has been thinning out instead of expanding for decades; some yards have been in operation for a whole century.[57] In the United States, railways were once a point of pride, but the country became the land of cars and planes in the twentieth century. Matters are much the same in Latin America. The elaborate railway systems put in place in the nineteenth century have suffered neglect, especially because of cuts favoring private bus companies in recent decades. In their quest for new markets, European and North American companies, assisted by the World Bank, urged governments to invest in roads.[58] In the 1930s, Argentina had forty-three thousand kilometers of railroad tracks; bananas, saltpeter, sugarcane, ore, and workers all traveled by train. Today, only one-fifth of the network is still in use. On occasion, services that have been revived for tourists—for instance, the Tren a las Nubes, which travels to a 4,144-meter-high viaduct in the Andes and back again—are pelted with stones because local communities derive no benefit from them.[59]

Since the 1990s, many of the state-owned railroads in Central and Latin America have been deregulated and privatized. Such measures were introduced, among other reasons, to curb the influence of unions. Moreover, the powerful shipping lobby has blocked further investment in railways; left-wing critics call neoliberal governments *ferrocidas*, or "train murderers." And in the rare case when railroads are reactivated, routine, everyday matters seem to have been forgotten. According to a 2010 report from Peru, vendors in Juliaca rushed to remove their stalls from the tracks whenever a train was coming. Costa Rica, in turn, witnessed repeated accidents on a line to suburban San José because residents didn't realize that the railways were still in operation.[60]

7

ACHILLES' HEELS: THE VULNERABILITY OF LARGE-SCALE NETWORKS

UNOBSTRUCTED OPERATIONS

A "little manual for popular decency" (*Anstandsbüchlein für das Volk*) published in 1894 devoted a chapter to "conduct on the street." Observing that everything one does here is "exposed to public attention and judgment,"[1] the author, Franz Vogt, offered practical advice. Among other things, he instructed readers to consider social status. "If you are walking with a more distinguished person, or one higher in standing, always yield the place of honor."[2] The same rule applied to verbal exchanges. "Conversation on the street should be brief; people should only say what is absolutely necessary. The sign of farewell must be given by the party of higher standing."[3] Needless to say, age was also a factor: younger people should defer to their elders. Such counsel rested on the principles of bourgeois propriety; Vogt was elaborating a code to make the gears of metropolitan traffic turn with as little friction and as few sparks as possible. In this regard, the behavior he advocated no longer derived simply from class-based notions of rank but also from functional logic: circulation should flow without interruption, and pedestrians should draw minimal attention to themselves.

Two things were necessary for streets to afford any sense of safety: an appropriate design of space and suitable instruction for pedestrians and vehicle operators. As early as 1902, the president of Arnsberg, an administrative district in western Germany, declared that schools needed "to offer classroom instruction about the causes of accidents and urge parents to supervise their young and feckless children carefully." "In particular," a contemporary author advised,

> a remedy must be found for the deplorable habit of school-age children crossing the tracks at the last moment out of sheer wantonness, in spite of warning signals from streetcar drivers. . . . The methods involve instilling concern for physical integrity in accordance with God's fifth commandment, making clear the far-reaching consequences of accidents caused by thoughtlessness, and establishing precise rules of conduct for relevant situations.[4]

No sector of infrastructure offers as direct a parallel between development and educational—indeed, disciplinary—measures as traffic and transportation. Here, the twin considerations of free mobility and safety stand in manifest contradiction. Neglect or failure to observe routines and convention can bring injury and death. All the same, risks are accepted, and children are taught as early as possible to exercise caution in this man-made environment.

As early as the nineteenth century, debates concerned the relationship between habitat and habitus, the setting in which people live and the ways they react to it. The preceding quotations from the "decency manual" make clear that attempts to control behavior and instill a salutary sense of danger have changed since then.[5] Cultivating responsible forms of movement and communication has long played a key role in psychology, pedagogy, and projects for social morality, even if the particulars have changed. The undertaking was, and is, vast. Vehicle operators and pedestrians needed to adopt patterns of conduct suited to mass movement. The overall balance was shaky because local cultures could vary significantly.[6] Legend holds that cyclists proved especially resistant to observing rules. The drunk, or otherwise incapacitated, driver represented a dangerous exception.

New regulations and increased circulation conflicted with established routines. People who had spent their entire lives in a given area often resisted rules imposed by anonymous officials and continued to act the way they always had. By the same token, they considered the outsiders using their streets to be suffering from a disorder ("gasoline rabies," in one memorable phrase).[7] Any European or North American who has tried to anticipate the behavior of pedestrians at a crosswalk in, say, Rio de

Janeiro or New Delhi will appreciate how much expectations and conduct vary from country to country and culture to culture.[8]

Travelers accept traffic signs and rules for the self-evident reason of avoiding mishap. Readiness to expose oneself to hazards has developed in parallel to the street—formerly an environment for living—transforming into a conduit to get somewhere else.[9] Force of habit and reflexes make behavior more or less predictable. For some time now, orderly circulation on roadways has offered a model for social equilibrium beyond what authorities are able to regulate and monitor from above. Reliability promotes a sense of security.[10] Trust and shared expectations enable spaces not only to function but to improve.[11] All the same, more than a million people die in traffic accidents every year, and road safety qualifies as a global challenge.[12]

Over a million now travel by air on any given day. Even though statistics confirm the safety of flight, it does seem surprising that passengers feel comfortable using it. More remarkably, perhaps, there were people willing to try out the new technology in the first place. Early travelers by rail said they felt like "piece[s] of freight."[13] When, after World War I, cabin airplanes came into use, a fur coat, aviator goggles, and an insulated leather cap were sometimes issued to those on board.[14] Supplying alcohol, granting permission to smoke, and more recently, offering all manner of electronic distractions have served to provide reassurance and promote calm.

"Sky girls" made their debut around 1930. At the time, airplanes had to make do without regulated pressure, and these women would serve as nurses, should the need arise. Stewardesses are still in charge of seeing that safety standards (which have only increased) are observed. With a consistently friendly demeanor, they perform valuable emotional work to help antsy passengers overcome anxiety.[15] In contrast to automobiles, airplanes offer travelers no room to take charge of the situation. Knowing, in the abstract, that flying is relatively safe is not easily reconciled with awareness that accidents do happen. Every case of something going awry fuels the fires of imagination. It seems unthinkable, but efforts have been made to conceal mishaps; as late as December 1986, East German authorities tried to hush up the crash of a Soviet Tupolev 134 near Berlin, which claimed seventy-two lives.[16]

The maintenance and safety work that airlines are required to perform represents a complex game of transparency and camouflage. Mass checks and clearances have yielded a highly professionalized field that relies on a vast array of confidence-building measures—for example, the security gates that were introduced in the early 1970s, following a series of hijackings. However, no new step can resolve problems once and for all. The effectiveness of changes is subject to ongoing debate.[17] The phrase *security theater* has been coined for procedures that contribute to a sense of safety without, in fact, enhancing it in any real way. Measures include the ubiquity of uniformed personnel, set questions and check-in rituals before passengers board planes, and instructions provided by crew members before departure.[18] From the time travelers arrive at the airport until their plane takes off, their freedom of movement has become more and more restricted over the last few decades. Increasingly elaborate, and profitable, commercial spaces in waiting areas distract them with assorted goods and services. In some instances, complaints have been voiced (especially about checks perceived as invasive following the September 11, 2001, attacks)—but resistance proves slight on the whole.[19]

Still, basic confidence in infrastructural reliability remains precarious. Without trained personnel available to personalize operations, it takes time and effort to cultivate trust. Staffing has continued to decline because technology can perform many functions previously done by human beings. Another incentive to reduce staff is that employees have the power to halt operations. This potential represents a core vulnerability of industrial society—especially when workers go on strike. Disruptions and added expenses, which increase the pressure to negotiate, amount to the dialectical complement of the conveniences otherwise enjoyed.

From the outset, strikes have represented sites of conflict and solidarity. Thus, on May 19, 1900, the employees of the Große Berliner Straßenbahn AG halted operations to demand higher wages and better working conditions. The majority of customers sided with them, throwing stones at streetcars driven by strikebreakers and lifting vehicles off the tracks. Angry strikers unhitched animals from horse-drawn streetcars, and drivers blocked roadways.[20] In November 1932, National Socialists and Communists even conducted a joint strike against wage cuts by the Berlin Transport Company (Berliner Verkehrsgesellschaft).[21] Strikes in South Africa that hit transport, supply, and waste disposal facilities in the beginning of the twentieth century also met with popular support, developing into a tradition for anti-colonial and anti-racist social movements. The symbolism of connecting some people while excluding others was just too obvious.[22] Opponents allied with business countered with dramatic claims that strikes affecting public services damaged the economy and threatened society as a whole; indeed, they might even lead to deaths. On principle alone, the argument went, strikes should not be tolerated. As early as 1917, the American Anti-Boycott Association demanded that walkouts be prohibited by law.[23] One way of doing away with the right to strike is to make vehicle operators civil servants.

Historically, workers have been able to rely on public understanding and even support. Temporary disruptions have met with acceptance, and people affected improvise other arrangements because strikers tend to be poorly paid for thankless jobs, especially waste disposal. But when the public has the impression that it is being held hostage to special interests, solidarity decreases. This has often been the case for pilots, air traffic controllers, and members of other professions who are prominent and well remunerated. Time and again, claims have been made that labor disputes do not harm employers so much as consumers who need vital services; accordingly, the argument goes, stricter standards should be observed to determine whether a strike is legitimate.[24] All the same, the European Court of Human Rights has established the right to strike for civil servants. At least in principle, "servants of the state" now enjoy the same rights as those employed in the private sector.[25]

It is generally assumed that infrastructures will run smoothly. Allowances are made for interruptions that, depending on the affected parties' dispositions, occasion humorous remarks or grumbling. Consumers have learned to take the promises of advertising with a grain of salt. Still, the gap between expectations and what is actually provided can be shocking. In the 1960s, the German Federal Railway (Deutsche Bahn) employed the slogan "Everybody talks about the weather. Not us." To this day, commuters cite the phrase with a smirk when weather-related disruptions occur—icy switches, downed overhead lines, failed air-conditioning systems, and so on. Most of the time, disruptions can be remedied quickly because special equipment and skilled personnel stand at the ready—removal services, teams of construction workers and electricians, and in extreme cases, relief organizations. Such organizations rarely make an appearance in

7.1 To stop streetcars driven by strikebreakers, rails were blocked in Berlin-Schöneberg in November 1932. *Source*: Bundesarchiv (Federal Archives)

day-to-day life because planning has minimized hazards.[26] Contingency plans exist for emergency situations—for instance, evacuation plans for cities and buildings. Regulations on the books prescribe how space should be organized so rescue and firefighting vehicles can get to sites of trouble.

The watchword is *resilience*. This term, put into circulation in the 1970s by psychologists investigating how children cope with trauma,[27] names a key category for assessing whether large-scale socio-technical systems will prove flexible enough to survive stress. Specialists research everything that can go wrong. Such information is handled with discretion and does not leave professional circles; after all, if safety and calm are to prevail, it is important that not everything be known. (Pest control experts in Hamburg, for example, keep details about managing the city's rat problem secret so that residents will not panic.[28]) Many things stay under wraps because knowledge could be misused for terrorist purposes. It has long been clear that infrastructures harbor points of vulnerability. Moreover, levels of interconnection and interdependence are only growing, which means heightened risk. In this context, large-scale systems can amplify instances of malfunction, causing problems to multiply. The record shows that defects do not cause the greatest damage; cascading effects that spread to adjacent systems are the real source of trouble.[29] Harbors, airports, railway stations, power plants, waterworks, and communications centers—that is, connective infrastructures—are especially liable to danger.[30] Special security forces (rail and harbor police, customs agents, and so on) work to ensure their safety by preventing unlawful activity. But connective infrastructures can also suffer damage as the result of accident. For example, in 2011, a seventy-five-year-old retiree looking for scrap metal allegedly cut off electronic communications between Armenia and Georgia with an electric saw. "I have no idea what the Internet is!" she tearfully protested.[31]

Environmental events qualify as disasters when Mother Nature overwhelms human-made nature. Under normal circumstances, the two realms are viewed as separate.[32] Trouble begins when vital networks of supply or communication are not up to the challenge posed by environmental forces, targeted attack, or systemic failure—that is, when they are not sufficiently resilient. Disaster can also result from weaknesses and defects in planning and operations, which is why experts speak of incubation periods. In such cases, seemingly insignificant errors build up and, in the aggregate, lead to breakdown.[33] Thus, for the Western world, periodic blackouts were all but a matter of course in the early days of gas and electricity; in many parts of the globe they remain regular occurrences. The parties affected cope as best they can by falling back on earlier technologies and practices; "obsolete" equipment can be a lifesaver in an emergency.

One of the perils of today's elaborate networks is that people, especially younger ones, are unlikely to be familiar with improvisatory measures such as repairing old devices or even making a fire.[34] And so long as systems are not visibly impaired or threatened—by war, for instance—readiness for contingency situations decreases even more. Hence the vulnerability paradox: the better that services function in a given society, the greater the disruption if they fail.[35] Experts note the mounting perils facing infrastructures; consider, for example, overlapping, interdependent networks, which pose risks that will only increase. In particular, smart grids—digital systems for controlling the generation and consumption of power—represent a source of concern; the consequences of a blackout will prove more severe if water supply and communications are affected, too. Additional dangers exist when earlier technologies and modes of living cannot be reactivated or resumed.[36] Professionals have drawn up scenarios; an electricity outage of long duration, for instance, would cause everyday routines to dissolve, to the obvious detriment of social order.[37] Events such as the power failures that occurred in New York in 1965 and 1977—which lasted longer than anticipated—represent the problem in miniature.[38]

In recognition of such dangers, networked societies have come to rely on measures for safeguarding conditions of normality, which means accepting restrictions on individual liberties. A constant topic of debate throughout the twentieth century was the extent to which governmental surveillance could be accepted. Threats, whether real or imagined, increased popular willingness to accept checks on freedom and monitoring. The questions persist. Surveys suggest that most citizens agree with security measures not only in cases of manifest threats but also, increasingly, to simply keep the infrastructures on which they have come to depend operational. But they usually embrace a return to the status quo as soon as disturbances are over; as if nothing had happened, everyday routines are resumed. Such compartmentalization, or oblivion, is telling both in terms of individual psychology and on the level of social psychology.[39] In the wake of terrorist attacks in the twenty-first century's first two decades, it has become a standard refrain to declare that life will go on as before.

In sum, the dialectic between expanding and restricting freedom represents one of the most remarkable aspects of infrastructures. Gray areas in systems of supply, disposal, communication, and transportation—especially when they are the object of scandal (say, mismanagement in the health sector, lack of proper governmental or corporate oversight, or accidents that should not have occurred)—provide the stuff of infinite discussion and debate. The resulting disagreements show just how easy it is for the public to lose faith in the systems it has been trained to accept and rely on from childhood. The longer one has enjoyed the peace of mind that infrastructures promote, the greater the distress when they cease to function. Disregard for proper conduct on roadways and in public spaces is a constant source of irritation—and sometimes the least of one's worries. To be sure, the standards that now prevail are defined differently than when Franz Vogt published his "decency manual" over a hundred years ago. Inconspicuousness and moderation set the tone more than ever. These priorities may promote smooth operations, but they still do not guarantee safety.

NETWORK CRITICALITY

The infamous *Protocols of the Elders of Zion* was concocted around 1898, most likely in Russia. The work became one of the most effective—and fatal—pieces of propaganda in the twentieth century. Anti-Semitic through and through, the screed provided a blueprint for later conspiracy theories claiming to expose how a cabal of sinister masterminds has orchestrated the workings of our interconnected world. Infrastructures supposedly play into the hands of agents who exploit operations to their own ends. All the diffuse, subliminal fears that modern society can inspire come together in an exchange that the *Protocols* pretends to document.[40] Pulling the strings from behind the scenes, the "elders" gloat about "throwing into action such a formidable force as will make the bravest of men shudder": soon, "metropolitan railways and underground passages will be constructed in all cities. From these subterranean places we will explode all the cities of the world, together with their institutions."[41] These words allude to the Paris metro, which began to be built in 1897 and opened in 1900. The *Protocols* rakes the coals of deep-seated fears about strange, new modes of transportation while conjuring up medieval legend—fantastic tales about how Jews poisoned wells and spread the plague in the fourteenth century.[42]

In fact, subways saved lives by providing shelter during World War II. All the same, these enclosed spaces can prove deadly. In March 1995, Ōmu Shinrikyō, a Japanese doomsday cult, carried out an attack in Tokyo. Members of the group used the tips of umbrellas to puncture plastic balls filled with a toxic liquid, and the vapors spread through fifteen subway stations. Thirteen people died and thousands were injured. The novelist Haruki Murakami wrote of the event as an "underground war."[43] Two decades earlier, in 1977, Armenian terrorists struck the Moscow metro; Chechen separatists, it seems, did the same in 2004 and 2010. In 2004 and 2005, Islamist bombers carried out attacks on subways in Madrid and London. Overall, the incidents did not diminish the use of public transportation, but many subway stations got rid of garbage cans as a precaution.

An explosion that blew up the north–south tunnel of the Berlin S-Bahn under the Landwehr Canal, which resulted in the massive flooding of the U-Bahn, goes back to the most notorious reader of the *Protocols of the Elders of Zion*: Adolf Hitler. Facing imminent defeat, the dictator issued the Nero Decree on March 19, 1945:

> It is a mistake to believe that when we win back the lost territories we will be able to retrieve and use these transportation, communications, production, and supply facilities that have not been destroyed or have been temporarily crippled; when the enemy withdraws he will leave us only scorched earth and will show no consideration for the welfare of the population.
>
> Therefore, I order: . . . All military, transportation, communications, industrial, and food supplying facilities, as well as all resources within the Reich which the enemy might use either immediately or in the near future to continue the war, must be destroyed.[44]

Fortunately, the Nero Decree did not go into full effect. For reasons debated to this day, Albert Speer, minister of armaments and war production, refrained from carrying out the directive.

During the US Civil War, some railroad bridges were demolished and rebuilt seven times in the back-and-forth of combat.[45] Geopolitical strategists saw the writing on the wall. Since then, they have constantly pointed out weak spots in supply, military installations, lines of communication, and transportation routes. As the geographer Friedrich Ratzel observed, "Since internal traffic routes are organs for internal cohesion . . . , wartime efforts aim primarily at rupturing parts of [the enemy] state and shaking its integrity."[46] His reflections focused on civilian facilities as potential liabilities for military security. Even the axiom of Carl von Clausewitz, the early nineteenth-century theorist of warfare, was revised. In 1905, the Prussian officer Willibald Stavenhagen wrote, "As a policy continued with more forceful means of expression, war, in essence, represents a particular, though not exactly polite, and rather violent form of interaction."[47]

The Hague Conventions of 1907 concerning land warfare called for limiting destruction to what was necessary for military purposes. Twentieth- and twenty-first-century conflicts have often witnessed the exact opposite. Military logic and the escalating violence that follows amount to a race to protect one's own facilities and destroy those of opponents as quickly and as efficiently as possible.[48] Modern war is not just a matter of mobilizing soldiers to replace words with deeds. It means that channels of traffic and communication previously shared now have to be expropriated or put out of commission entirely: cutting submarine cables, tearing down radio masts, seizing broadcast stations, and so forth.

World War I demonstrated the need to reassess inherited ideas and practices, at least in Europe. Belligerents got bogged down at the front and had to focus logistical efforts on getting supplies to the trenches. The theaters in Africa and the Middle East proved to be considerably more dynamic and open-ended. Thus, Thomas E. Lawrence (Lawrence of Arabia) found expedients to remedy the fighters under his command being outnumbered and having inferior equipment when he attacked the German-built Hedjas Railway, water supply systems, and port cities. The same tactics were employed—and refined—by countless guerrilla forces and partisans over the course of the century.[49]

In turn, World War II was decided in large part by communication and transportation, which both sides tried to block the enemy from carrying out at all. The first—staged—action was undertaken by German operatives in September 1939 at a radio station in Gliwice and the post office in Gdansk. The Japanese employed a similar ruse to justify invading Manchuria in 1931 (which, from a broader perspective, can be seen as the actual beginning of global conflict). In the Mukden Incident, Japanese officers faked an explosives attack on the South Manchurian Railway, for which they blamed the Chinese.[50] Lightning strikes to overwhelm and paralyze the enemy became more important than ever. The Allies' strategic bombing of German transportation facilities represents a case in point: railways and junctions of central logistical importance, train stations and repair yards, canals and locks—and then, finally, cities themselves. Later accounts magnified the heroism of figures such as Barnes Neville Wallis, the aircraft engineer who developed the Dambuster bombs that rained down on facilities on the Möhne, Eder, and Sorpe rivers, in addition to other devices for destroying infrastructures. After the war, the British and Americans studied the effectiveness of air offensives on supply and resupply systems, as well as their impact on the morale of the population. The results were sobering. Infrastructural systems and German spirits had been broken only by massive losses and, ultimately, fighting on the ground.[51]

Much of this knowledge remained a military secret.[52] With the arrival of the atomic bomb, however, it lost relevance. No matter how stable infrastructures and morale were, no one could remedy the effects of a nuclear strike. If the Cold War turned hot, the first and, presumably, most disastrous site of attack would be Central Europe. Nevertheless, opposing sides prepared for conventional warfare. Measures included bunkers, of course, but also preparations for preventing the enemy's advance. Thus, pioneer troops trained to blow up bridges and roads in their own countries; concrete barriers would seal off tunnels, and nuclear mines would make stretches of land impassible.[53]

Various scenarios reckoned with escalation on an unpredictable scale. This state of affairs led to conflicts being carried out in proxy wars, outside Europe, where they combined with decolonization efforts, national liberation movements, and partisan tactics. For the most part, the latter aimed to hit the technologically superior opponent at the most sensitive points. A strategic strike would not just deal a blow to enemy systems but also detonate a symbolic charge. Infrastructures were key in operations. War erupted when an electrical plant in Hanoi was destroyed on December 19, 1946. Neither France nor, subsequently, the United States could gain the upper hand. The fighting ended only in 1975; a helicopter that evacuated American agents and Vietnamese refugees from the roof of a residential building used by the Central Intelligence Agency in Saigon has become the iconic image of hasty foreign retreat.

Mao Zedong famously declared that the partisan should move among the masses "like a fish in water." He meant that the anonymous structures of the modern world offer underground fighters opportunity to change their positions and tactics as circumstances require. Such maneuvering applies to guerrilla methods in a wide array of contexts; it has also been adopted by those who employ terror to unsettle majoritarian society and the ruling classes. The attack that Comandante Ernesto "Che" Guevara and two dozen others made on an armored train belonging to the Cuban leader Fulgencio Batista has become the stuff of legend. On December 29, 1958, the forces under Guevara's leadership used a bulldozer to destroy the railroad tracks at Santa Clara. The train was derailed, and Molotov cocktails thrown by rebels sent the three hundred soldiers onboard running for safety. The insurgents looted the supply of weapons and ammunition and used them to take over the city. Shortly thereafter, on New Year's Day, 1959, Batista fled the country.[54] The legend of the Cuban Revolution was born; since then, imitators have employed guerrilla tactics to prevail in comparably asymmetrical constellations of power.

Since the emergence of modern terrorism in the mid-nineteenth century, tactics have changed far less than the technologies involved.[55] For the most part, attacks have targeted people who occupy positions of symbolic significance. Often, they aim at important places and buildings. Events on September 11, 2001, brought together systemic and symbolic criticality in horrific fashion,[56] trumping all disaster movies about airplanes, subways, and boats. Terrorists do not necessarily want to maximize the number of victims, even though a certain morbid competition may sometimes be at work. Instead, the priority is to amplify the attack by raising social insecurity to a fever pitch.[57] The importance of infrastructures—and their power to activate latent fears—was underscored when al-Qaeda fighters were found in possession of construction plans for dams and nuclear power plants; they had searched the Internet for information about networks controlling electricity, traffic, communications, and water.[58] Western interpreters declared that terrorists were launching an "attack on modernity" itself, and they called for system architectures to be made more flexible in response.[59]

Since 2001, by the latest, measures to ensure public safety have superseded the desire to preserve individual liberties in Europe and the United States. The very same greater networking and interconnection in transportation, supply, and communication that softened political borders—or even eliminated them, as in the European Union—now, *sub specie securitatis*, are a problem. Effects include the ascendancy of decentralized networking, which formed the basis of Arpanet and, later, the Internet. This paradigm first emerged in a military setting. During World War II, telephone lines in London were networked to minimize the possibility of interference from hostile agents.[60] Similarly, "meshing" electricity networks and overhead lines sought to make the main arteries of energy supply less vulnerable and facilitate the development of more resilient structures.[61]

But no matter how many security measures are taken, terrorist attacks will continue to shake public morale. After 9/11, President George W. Bush identified popular support of the government itself as a "critical infrastructure."[62] The President's Commission on Critical Infrastructure Protection had been founded in 1996 to safeguard the expansion of the Internet, the information superhighway advocated by Vice President Al Gore. Since the 1980s, antagonistic exchanges between software developers and hackers had given rise to a digital infrastructure that expanded rapidly from a closed system into an open, global network.[63] The ongoing challenge was to keep data safe from misuse and cyberattacks. The Internet achieved a real breakthrough when business interests recognized the value of convenient e-commerce (especially products and services of an erotic nature).[64]

Ever since, political, administrative, economic, and technical processes have grown only more dependent on the Internet, which merges previously discrete information flows. The preconditions have been met for intelligent logistics systems: smart homes, cars, factories, and cities. Such arrangements produce vast quantities of data by means of constant monitoring, measuring, and calculating—which can, in turn, be misused or attacked. Experts warn of threats to personal information and cybersecurity.[65] Most people now live in cities. Urbanization presents both opportunities and risks, requiring habits and practices far beyond measures adopted for road safety and dealing with traffic.[66]

In 2004, the German Federal Ministry of the Interior defined "critical infrastructures" as "organizations and facilities of (vital) importance to the state community, the failure or impairment of which would result in lasting problems of supply or other dramatic consequences for large sections of the population."[67] In the 1960s, in response to student protests, emergency laws had been passed—which only compounded protests' frequency and volume. In the 1980s, when a census was conducted and supposedly forgery-proof identity cards were introduced, large parts of the population contested the measures. Today, in contrast, when steps are taken to improve and expand infrastructures, debate arises about the details but not the underlying principles. Questions about analogue practices—for instance, whether it is legitimate to shoot down a hijacked airplane likely to be used as a weapon—prove more controversial than surveillance systems in the digital world.

Indeed, instead of prompting discussion, issues provide entertainment for a public increasingly accustomed to spending time in virtual realities. Take an exchange that took place in 2002 on the forum for *Civilization*, a computer strategy game. With minor adjustments for grammar, it reads,

GRUMPYBEAR: I'd be interested to know the effect of destroying enemy infrastructure in your experience. Since destroying well-fortified cities with naval and air units doesn't produce the desired results, . . . I wonder if you can reduce the population of enemy cities by eliminating agricultural and rail areas. Or do you leave enemy infrastructure untouched so you can make the most of it after taking over?

SOERENS: Personally, I rarely destroy the enemy's infrastructure, except for resources . . . , so the rest is left over. . . . If I wind up taking the cities, I'm glad the infrastructure is there.

SCAREDYRABBIT: Always war! These computer civilizations keep coming up with troops even without mines. So why wreck everything? I almost always lose anyway.

FOSSIBEAR: Well, once I get my own continent, I'm only interested in cities on other continents with luxury goods and resources. And since I don't want the cities themselves, I just bomb everything to pieces and send the enemy continent back to how it was in 4000 BC.[68]

8

THE COLD PERSONA: USERS AND OPERATORS OF INFRASTRUCTURES

TRAFFIC FACE

The almost-daily suicides in the Tokyo subways, when people throw themselves in front of trains, are called *jinshin jiko*, or "human accidents." The phrase turns the life, and death, of a human being into an infrastructural disruption; usually it takes forty minutes to "clean up." Such intervals can represent a long time for passengers squeezed into cars at double capacity. Most of them react with composure and, via cell phone, tell those waiting for them that they will be late. Commuters often mention that they consider the action of the deceased antisocial: they have forced millions of people on a tight schedule to shoulder the consequences of their personal problems. Subway companies have responded by increasing the amount of light on platforms and in trains, to help dispel gloomy thoughts. They have also threatened to hold relatives liable for costs. Neither measure has produced satisfactory results.[1] Retooling logistical arrangements as a whole is more effective.

Since the 1920s, starting in the United States, it has been standard practice worldwide to monitor trains and schedules from a central office; the control room features prominently in disaster films (e.g., *The Taking of Pelham One Two Three* [1974]) because all information converges here. Beginning in 1996, the Tokyo subways have also implemented the Autonomous

Decentralized Transport Operation Control System, based on the ideas of engineer Kinji Mori. Instead of the high-modern practice of modeling infrastructures on an organism with operations controlled by a brain, this design is meant to react to technological and social disturbances in the manner of an ecosystem, which is constantly changing.[2] Subsystems deal with unpredictable events as independently and decisively as possible. In this framework, incidents such as suicides do not represent a state of exception that compromises all operations; they can be viewed in terms of a functional logic that, although always precarious, admits stabilization through *distributed autonomy*. Now, Tokyo subways no longer adhere to strict schedules (which never work, anyway); however, they do follow a reasonably reliable rhythm. Passengers are informed of disturbances or cancelations via display panels or text messages so they can decide, independently, how to arrive at their destination.

When *crisis* means normal operations, the system corrects itself autopoietically. But conditions change. In 1984, Bernhard Hugo Goetz fired at four youths on a New York subway because he felt threatened. The violent act met with a fair amount of public approval; the incident offered striking parallels to *Death Wish*, a film from ten years earlier in which a character played by Charles Bronson shoots two petty criminals for provoking him on a train. Many New Yorkers had a sense of helplessness about high crime rates—especially on subways, where they felt trapped. In their eyes, Goetz's vigilantism was self-defense.[3]

But the habits that emerge around infrastructures do not usually generate callousness, much less fits of brutality or taking justice into one's own hands. Nor, for that matter, is "psychosocial conditioning of users" the force behind the unfortunate incidents that do occur.[4] Most people learn how to deal with communication and transport arrangements in a way that allows things to run smoothly and with little danger. On the whole, the growth of infrastructures has promoted circumspect bearing and flexibility. All the same, critics maintain that people who spend their time in the no-man's-land of traffic jams, train stations, airports, and hotels are losing their individuality; conformity, they say, is rising in proportion to the anonymity of modern life.[5] As interaction becomes more and more fleeting, people need to learn how to behave in the first place. The history of infrastructures involves new codes of politeness and respect, from telephone protocol ("Keep it brief! Other people are waiting") to netiquette.[6]

For well over a century, writers have been exploring processes of standardization and uniformization. Most of the time, this involves juxtaposing actual human beings with abstract systems that seem to expand all on their own. "The Berliner is a slave to his apparatus," wrote Kurt Tucholsky in 1919, feeling himself to be more of a ticketholder than a real person.[7] At the other end of the spectrum, a luckier party might feel like Phileas Fogg, the protagonist of Jules Verne's famous novel who manages to exploit intermodal networking to travel around the world in eighty days. Franz Kafka's K. occupies the space between, bumbling from one bureaucratic entanglement and waiting room to the next in a never-ending odyssey down the corridors of modern circulation and progress.

Circulation, especially the motion of bodies in space, gives rise to unexpected encounters. Chance has become inescapable, a matter of fate. In 1902, the psychologist Willy Hellpach offered an exemplary account of the challenges to psychic equilibrium this state of affairs presents:

> I step out of the house, just as the electric train is pulling away. I have to wend my way over the sidewalk; a few people from a small town are blocking the flow; a cab comes around the corner at breakneck speed. I have to wait for the next streetcar. There's barely anywhere to stand on the upper level; the man next to me is smoking a foul cigar, and the soot flies onto my fresh linen collar. The trolley soon is flying ahead at a furious pace, so that everyone and everything bumps together; just as abruptly, it comes to a stop because a truck is blocking the track.[8]

In such an environment, sensory impressions and discontinuous movements condense into an overcharged mass of stimuli. In the city, the novelist Robert Musil noted in 1927, "the only speed one actually still feels" comes from "trying to make connections, the hurry to change trains and uncertainty about being on time."[9] An earlier culture of leisurely strolling had transformed into a state of drifting and surfing. This mode of being required highly selective perception and intuitively deciphering fleeting signals. As early as 1903, the sociologist Georg Simmel observed that metropolitans, overburdened by the constant influx of impressions and increasingly nervous, had developed a blasé attitude toward their fellow human beings.[10] The "unfocused interaction" (Erving Goffman) that prevails in the city requires exercising discipline and self-control not to look at others or otherwise engage with them (say, by eavesdropping on conversations). Along the same lines, the journalist Walther Kiaulehn examined how city residents, when leaving their apartments, adopt a masklike, impersonal, tense yet absent "traffic face" (*Verkehrsteilnehmergesicht*) to avoid social entanglements.[11] Using a hidden camera, Walker Evans took countless pictures between 1938 and 1941 on the New York subway to capture its variations; passengers often looked strained as they tried to avoid being distracted by others.

Anonymous but shared experiences have not always received a negative judgment. The American poet Walt Whitman had both a sociological and an aesthetic interest in the ways that infrastructures enable large-scale movement in space; chiming bells, timetables, and rituals of travel are the pulse of communal experience.[12] Contemplating the Hudson River, he observed, "I have always had a passion for ferries; to me they afford inimitable, flowing, never-failing, living poems."[13] Whitman was fascinated by how moments of pause, rest, and idleness create a collective; he believed that people in motion share a world of sensation, even if they do not speak of it to each other. For Whitman, the diverse *crowd* generated by infrastructures symbolized a collective: US society as a whole.[14] The poet struck on a point essential to our purposes here: the formative role of social systems and networks for individual and group consciousness. By detaching people from familial and local attachments (and the impositions they entail), these systems engender isolation, but this isolation is remedied by new relations that are matters of choice. The existential bearing that results represents a gain in autonomy, even if it requires acknowledging new restraints.

In this context, sociologists have described the human animal as employing "radar" (David Riesman) to follow cues and staging "theatrical performances" (Erving Goffman) for others.[15] Consumers and passengers, the argument goes, are controlled by external signals.[16] On the whole, then, and to borrow a phrase from the cultural historian Helmut Lethen, the conditions of modern life favor "cool conduct" in lieu of familiarity and deeper connections.[17] Very few theoretical perspectives on personal and social dynamics refer explicitly to the technological and administrative basis for such developments. However, environment is precisely what made behavior that prioritizes calculation and coordination possible—and necessary—in the first place.[18] Distance, not proximity, is the watchword. Some features of "inner urbanization" such as telephones and televisions create new forms of privacy; others, like mass transportation or the cinema, reshape the public sphere.[19]

Public transportation has given rise to spaces defined by openness and narrowness at once. Travelers learn to filter and block out elements of their surroundings. Urban residents immediately recognize "country folk" by their different bearing—for instance, looking around and making eye contact. Beggars, pickpockets, and sundry opportunistic types find a great resource here. But even seasoned passengers can at times be overstimulated—for instance, by the erotic allure of a stranger or the defiance of a manspreader. (In 2014, the Metropolitan Transportation Authority in New York even launched a campaign against this expression of entitlement, which is often unconscious.[20]) Deviations from the norm are usually obvious.

It is open to debate whether conditioning is really turning human beings into abstract entities devoid of character. Perhaps people are proving more adaptable, and their habits are becoming increasingly elastic. It is certain, however, that infrastructures have promoted greater awareness of timing, a quick and intuitive knack for orientation, and a predictable bearing under conditions of heavy traffic. A sixth sense for navigating social space has developed, as well as new codes of etiquette for interactions—in brief, "implicit knowledge," which works even if no one can put it into words.[21] Activities that fall under this heading combine routine and reflex operating just at, or below, the threshold of self-awareness.[22] Safely navigating city streets, riding a bicycle, traversing busy crowds, observing decorum on the telephone, consulting schedules and maps, remembering passwords, and much, much more—the list of practices that ease life in isolation but give rise to stress in the aggregate admits endless expansion.

Ideally, knowing and doing should constitute an integrated whole.[23] However, a seemingly new form of distraction has emerged: multitasking. In fact, the overdistribution of attention occasioned discussion in the nineteenth century, when books ceased to be luxury items; the practice of reading numerous works in scattered fashion became widespread. As the novelist and critic Friedrich Spielhagen noted in a 1898 discussion of epic poetry and modern transportation, "the paperback, bought at the station for just a mark, reigns supreme," because it "fits conveniently in a pocket and can be flipped through between the start and finish of a journey."[24] New media and infrastructures form a continuum. Today, radio dials, remote controls, and other technologies enabling users to pick and choose have shaped what has come to be known as the "attention economy."[25] Of course, not everyone is up to the challenge. Modernist novels draw on the divide between characters' inner rhythms and the demands of the outside world. Thus, Alfred Döblin's *Berlin Alexanderplatz* (1929) tells how the roughhewn protagonist, Franz Biberkopf, is beaten into submission by the metropolis. After years at odds with the social order, first as a criminal and later as a patient in a psychiatric institution, he finally learns to accept his station in a world of statistics and averages and becomes a doorkeeper—in other words, a disposable "human resource."[26]

The basic goal, it seems, is to fit in and go with the flow as proactively and adaptably as possible. A key role is played by conditioning and training, ranging from traffic safety classes for children to media and technology workshops offered by employers.[27] Of course, cultural and historical differences do exist. Today, for instance, the shocks to patterns of perception on railroad travel that nineteenth-century observers remarked seem like no big deal.[28] There is no comprehensive assessment of the challenges and consequences of inhabiting a world teeming with networks of transportation, communication, supply, and disposal—let alone infrastructural arrangements in the social and cultural spheres. Basic optimism or pessimism—that is, personal disposition—is vital. Early adopters have an affinity for the latest and (supposedly) most advanced technologies. Such people exhibit confidence; they like to think of themselves as pioneers and acquire new products in this spirit. Early adopters seize opportunities where others, of a more reluctant cast of mind, tend to hesitate, and they pull them along. They are more likely to keep track of technological advances and trust themselves to notice potential disadvantages.

Generally speaking, infrastructural developments favor curious and creative types. Traditionalists stand at the other end of the spectrum and wait for products and services to prove their utility and value. In their eyes, innovations can impair established activities and routines. Conservative by temperament, they regard early adopters and their hangers-on as snobs, fashion victims, and trend surfers—and note the potential for addiction.

People inclined toward criticism have always suspected that their contemporaries spend too much time reading, listening to the radio, watching TV, talking on the phone, or poking around on the Internet, and they lament the one-sided personalities that have developed as a result.

These varying attitudes toward public institutions generate tension. Some people readily handle the demands associated with infrastructures—standardization, regulation, and the corresponding mindsets.[29] In extreme cases, they succumb to a mania for all that is new and inflict their superior understanding of things on others. Conversely, those who like to swim against the tide or fancy themselves gadflies engage in provocative behavior, especially with regard to institutions that symbolize majoritarian society. Graffiti is the textbook example. On the scale of public nuisance, it occupies a place somewhere between pranks (say, children ringing doorbells and running away before occupants can open the door) and vandalism (for instance, damaging telephone booths—back when they still existed). Hackers and others who bend the rules occupy an intermediate position; their actions can prove beneficial under authoritarian regimes—for example, intercepting police transmissions, setting up pirate radio stations, and subverting state-controlled media—but they can also be purely antisocial.

Infrastructures qualify as such only to the extent that they find everyday use. The history of bicycles provides an instructive example. Even though cyclists embody the ideal of personal mobility—at a low ecological cost, moreover—they have received fairly shabby treatment from civil engineers and planners. In response, and especially in cities, activists have adopted semi-illicit strategies combining the advantages enjoyed by pedestrians, on the one hand, and drivers, on the other, to navigate the streets as quickly and as effectively as possible. This means following some rules while disregarding others. Their freewheeling, anarchic behavior often provokes strong reactions from others, from vehicle operators to municipal authorities. The resulting complications and concessions make it clear that infrastructures are dynamic arrangements subject to change.[30]

It also matters whether urban modes of life are internalized or, for whatever reason, rejected. That said, the distinction between city and country is not tidy; many infrastructures have spread quickly beyond the urban sphere, and reactions vary from one place to the next. Many new technological offerings are adopted without delay, but not all. The booming megacities of China still display instructional placards to advise a predominantly rural population about what to do and how to act in uncustomary surroundings. Conflicts have arisen in Israel for decades because Orthodox Jews wish their countrymen and countrywomen to observe religious commandments (avoiding travel on the Sabbath, for instance).[31] Every conceivable variant occurs on the spectrum. Time and again, infrastructures and their operators have adapted to cultural sensibilities and practices. In the nineteenth century, railroads introduced a fourth-class category of travel (known, in the day, as wooden class because of the spartan appointments) so the poorest folk could use the train. A more recent phenomenon is the Kids on Tour service for children going back and forth between divorced parents.[32] Many cultural developments could hardly have been foreseen when arrangements first were planned. For some time now, the United States has been home to *shoefiti*—footwear hanging from utility lines to mark gang territory; it's also unlikely that maintenance personnel and regulatory authorities ever expected so many people to fasten "love locks" to the railing of bridges.

Many practices have emerged that run counter to the infrastructural ideal of orderly circulation. Homeless encampments, which amount to parallel universes, pop up in parks, under overpasses, and in subway tunnels. Their residents' lives contrast sharply with those of the general population, who enjoy comfort and convenience. In the 1990s, journalists began documenting the mole people, who made their homes in the shafts for gas, electricity, sewer, and railroad services beneath New York. It turned out that more than five thousand people were dwelling there, like the unfortunates in the bowels of Paris described by novelists Eugène Sue and Émile Zola.[33] Journalists sometimes numbered the mole people among the homeless indiscriminately; perceptive observers recognized individuals who had fallen through the cracks or chosen to take a different path in life. They included Vietnam veterans, macrobiotic hippies, drug addicts, Cuban refugees, convicted murderers, computer nerds, philosopher-hermits, and fugitives from the law.[34] Unconnected to public service networks, these people defied identification in terms other than what they said about themselves.

Meanwhile, the world above has discovered the value of detachment—and capitalized on it. Since 2014, Café Seymore+ in Paris has offered an offline hideaway: a technology-free zone without the sockets or smartphones that otherwise dominate professional and private life. *Mindfulness*—that is, attention and concern for oneself and others—is the new buzzword for digital junkies who don't know where to turn.[35] One wonders whether this enlightened mindset makes room for empathy with people who throw themselves in front of the subway or otherwise fail to keep up with the times. A 2014 incident in Zhengzhou, China, underscores the conflict between traditional social mores and the functional logic of infrastructural systems. On a public bus, an elderly passenger punched a younger man in the face several times when the latter refused to give up his seat; instead of fighting back, he got off at the next stop. The old man was so upset by the lack of respect he had been shown that he suffered a heart attack and died. The matter was hotly debated on Chinese Internet forums, with many young people declaring that they worked so hard they no longer saw the point of accepting a position inferior to that of their elders who, after all, had been useless for some time.[36]

SERVICE AND SELF-REGULATION

In 2015, a team of computer scientists in Bonn declared that smartphones lead to digital stress. Analysis of data from sixty thousand users indicated a high level of fractured attention and permanent multitasking. On average, participants in the study used their cell phones eighty-eight times a day, or every eighteen minutes, for a total of almost two and a half hours; young people reached for devices more than other age groups. On the whole, researchers concluded, there were clear signs of behavior detrimental to social productivity.[37]

The smartphone—like the automobile, earlier on—epitomizes achievements charted since the eighteenth century (that is, the Enlightenment, Industrial Revolution, and bourgeois liberalism). It stands for independence from spatial and social restrictions, access to information and data streams, membership in a global community, consumer choice in an international market, a certain level of equal opportunity in education and work, civic responsibility, and room to determine the course of one's own existence. With a slight change of perspective, we can hear echoes of slogans and campaigns propagated for decades now, at least in Western countries, by politicians, business concerns, managers, technicians, and scientists eager to promote individual autonomy and self-reliance.

Infrastructures enable exchanges within communities and between human beings and nature. But it is meaningful to speak of freedom of choice only if each and every person has access to services of supply, disposal, communication, and physical movement; democratic engagement; broadened horizons; an open society; universal rights; and so on. In other times and under other social conditions—which still prevail in most parts of the world—decisions have fallen to families, clans, groups and associations within the larger community (fraternal or volunteer organizations, for example), political parties, churches, congregations, and so on. Taking advantage of opportunities presupposes a strong and sovereign personality, which in turn rests on the premise that one is comprehensively informed and ready to choose among options.

In fact, sovereign individuality evolved over time and was long reserved for the few. Services were based on personal acquaintance; only gradually did they yield to impersonal, market-based systems with replaceable personnel. For centuries, privileged groups employed servants to avoid the bother of everyday affairs, but all—manorial lords, ladies, and their staff—lived in mutual interdependence.[38] Occupations such as cloakroom attendant, courier, street sweeper, lamplighter, coal carrier, switchman, streetcar conductor, and stagecoach driver emerged when the social pyramid became less steep. Earning a livelihood has become more precarious, and technological advances have done away with most of these jobs. Now, one sees a human being directing traffic only when electrical systems are out of order. In the nineteenth century, thousands of kilometers of pipes laid in Paris to supply drinking water put twenty thousand water carriers out of work in one fell swoop.[39]

There is pride in old-fashioned lines of work; not just yesteryear's equipment but also professions occasion fond recollection.[40] Personnel with highly specialized skills model reliability and integrity. Their station is characterized by a certain bearing; often, a uniform marks their status. In the past, it was clear who served as a canal and river boatman, conducted trains, or worked as a postal clerk or traffic policeman. Positions passed from generation to generation within families, which could become veritable dynasties with a specific ethos and pride. Much the same held for employees of large enterprises, who shared a corporate identity. "Neither snow nor rain nor heat nor gloom of night stays these couriers from the swift completion of their appointed rounds." These words, from the Greek historian Herodotus, stand above the portico of the James A. Farley Building in New York, former home to the city's main US Postal Service branch, reminding the public of the dependability of mail carriers, even under unfavorable conditions.[41]

The same aura of permanence also had a forbidding aspect. The *Zeitschrift für Kleinbahnen* declared in 1916, "The natural adversary of the stowaway is and will remain the sharp-eyed conductor who manages to collect fares even against passenger will!"[42] Kurt Tucholsky described such diligence from a less appreciative standpoint. The passenger whose ticket is checked "feels caught"; "confronted with what is holiest to Germans—the uniform—he is suddenly made small" and no longer seems like "an adult . . . who perhaps has a wife, children to bring up, and employees to bark at."[43] Lenin's famous remark that Germans would never carry out a revolution because they would all buy tickets before storming a train station conveys the same idea, but with a different emphasis. In many respects, the ideal of commitment and cooperation calls military structures to mind. It is no coincidence that the Nazi propaganda minister Joseph Goebbels praised German railroad workers in terms sounding as if meant for soldiers, extolling their "superior improvisational elasticity," which kept "the arterial system of our entire war effort in motion."[44]

Employment in the infrastructural sphere in the past offered opportunities and freedom from predetermined social positions. Examples include the young women working at telephone exchanges, beginning in the late nineteenth century, who were preferred at switchboards because their voices were thought to have a soothing effect.[45] Like letter carriers or counter clerks, they lent a personal touch to an anonymous institution. Only in the 1960s did automated dialing systems and recorded announcements take their place. During the world wars, women conquered other masculine preserves, working as conductors and even drivers. When the fighting ceased, social pressure forced them to relinquish these roles. Some lines of work are dominated by men to this day—flying airplanes and piloting boats, for instance. Professions with a lower social profile—nursing and hospitality, among others—are still largely the realm of women.

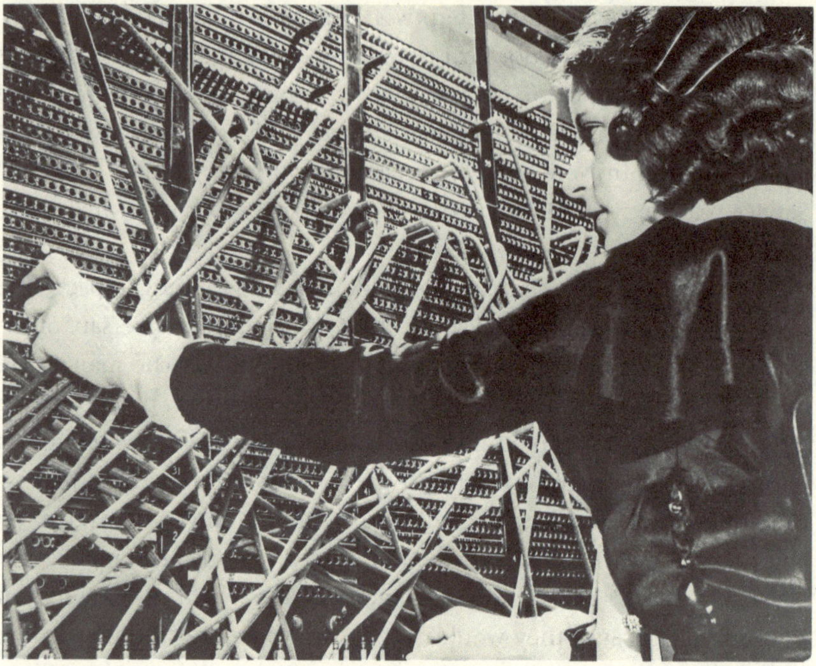

8.1 Before services were automated, many young women achieved independence by working on telephone switchboards (photograph from the United States, ca. 1930). *Source*: Getty Images.

What is most remarkable is just how many jobs have disappeared altogether as customers have taken over work formerly performed by someone else: purchasing and validating tickets, filling out and mailing forms, reading meters, and so on. (Such monotony is memorably captured by Serge Gainsbourg's 1958 song "Le Poinçonneur des Lilas," from the perspective of a lonely man who punches tickets in the Paris metro; his dead-end job amounts to long-term suicide.) Kindly measures to spare employees from drudgery do not explain the development.[46] Nor have service personnel been reduced to keep criminals or politically compromised individuals from finding a hiding place (as in Bernhard Schlink's novel *The Reader*, in which an ex–concentration camp guard works as a streetcar conductor after the war).

Strong unions, on one side, and resolute management, on the other, have played a decisive role. Effective worker advocacy led to rising salaries and handsome pensions; from the perspective of employers, such gains amounted to additional expenses justifying measures to scale down the workforce. Commercial enterprises and state agencies have continued to relieve themselves of as many tasks as possible. Studies are constantly conducted to analyze customer behavior, yielding a comprehensive profile of the average user. To justify changes, business and government declare automated operations and self-service arrangements provide opportunities for greater choice and convenience; inconvenient opening hours and long lines no longer pose an obstacle, the argument goes. Furthermore, it is claimed, the new setup is more reliable in technical terms. (The disadvantages for consumers of figuring out tariff structures, how to pay for tickets electronically, and remembering passwords are conveniently disregarded.)

In consequence of downsizing, many infrastructural services no longer present a human face at all. They must simply be trusted to keep working. The few people who are left count as parts of a machine—and often receive commensurate treatment. Thus, at the beginning of the 1980s, municipal authorities in Vienna found it necessary to mount a campaign to remind the public that street sweepers are also human beings.[47]

Across the board, the process has been more or less the same. Institutions that once employed a sizable workforce yielded to operations with a reduced but permanent and respected staff; in turn, large-scale automated systems took over. In the most recent development, human beings perform only monitoring services or have jobs ensuring "customer satisfaction." Actually talking to someone comes as a surprise. Relations between users and operators have cooled and become increasingly matter-of-fact. The personal dimension has diminished. Indeed, anti-corruption laws and similar ordinances now prevent garbage collectors from ringing doorbells to receive the modest holiday bonuses that once were the norm; these tips had formerly offered the occasion for friendly greetings and adding a face to essential services.

The objective at the core of much infrastructural strategy, to eliminate the ethos of *do ut des*, now sets policy from Karachi to Cologne.[48] In the name of equal treatment for all, bureaucracy has made exchanges anonymous. However, tiered offers have been introduced: people who pay more have access to better services. This arrangement, which is particularly widespread in the digital realm, undermines egalitarian ideals. Whereas privilege previously depended on rank and social estate, now it is a matter of prices and tariff systems. More than ever, social standing is measured by the degree to which one is connected to goods and services.

Questions of distinction represent a constant in the history of infrastructure.[49] They include dividing railroad cars into different classes, reserving certain elevators for clients and others for personnel, and exclusive waiting lounges. Although such arrangements are supposedly tailored to individual needs and preferences, they are possible only because research has calculated a typical "passenger," "customer," or "end consumer"—a generic nonentity, in other words—and introduced what engineers call "singularization systems" (turnstiles in subways are the classic example). In consequence, the culture of self-service,[50] which alternates between community-oriented and self-interested behavior, has mobilized counterforces: activists who appeal to the classical notion of infrastructure as universal connectedness.

For most people, high-mindedness represents a matter of secondary interest, at best; they are too busy keeping up with technological developments, sorting through packages and plans, and maybe, occasionally, thinking about environmental risks. Personalized services require a continuous learning process. They promise to enhance life but demand that consumers survey options, make appropriate choices, and familiarize themselves with new arrangements. In some cases, sound intuition is enough. But often trouble lies in store, if only because one needs to update software and equipment time and again (consider bundled technologies such as multifunction printers).[51] We have noted the enduring relevance of Mary Shelley's *Frankenstein or Prometheus Unbound*; technological advances abound in unintended consequences. For the foreseeable future, it seems there will be no escape from the perils of innovation.[52]

New technology can change life from top to bottom. In the nineteenth century, one of the first personalized devices to hit the market was the pocket watch, which introduced new expectations of punctuality. Innovations such as sewing machines and motor vehicles, among many other things, likewise instilled a regime. Learning how to use machinery in the first place took time, and operations set the pace for daily life.[53] In turn, radios and televisions, which are pretty easy to use (but not to repair), provided an intermission of sorts. The computers and smartphones that have become ubiquitous in the last few decades are another matter. Keeping up with technology turns some people into enthusiasts, and others into curmudgeons. Most people manage to acquire just enough familiarity to get devices functioning, and then they carry on as normal.[54]

The history of the telephone offers a vivid illustration of the dynamic. The starting point for telecommunications was the nineteenth-century postal system, with which the public was relatively content. As telephone use expanded, companies experimented with offerings to attract customers in keeping with technological developments. Initially, operators made connections; in turn, dials and then keypads enabled communication that was even more direct—and personal. Longer cables and, later, wireless services allowed users to withdraw into privacy.[55] If the pushbutton heralded a change, the more recent advent of portable media has reinforced the dialectic of (seemingly) individualized operations and (real) dependency on infrastructural arrangements.

Developments have culminated in the smartphone, which promises complete autonomy.[56] Although speaking loudly on the phone is viewed as just as intrusive as filling the air with cigarette smoke,[57] the device presents an overall picture of quasi-divine sovereignty. By pointing, swiping, or performing some other elementary gesture, things can be made to appear or vanish at will. The smartphone represents the power to shape one's world. It is no surprise, then, that it serves as a kind of personal remote control for the whole of networked society, enabling—or disabling—what is heard, seen, said, written, arranged, examined, used for work or entertainment, consumed, or shared with other people. Its use has become so widespread that it has reached parts of the world that earlier infrastructures never penetrated—for example, countries in Africa without an electrical grid.

As the site where varied spheres of life intersect, the smartphone has ushered in connectivity on an unprecedented scale. Still, local variations exist. In Tokyo, for instance, it flourished earlier than elsewhere because of the virtual space it opened on congested mass transit systems.[58] The device offered something like a home, a guarded room where visitors were admitted at personal discretion—especially when someone else's act of desperation, in the real world, had brought traffic to a standstill.[59]

CONCLUSION: DOWN THE DRAIN?

PIPE SYSTEMS AND DATA HIGHWAYS

Almost thirty years ago, a German reporter made fun of Al Gore for declaring that all information should be available to all citizens at all times. When the vice president made this pronouncement, in 1995, a system that was coming to be known as the Internet connected just about 3.5 million computers worldwide. Getting online was still a relatively complex procedure; modems converted digital information into acoustic signals that were sent down telephone lines. Gore wanted these wires to be replaced by more efficient fiber-optic cables. It was unclear to the foreign correspondent what good such data highways would do, but he noted that people in Santa Cruz, California, could now order pizza with their personal computers. "Whether anyone would ever want to do that," he wrote agnostically, "it does work." Presumably, the article concluded, people would one day have flat, large-screen devices—a kind of "telecomputer"—combining the functions of televisions, desktop computers, fax machines, telephones, and video players.[1]

In the sixteenth century, the literary genre of the utopia inaugurated a tradition of thinking about optimal social organization by means of technology. What has become reality since then, however, is the result of incremental progress and knowledge obtained after the fact, not planning beforehand. The history of infrastructure includes many relatively brief phases of development when transportation, communication, supply, and disposal changed rapidly. In our own time, digital technology is ascendant; many older infrastructural arrangements are being absorbed into the "universal machine"—the computer.[2] The Internet, it is often claimed, is at the heart of the "information age"; now, data can be disseminated and retrieved as a matter of course the world over. Although this assessment may be true, its origins started 150 to 200 years ago.[3]

Celebrants of the Internet and the superiority of modern technology overlook—or are unaware of—the many ways that digital infrastructures reproduce analogue ones. In 2006, for example, the blogosphere jeered at Senator Ted Stevens, chairman of the Commerce, Science, and Transportation Committee, for declaring that people shouldn't just dump anything and everything online; too much data would clog the system. The comparison was anything but misguided, but because Stevens hailed from Alaska, he was assumed to have waded out past his depth.[4] In June 2013, the German chancellor Angela Merkel met a similar reaction when she spoke about the Internet as "uncharted territory"; commentators sneered that her words revealed her provincial origins in the German Democratic Republic.

In fact, comparing newer infrastructures with older ones makes good sense. In 1996, the science fiction novelist Neal Stephenson wrote an article about using fiber-optic cables for global data highways, "Mother Earth Mother Board." The piece, which has come to be regarded as a classic, compared the project to undertakings in the 1850s, when the first transatlantic telegraph cables were laid.[5] Stephenson also noted that information sources requiring in-person visits, such as libraries, were becoming unpopular. He urged the community of "hackers" to examine the "analog" nodes of nascent network society; otherwise, it would be impossible to prevent a handful of corporations from monopolizing the master switch.[6] His warning went unheeded; evidently, the ready availability of data (secure or not) was just too distracting.

Analogue infrastructures continue to hold their own, but something qualitatively new is emerging. It is not a revolution so much as a transition along a continuum. A large part of the infrastructures in place since the end of the eighteenth century—and in some cases since well before then—are still in operation because they have been regularly updated and connected to new networks. Thus, Deutsche Bahn is still replacing mechanical signal boxes, some of which are more than a hundred years old, with electronic devices.[7] As a rule, routing systems are maintained once they are established, and they become permanent channels of traffic or communication; new branches are simply added. A few means of communication—optical telegraphy and carrier pigeons, for example—have disappeared, but in principle they could be reactivated at any time. Many arrangements have ceased to be overseen by the state and have passed into private hands. Examples include systems supplying fresh milk, which began at the end of the nineteenth century in many European countries.[8] Until quite recently, milk's reputation was as a cure-all. There were communal milk yards, deliveries by milkmen, school milk for children, and milk bars for adults. Today, however, no one would speak of a lactic infrastructure; extended shelf life and better systems of distribution have made milk so ubiquitous that producers have to receive subsidies.[9] Operations have been privatized. Much the same holds for other services that once were deemed matters of general interest and the public good. In the 1990s, for example, package delivery systems seemed to have entered a period of stagnation—until online ordering triggered a boom that has lasted to this day. Likewise, demand for services such as cleaning and yardwork was waning; the Internet, by facilitating transactions between companies and homeowners, reinvigorated the market.[10]

Economics and prosperity are no longer conceived in terms of production, as they were 150—or even 50—years ago, but in terms of consumption. International commercial logistics is now so cost effective that goods can be sent almost anywhere in the world.[11] Since the 1950s, containers are the red blood cells of globalization, reducing expenses and improving security. Bigger and bigger ships are transporting them. The largest was commissioned in 2015; *Oscar*, a boat nearly four hundred meters long owned by the Mediterranean Shipping Company, can carry almost twenty thousand containers at one time, with a crew of just thirty-five.[12] Modes of transport are in constant competition for the highest speed and the lowest price. Political and economic conditions make one form preferable in some cases and another in others. Accordingly, freight can travel by rail, sea, or air. The flow is constant—both for merchandise and for people.

A common impression is that we're living in a world that is a "commuter republic"—that there's no way to escape constant travel. In Germany at the beginning of the twentieth century, only one-tenth of the population even left their place of residence to get to work. By the 1950s, the figure had risen to one out of every four people. Today, 60 percent of employees leave their neighborhoods to get to their jobs. Half of them—8.5 million people—spend more than an hour in transit each way, and that's not including weekend commuters.[13] Coordinating the challenges of this rhythm has come to define countless lives. It is vital, both literally and figuratively, not to miss connections.

By the same token, mere communication is taking up more and more time. The meditative age of epistolary correspondence (which started in the sixteenth century with the Thurn und Taxis postal system) seems to be over and done. But people are still sending messages, now with images and sound attached. High-tech communications run on a parallel track with enhanced mobility. As more and more ground is covered physically, information is being exchanged faster and faster. The end result is that the practices and processes governing daily life are increasingly removed from face-to-face encounters; even when occupying the same building or room, people use technology to interact. Although line systems continue to be employed, radio networks and, since the 1950s, satellite systems have played an ever-larger role, at least for terminal devices. Users—who received this designation in the 1980s—are enjoying greater freedom of movement at every turn.[14] Thousands of positioning and transmission satellites orbit the earth—so many, in fact, that the European Union found it necessary to issue a *Code of Conduct for Outer Space Activities* in 2012. Steady monitoring aims to prevent the eleven thousand satellites in operation from colliding with each other (or with the growing amount of space junk also in orbit, twenty-one thousand pieces so far).[15]

Digital databases, democratized network access, news in real time, and similar provisions mean that information is more readily and fully available than ever before. As early as 1900, the Belgian author and activist Paul Otlet came up with the idea of a universal library. In 1945, the American engineer Vannevar Bush wanted to build a giant computerized brain he called Memex, short for "memory extender."[16] Their visions came much closer to reality in the 1960s with the advent of electronic data processing.

The Internet, search engines, Wikipedia, and so on represent the fruits of such "information euphoria."[17] At every step of the way, today more than ever, battles rage for who gets to decide what enters into circulation, who has access to it, and who makes money from it. The questions surrounding big data—on monopolies, surveillance, and privacy, among other things—are burning social issues. Unbounded cycles of information, goods, and finances have given rise to concerns about not only competition and solidarity but also copyright and intellectual property.[18] Since the early 2000s, interest groups have rallied around the catchphrase *open access* to demand the ready availability of information and social resources—including a basic income, in some instances; indeed, many activists couch their calls in terms of human rights.[19] For a spell, the Pirate Party in Germany and similar organizations elsewhere dominated political discussions; debates still have not been settled.

The most significant development concerns energy production, which has taken a new direction in light of climate change. A 2010 report by researchers in the United States concluded that emissions targets can be met only by drawing on infrastructures already in place.[20] The environmental consequences of burning coal, oil, and gas can no longer be shifted somewhere else or deferred to a later date. The massive human interventions that have occurred since the beginning of industrialization, but especially since the 1950s, have inspired a new designation for this epoch of the planet's history, the Anthropocene.[21] Infrastructures, cities, and manufacturing centers did not directly cause the conditions that have emerged, but they have played a major role and, indeed, epitomize the processes at work. The new forms of energy that are being developed to ensure a carbon-free future often draw on resources and practices of long standing—for instance, solar and wind energy, biomass, and dams. However, employing these technologies on a broader and intensified scale requires new storage and distribution systems; problems arise wherever a high level of infrastructure already exists and new loads cannot be readily accommodated. The obstacles are not so much technical as inertial; factors include goals of economic independence and political efforts to preserve jobs. On this score, an "infrastructure curse" prevails.[22]

What holds for the transportation sector holds for the energy sector, and to an even greater degree. Political and economic factors suggest one course at some times, and a different course at others. Individual countries' dependency on global energy flows has made them the object of strategic calculation.[23] International tensions emerge when pipelines cross borders. Conversely, price increases heighten the risk of domestic trouble. In premodern times, the cost of bread was the index for governmental success or failure; now, infrastructure prices are the yardstick. Investment in infrastructure is a measure of confidence.[24] Poor maintenance suggests an unfavorable state of society as a whole. Unrest surrounding tariffs (for instance, events in Brazil in 2013) can be observed worldwide. Matters of social justice take center stage when citizens object to the commercialization and privatization of basic services such as garbage collection and water supply. Better policy, the reasoning goes, could be implemented instead of handing operations over to business interests.[25] In societies accustomed to abundance and constant availability, periods of stagnation, shortages, and price hikes quickly take on an existential dimension.

Basic provisions for supply and disposal, even on a global scale, are better than ever, however. A growing portion of humanity has connections to water and electricity. Hygiene presents a dramatic point of contrast, however. According to estimates of the Bill and Melinda Gates Foundation, 2.5 billion people lack access to sanitary facilities, and four-fifths of illnesses in developing countries are caused by contaminated water.[26] In 2014, the Indian government constructed public toilets in a campaign to prevent people relieving themselves in fields (where numerous sexual assaults also take place). In 2014, the World Toilet Organization declared November 19 World Toilet Day—an occasion recognized even by the United Nations.[27]

In countries at the prosperous end of the spectrum, infrastructural arrangements represent a major factor in where people choose to live; this has long been the case when companies consider where to locate operations. The level of services relating to quality of life, transportation, and childcare can prove decisive for transnationally mobile classes, especially for creatives. And despite countermovements, globalization is expanding.[28] If everything remains flexible, the reasoning goes, individuals will find the most suitable conditions for themselves. Intermodal connections among networks of transportation, communication, supply, and disposal have made immense progress. The downside is that systems are susceptible to disruption, and battles fought over technical and organizational standards will set the course for the foreseeable future. Points of incompatibility—for instance, different voltages, track widths, and frequencies—mean that further improvements are necessary.

Across the globe, urban hotspots for infrastructure development have emerged. Major companies set up operations in "alpha cities," which have busy airports, good basic provisions for communication, and a high rate of foreign investment.[29] Unsurprisingly, they attract mobile, creative segments of the world population. Elsewhere, little new infrastructure is being created, and some established systems are being dismantled because of declining populations.[30] In other respects, spatial and social disparities are leveling out, especially in terms of communication. Digital contact does not have to meet as many conditions as face-to-face encounters. If corporations were nation-states, Facebook or WhatsApp would be the largest countries in the world today.[31] Indeed, they are sometimes considered to have taken over functions performed by governments—including social services.[32]

The high-modern infrastructure ideal—universal access, standardized services, and monopolistic structures—has faced mounting challenges since the 1970s. Networks such as the postal system and telecommunications have become less rigid in organizational terms. The competitive and lucrative components of infrastructures have often been privatized, while parts prone to financial loss remain publicly owned. This state of affairs has resulted in a spectrum of prices and quality, with privileged "spaces of flows" (Manuel Castells) moving closer and closer to one end and less attractive options crowding the other.[33]

Megacities of the Global South sharply diverge from the bacteriological and infrastructurally networked cities of high modernism. Indeed, they are so fragmented and their functionality so incalculable that they might merit another name altogether. Ideally, the modern city provides universal services, and relatively homogeneous norms, expectations, and values prevail in them. An interconnected and cooperative citizenry holds everything together under the auspices of a government that monitors and promotes general welfare. This is not the case for the urban agglomerations in question, where infrastructures emerge and function in an uncoordinated manner; who has access depends largely on chance. Consequently, relationships between users and networks in the Global

South are completely different from those in the Global North.[34] Only time will tell whether these conditions represent an intermediate step toward alignment with standards and services elsewhere, or—as is more likely the case—they will give rise to more flexible models from which developed countries will learn. Infrastructures will persist, but not the concepts that have guided their development in Europe since the end of the eighteenth century.

WHAT NEXT?

Minor shifts often herald larger changes. The television show *Wünsch Dir was* (Make a wish), which aired in Germany, Austria, and Switzerland in the late 1960s, represented an early experiment in audience participation. Families from different countries had to demonstrate their cohesiveness in games that placed various kinds of stress on their members. At the end of the broadcast, a preselected segment of viewers voted on who passed muster by flushing the toilets in their homes according to a certain pattern. Local water companies then reported the rise of consumption to the studio, which announced which family had won. With the heightened environmental awareness that developed during the early 1970s, the procedure changed. Voters were asked to turn off electrical appliances, and the dip in energy consumption was the measure.[35]

All this sounds dated in an age prone to rhapsodies about the hive mind, e-democracy, and the like—especially because the game shows of yesteryear presupposed that the nuclear family still set the norm. Nevertheless, the intermodal connection and voting procedure pointed to the road ahead, as did the ecologically minded reorientation that took place a little later. Different spheres of life cannot be neatly separated from each other, and no infrastructure can simply subordinate "first nature" (or Nature writ large) to "second nature" (that is, culture). Centralized, linear thinking that prizes constant growth has given way, more and more, to models that incorporate cycles and interconnectedness.

Industrialized nations have come to recognize that the Global North cannot continue to consume resources at the current level. Because the world is a closed system, not an open one, it is impossible for particular regions to produce and consume beyond their means, damaging the environment, without harming themselves and others. But the levels and standards that prevail in these countries cannot be obtained worldwide without ecological damage. The challenge, then, is to change conditions without imperiling comforts and conveniences that form part and parcel of everyday life. Doing so calls for solutions that increase the efficiency of systems already in operation while holding the consequences in check—a process often described by the vivid phrase *end-of-pipe technology*. The only *real* change is supposed to be emissions.

Since the 1970s, more and more appeals have been made to individual responsibility, moderation, and environmentally conscious practices. In other words, the public, which is used to thinking and acting within a high-modern framework, should anticipate future conditions that require changes to habits and routines along sustainable lines. Some people are proving more adaptable than others. Thus, in 2013, members of the younger generation in Berlin struck on the idea of setting up water fountains that could be located by means of smartphone apps. The goal was to counter the corporate interests that sell some ten billion liters of bottled water per year (at a significant markup), when what comes out of the tap is usually perfectly fine.[36] Similarly, slow-food and buy-local movements discourage shipping goods from far away and advocate following the rhythms of traditional agriculture; an apple from a neighboring field is just as appetizing as a mango from overseas. Proposals in other areas advocate deceleration and rediscovering a mode of life more in keeping with human rhythms than the revved up logistics of constant production and consumption. The stakes are much higher than the temporary remedies sold by the booming wellness industry.

Fascination with automobiles and personalized travel, which lasted for more than a century, seems to have passed its peak, and younger people demonstrate considerably less enthusiasm for cars than their elders. An instrumental and matter-of-fact attitude toward transportation is now the norm; bicycles and public transit, combined with the sharing economy, are experiencing a heyday. At least some members of the new generation also have a critical view of long-distance travel and other resource-intensive forms of social distinction. Expansive globalization that has tourists and commodities circulating more and more broadly is no longer a sign of necessary (much less desirable) modernization.

In turn, the ecological paradigm valorizes decentralized infrastructural arrangements adapted to local settings. It considers environmental implications from the outset to avoid compounding the "denaturalization" of time and space.[37] Developments include not only emissions trading (which has proved problematic) but also counterbalancing the space claimed by new buildings and facilities with a comparable surface area devoted to ponds, plantings, and the like; this practice is observed today in Germany, whereby requirements also exist to preserve sites of archaeological importance and monuments.[38] (Together with increased levels of public self-confidence and organization, such provisions to protect land, biodiversity, and cultural heritage are the stuff of technocratic nightmare.) There is even talk of green infrastructure. In 2010, the European Commission agreed on biodiversity targets to reestablish green belts and ecosystems in areas that have been fragmented by roads, railways, pipelines, and so on. Bridges for vegetation over alluvial paths will connect untouched areas, allowing habitat for flora to spread and fauna to roam.[39] Such renaturalization will also counteract the eyesores of high modernity scarring the landscape (which often is green only on a superficial level).[40] A glance from any airplane window confirms that infrastructures have carved up the earth into grids.

Yet despite heightened ecological awareness, infrastructural development remains focused on the abstract individual untrammeled by spatial, temporal, and natural constraints, who is constantly making self-aware and informed decisions. All the trends mentioned in the preceding will not change much in the short term. Competition to reduce margins of time, space, and speed will continue among producers, middlemen, consumers, and commuters. Flowing, surfing, switching among connections, avatars, and digital identities represent the predominant mode of existence in a society of "liquid modernity," which has set everything in flux.[41] Infrastructures have accelerated activity and shortened distance—which means that much of what used to take time (or not happen at all) now occurs simultaneously. What is gained by faster travel or better organization is spent on unplanned activities. Some believe that these conditions give their lives greater intensity. The public intellectual Roger Willemsen denominated them "second-screen people"; they cannot abide a world in which multiple events are not occurring in parallel.[42] Others feel like proverbial hamsters in a maze they cannot escape, that anonymous forces control their lives.

Now as ever, reactions can be more optimistic or pessimistic. Those who see the brighter side stress the benefits of individual autonomy, social contact, convenience, and cultural horizons that globalized transportation and communication have brought. Their pessimistic counterparts voice concern about social alienation and isolation, technological dependency, and constant monitoring; they fear that patterns of consumption, social interaction, and opinions are becoming uniform in tides of (manipulated) data. But they also fear missing out. Max Scharnigg, a writer born in 1980, notes that members of his generation feel obliged to stay informed and try out everything new, lest they fall behind the times.[43]

Being enmeshed in digital technology gives rise to new forms of interaction. Current notions must be constantly revisited. Attendant disputes involve what Max Weber, at the beginning of the twentieth century, identified as "*stahlhartes Gehäuse*" and Martin Heidegger, a few decades later, "*Gestell*" (respectively translated, somewhat gracelessly, as "iron cage" and "enframing"—in other words, constricting conditions of life). The judicial and administrative spheres must figure out, even as developments are unfolding, the new possibilities and risks at play and try to hold them within acceptable limits. Superstructures emerge, which, like infrastructures, have a normative and standardizing effect, albeit at a different level. For example, the World Wide Web harbors potential that has been, and still is, used for crime, defamation, and political manipulation. The algorithms employed by search engines and data corporations steer Internet users toward information and services that reinforce habits and prejudices.[44] Abuses provoke belated regulatory intervention. Dependency, especially when amplified by technological means, proves difficult to escape.

Significantly, the same process unleashes subversive forces. As in the days of amateur radio operators, pirate broadcasts, and free cable TV stations, actors intent on undermining control mechanisms advocate, among other things, that large Internet providers be countered by independent "citizen networks."[45] These nonconformists, who often have anarchistic sensibilities, remind us that infrastructures can be steered in several directions and used for different purposes. Creative potential can take unexpected turns, giving rise to an opposition or resistance movement. In practical terms, their activities demonstrate that it is possible to break through material, institutional, and mental regimes and open up space for alternatives. It is worth noting that infrastructures demonstrate a certain inertia.[46]

Twenty years after coining the term *Daseinsvorsorge* in 1938, Ernst Forsthoff advised governments to focus on "securing basic conditions of existence for citizens to the extent possible"—that is, not to cater to the entitled consumers who had emerged in the interim.[47] No scheme has been developed that would take care of all the complications that resulted from the late nineteenth-century bacteriological city, the mid-twentieth-century model of governmentally administered services, or the open access promoted more recently. Deregulation has given rise to a host of organizations that exist where the public and private spheres intersect. Almost every large-scale technical system can be considered either an infrastructure or a commercial operation, but the public tends to be reluctant about letting go of the idea that the state is responsible for satisfying basic needs.[48] Transitions are fraught with uncertainty.

Looking back at the last two hundred years reveals that facilities for communication, transportation, supply, and disposal have taken shape in keeping with a wide range of factors. Likewise, today, infrastructural networks represent the result of countless decisions in which technology, economy, politics, and other interests converge; necessarily, these constellations include elements of chance. In the digital era, participation is no longer reserved for a small number of television viewers who decide the winner on a game show by flushing the toilet. Systems have become simpler but also more complicated. On the one hand, matters are more straightforward because networks motivate behavior that no longer (or, at any rate, no longer primarily) depends on personal ties and site-specific actions. On the other hand, matters are more difficult because closing gaps in space and time tempts people to do more, and to do it while pursuing other activities.

Now, in the twenty-first century, turning off electricity or smartphones can send a message with broad-scale ramifications. Since 2007, people across the globe have turned off power to observe Earth Hour and make a political statement about the environment, in a gesture that recalls *Wünsch Dir was* and other past entertainments.[49] As noted earlier, this will hardly change infrastructures themselves. It does, however, reflect priorities that challenge habits and ideas deeply rooted in the nineteenth and twentieth centuries that are no longer self-evident.

RECAPITULATION

The preceding has sought to make visible a process of networking with infinitely more aspects than any one book can discuss. To be sure, it has not always been possible to draw a clear line of separation between material infrastructures in themselves, the devices connected to them (which often communicate with the commercial sphere), and the lived practices that rely on them. All these lines of investigation are interwoven.

However, it should be clear that the notion of infrastructure implies an understanding of modernity based on circulation and growth. The concept and attendant institutions have been, and continue to be, the vehicle for liberal ideas and ideals that originated in the Enlightenment and have continued to spread, albeit unevenly. Hereby, social spaces and orders have been recoded, material and immaterial borders have been transgressed, and new categories of commodities, both for domestic use and for export, have emerged. Indeed, infrastructural development has fueled the dazzling, if highly contradictory, process of globalization. Infrastructures have been hailed as instruments of transformation and derided as "tentacles of progress" (in the words of historian Daniel Headrick).[50] Political, economic, and cultural actors have fought over them. On the one hand, infrastructures have promoted communication and travel and, in so doing, integrated and connected once-isolated parts of the world. On the other hand, technology and attendant procedures have brought about standardization that proves unwelcome and, moreover, leaves systems vulnerable to disruption and even terrorist attack. Similarly, public utilities and waste disposal facilities have promoted a hygienic way of life far beyond urban centers, but they have also led to new social hierarchies.

Wherever services and connections are available, infrastructures offer a promise as well as a point of comparison with arrangements elsewhere. On this score, it is warranted to view their state of development as a measure of modernity. The very word *infrastructure* signals that there is a provider with obligations toward (or at least relevant to) the general public. Accordingly, infrastructures have largely retained the appearance of impartiality, even though—or for precisely this reason—they have served as instruments for influencing politics, economy, society, and culture.[51] This book has focused on the changes to everyday life and shifting interpretations brought about by infrastructural networking. One item of particular importance is the way such networking, which mediates between human beings and the natural world, has transformed spatial and social relations almost everywhere on the globe. Infrastructures have made it possible to specialize labor, extract material and immaterial resources, and produce and distribute not only mass goods but also information. They have promoted mobility for work, leisure, and even permanent migration. Infrastructures allow people to broaden their horizons, and they afford options for shaping life individually and collectively. Burdens have grown, however. Not only must one keep up with the possibilities on offer; one must also adapt one's thinking to large-scale technical systems, follow their operative logic, and conform to the regulations that keep them functioning smoothly.

Over and over, resistance has flared up because of the unwelcome feeling that things are going too fast, a mounting sense of estrangement from natural ways of life, the environmental cost of progress, and so on. All the same, most people submit to infrastructural regimes more or less willingly. The attraction exercised by supposed modernization seems to outweigh the drawbacks. Besides practical benefits, there are many other inducements: the desire to acquire personal or collective prestige, legitimize governmental institutions, compete economically or culturally, and achieve technological (or technocratic) ideals. Factors inhibiting infrastructural expansion include, especially in rapidly growing urban areas, the increased pressure caused by cramped living conditions, scarce resources, and health hazards. Apart from publicly celebrated prestige projects, initiatives have proceeded anonymously. A standard feature of infrastructures is that they permeate everyday routines quickly and without drawing much notice. Accordingly, they constitute something like a collective unconscious—the basis for the activities that *do* receive attention. As far as most segments of the population are concerned, overall social conditions, particularly economic and cultural aspects of life, depend on the extent to which these systems have expanded and now function on a quotidian level. Closer examination reveals that they are riddled with disparities and not uniformly accessible; infrastructural networks tend to benefit enterprising individuals and classes.

Infrastructures augment and amplify the senses. They form a "metabolism" sustaining the human and natural worlds. They have played a vital role in freeing people not only from environmental and meteorological conditions but also from local, regional, and national habits and traditions and from daily, weekly, and seasonal rhythms; under their pressure, even opening hours at places of business have changed. However, as new ways of life and patterns of perception develop, human beings grow dependent on the material practices and conditions that sustain them, and they become susceptible to shocks to the system, which can be caused internally or come from without. General ability to handle states of exception (scarcity or unforeseen events, for instance) has diminished. Connected societies face the troubles attending excessive abundance. Because of such dependency, temporary disruptions quickly turn into a nuisance—or even a threat.

Still, on the whole, infrastructure systems have proved resilient when accidents and attacks occur. Most of the time, systems are repaired or even rebuilt immediately. Security measures promote adaptable and modular arrangements with strong entry and exit controls. If life is to go on with a semblance of normality, everyday routines should not be shaken up too much, even though the anonymity inscribed in infrastructures—and, for that matter, human error—cannot be eliminated entirely. Infrastructures are built to last, if only because of the high costs of investment; older technical systems that remain available are a valuable resource, too. Stability is ensured by extrasystemic factors (mainly the routines of users) and by social, political, and economic incentives.

Ongoing debates about how infrastructures can be expanded and protected, and how costs and benefits can be fairly distributed, attest to the central importance of these institutions. Infrastructures open avenues of social participation, which in turn become matters of legal right, reconfiguring relations between groups. Today, connection to networks determines civic belonging. State socialism failed to realize this promise in full; access depended on hierarchy and could be revoked for political reasons. Market-based (neoliberal) systems have not managed to overcome unequal distribution, either, but socially inclusive, affluent societies come closest to achieving the ideal of universal access.

Although political and economic systems play a secondary role, infrastructures go hand in hand with certain values and standards. But they don't promote global standardization so much as hybridization. They have at times been built and operated by coercive means, excluded or marginalized certain groups, and served as vehicles for ideology. To appeal to, and be used by, the population at large, infrastructures must serve a range of interests and admit varied applications. By being open, they include options that run counter to the intentions of their creators; consequently, they have contributed to the growth of social formations that are plural, decentralized, and largely self-organizing. Rigid management and efforts at control have been repeatedly undermined, if only by parties (hobbyists, rebels, hackers, and so on) who are seemingly constitutionally averse to majoritarian practices and who reflexively challenge the boundaries of established systems.

Future developments cannot be predicted, then. This is especially the case for societies that are just beginning to deal with the concept and material reality of infrastructure—and under conditions that are quite different from those in the West. All the same, questions of varying levels of urgency, identified here, are likely to come up again and again. The first concerns organization, for which models are emerging that fuse public and private interests in a manner that needs to be negotiated locally (as in the case of public-private partnerships). Second, there is the question of visibility, concerning a tendency toward camouflage: infrastructures that are increasingly comprehensive in scope are being set up less and less conspicuously. The third issue is the cycle between aging arrangements and ongoing modernization; discussion about who is responsible and how cost burdens can be distributed equitably will continue. The fourth question concerns the vulnerability of networks and flow spaces, which necessitates trade-offs between security and freedom. Finally, there is the matter of whether the availability of infrastructures requires continued industrial growth, and if not, how to devise and implement more sustainable alternatives on a global scale.[52]

Precisely this constellation is what brings things back to ordinary people. History suggests that access to transport, communication, supply, and disposal, even when routine, does not instill a lasting sense of privilege. Deviations upward occasion envy, and deviations downward are perceived as backwardness. The elevator effect of infrastructures has led most people who are connected to become accustomed to certain levels, norms, and standards of comfort and convenience, which admit change only over the medium and long term. As has been the case for the past two hundred years, individual priorities and overriding concerns for the common good will need to strike a balance that everyone can live with. The history of infrastructures, then, holds many exciting chapters in store.

NOTES

FOREWORD

1. Emily Elhacham, Liad Ben-Uri, Jonathan Grozovski, Yinon M. Bar-On, and Ron Milo, "Global Human-Made Mass Exceeds All Living Biomass," *Nature* 588 (2020): 442–444, https://doi.org/10.1038/s41586-020-3010-5.

2. P. K. Haff, "Technology as a Geological Phenomenon: Implications for Human Well-Being," *Geological Society, London, Special Publications* 395 (2014): 301–309, https://doi.org/10.1144/SP395.4.

3. Geoffrey C. Bowker and Susan Leigh Star, *Sorting Things Out: Classification and Its Consequences* (Cambridge, MA: MIT Press, 1999).

4. Nathan Ensmenger, "The Environmental History of Computing," *Technology and Culture* 59 (4S) (2018): S7–S33, https://doi.org/10.1353/tech.2018.0148.

PREFACE TO THE ENGLISH-LANGUAGE EDITION

1. Emily Badger and Darla Cameron, "How Railroads, Highways and Other Man-made Lines Racially Divide America's Cities," *Washington Post*, July 16, 2015; Huub Dijstelbloem, *Borders as Infrastructure: The Technopolitics of Border Control* (Cambridge, MA: MIT Press, 2021).

2. Kregg Hetherington (ed.), *Infrastructure, Environment and Life in the Anthropocene* (Durham, NC: Duke University Press, 2019); Martin Meiske, *Die Geburt des Geoengineering: Grossbauprojekte in der Frühphase des Anthropozäns* (Göttingen: Wallstein, 2021).

3. See, for instance, Penelope Harvey, "Cementing Relations: The Materiality of Roads and Public Space in Provincial Peru," *Social Analysis*, 54.2 (2010): 28–46.

4. Archie Davies, "The Coloniality of Infrastructure: Engineering, Landscape and Modernity in Recife," *Environment and Planning D: Society and Space* 39.4 (2021): 740–757.

5. Gabriele Balbi and Andreas Fickers (eds.), *History of the International Telecommunication Union (ITU): Transnational Techno-Diplomacy from the Telegraph to the Internet* (Berlin: De Gruyter Oldenbourg, 2020).

6. For instance, Aparajita Mukhopadhyay, *Imperial Technology and "Native" Agency: A Social History of Railways in Colonial India, 1850–1920* (London: Routledge, 2018); Simone Abram, Brit Ross Winthereik, and Thomas Yarrow (eds.), *Electrifying Anthropology: Exploring Electrical Practices and Infrastructures* (London: Routledge, 2020); Timothy Moss, *Remaking Berlin: A History of the City through Infrastructure, 1920–2020* (Cambridge, MA: MIT Press, 2020); and Lyubomir Pozharliev, *The Road to Socialism: Transport Infrastructure in Socialist Bulgaria and Yugoslavia (1945–1989)* (Göttingen: Vandenhoeck & Ruprecht, 2021).

7. Anke Ortlepp, *Jim Crow Terminals: The Desegregation of American Airports* (Athens: University of Georgia Press, 2017); Stefan Höhne, *Riding the New York Subway: The Invention of the Modern Passenger* (Cambridge, MA: MIT Press, 2021).

8. Quoted in Norman Aselmeyer, "Capturing the Zeitgeist of Infrastructure." *TRAFO—Blog for Transregional Research*, January 31, 2019, https://trafo.hypotheses.org/17723.

INTRODUCTION

1. Daniel Deckers, "Unter den Wolken," *Frankfurter Allgemeine Zeitung*, April 5, 2013.

2. Paul N. Edwards, Geoffrey C. Bowker, Steven J. Jackson, and Robin Williams, "Introduction: An Agenda for Infrastructure Studies," *Journal of the Association for Information Systems* 10.5 (2009): 364–374, at 365.

3. Bram van Stolk and Cas Wouters, "Die Gemütsruhe des Wohlfahrtsstaates," in Peter Gleichmann, Johan Goudsblom, and Hermann Korte (eds.), *Macht und Zivilisation: Materialien zu Norbert Elias' Zivilisationstheorie 2* (Frankfurt am Main: Suhrkamp, 1984), 242–260.

4. Monika Dommann, "Alles fliesst: Soll die Geschichte nomadischer werden?" *Geschichte und Gesellschaft* 42 (2016): 516–534.

5. Siegfried Giedion, *Mechanization Takes Command: A Contribution to Anonymous History* (Minneapolis: University of Minnesota Press, 2013).

6. Stephen Graham and Simon Marvin, *Splintering Urbanism: Networked Infrastructures, Technological Mobilities and the Urban Condition* (London: Routledge, 2001).

7. The most far-reaching inquiry along these lines to date is Per Högselius, Arne Kaijser, and Erik van der Vleuten, *Europe's Infrastructure Transition: Economy, War, Nature* (New York: Palgrave, 2015).

8. "Infrastruktur," *Frankfurter Allgemeine Zeitung*, March 1, 1952.

9. Bernard S. Katz, "Infrastructure Ills: Or If You Can't Use It Correctly, Don't," *Advertising Age* 60 (1989): 23.

10. Till Briegleb, "Vor allem obskur: Was der Begriff Infrastruktur heute alles bedeutet," *Süddeutsche Zeitung*, May 27, 2013.

11. Renate Mayntz, "Zur Entwicklung technischer Infrastruktur-Systeme," in Renate Mayntz, Bernd Rosewitz, Uwe Schimank, and Rudolf Stichweh (eds.), *Differenzierung und Verselbstständigung* (Frankfurt am Main: Campus, 1988), 233–259.

12. William Rankin, "Infrastructure and the International Governance of Economic Development, 1950–1965," in Jean-François Auger, Jan Jaap Bouma, and Rolf Künneke (eds.), *Internationalization of Infrastructures: Proceedings of the 12th International Conference on the Economics of Infrastructures* (Delft, Netherlands: Economics of Infrastructures, 2009), 61–75. See also Dirk van Laak, "Der Begriff 'Infrastruktur' und was er vor seiner Erfindung besagte," *Archiv für Begriffsgeschichte* 41 (1999): 280–299; and Steffen Richter, *Infrastruktur: Ein Schlüsselkonzept der Moderne und die deutsche Literatur 1848–1914* (Berlin: Matthes & Seitz, 2018).

13. Harald Welzer, *Mentale Infrastrukturen—wie das Wachstum in die Welt und in die Seelen kam* (Berlin: Heinrich-Böll-Stiftung, 2011), 14.

14. Ernest W. Black, *Cursus Publicus: The Infrastructure of Government in Roman Britain* (Oxford, UK: Tempus Reparatum, 1995); Anne Kolb, *Infrastruktur und Herrschaftsorganisation im Imperium Romanum* (Berlin: De Gruyter, 2014); Margot Klee, *Lebensadern des Imperiums, Strassen im Römischen Weltreich* (Stuttgart: Theiss, 2010); Nico Ernstberger, *Die Versorgung Roms mit Wasser, Waren und Energie: Die Infrastruktur der Kaiserzeit* (Hamburg: Disserta, 2014).

15. Susan E. Alcock, John Bodel, and Richard J. A. Talbert (eds.), *Highways, Byways, and Road Systems in the Pre-Modern World* (Chichester, UK: Wiley-Blackwell, 2012).

16. Adam Smith, *The Wealth of Nations* (New York: Modern Library, 2000), 779.

17. Thomas Schulze, *Infrastruktur als politische Aufgabe: Dogmengeschichtliche, methodologische und theoretische Aspekte* (Frankfurt am Main: Lang, 1993), 80.

18. Christoph Asendorf, *Batterien der Lebenskraft: Zur Geschichte der Dinge und ihrer Wahrnehmung im 19. Jahrhundert* (Weimar: Verlag und Datenbank für Geisteswissenschaften, 2002), 43. See also Armand Mattelard, *Networking the World, 1794–2000* (Minneapolis: University of Minnesota Press, 2000).

19. Sebastian Giessmann, *Die Verbundenheit der Dinge: Eine Kulturgeschichte der Netze und Netzwerke* (Berlin: Kadmos, 2014).

20. Rolf-Peter Sieferle, *Rückblick auf die Natur: Eine Geschichte des Menschen und seiner Umwelt* (Munich: Landtverlag, 1997); Joachim Radkau, *Natur und Macht: Eine Weltgeschichte der Umwelt* (Munich: Beck, 2002).

21. For a contemporary account, see Bernard Forest de Belidor, *Architectura Hydraulica oder die Kunst, das Gewässer des Meeres und der Flüsse zum Vorteil der Verteidigung, der Festungen, des Handels und des Ackerbaus anzuwenden* (Augsburg: Kletts, 1766). See also Marie L. Allemeyer, *"Kein Land ohne Deich . . . !" Lebenswelten einer Küstengesellschaft in der Frühen Neuzeit* (Göttingen: Vandenhoeck & Ruprecht, 2006).

22. Jacques Le Goff, *The Birth of Europe*, trans. Janet Lloyd (Malden, MA: Blackwell, 2014); Peter Payer, *Die synchronisierte Stadt* (Vienna: Holzhausen, 2015).

23. Dieter Schott, *Europäische Urbanisierung (1000–2000): Eine umwelthistorische Einführung* (Cologne: Böhlau, 2014), 306.

24. Paul N. Edwards, "Infrastructure and Modernity: Scales of Force, Time, and Social Organization in the History of Sociotechnical Systems," in Thomas J. Misa, Philip Brey, and Andrew Feenberg (eds.), *Modernity and Technology* (Cambridge, MA: MIT Press, 2002), 185–225, at 185–186.

25. Jesse H. Ausubel and Robert Herman (eds.), *Cities and Their Vital Systems: Infrastructure—Past, Present, and Future* (Washington, DC: National Academies Press, 1988), 1.

26. Michael Flitner, Julia Lossau, and Anna-Lisa Müller (eds.), *Infrastrukturen der Stadt* (Wiesbaden: Springer, 2017), 2.

27. See, e.g., the call for papers published by the Center for Metropolitan Studies for the workshop Gefüge, Kollektive und Dispositive: Zum "Infrastrukturalismus" des Gesellschaftlichen, March 18–19, 2011, http://www.heike-delitz.de/Artefakte2011.html.

28. Dietrich Stein, *Instandhaltung von Kanalisationen* (Bochum: Stein, 1999), vii.

29. Schulze, *Infrastruktur als politische Aufgabe*, 84; Vincent Ostrom, "Adam Smith und öffentliche Güter," in Franz-Xaver Kaufmann and Hans-Günter Krüsselberg (eds.), *Markt, Staat und Solidarität bei Adam Smith* (Frankfurt am Main: Campus, 1984), 135–157.

30. Schulze, *Infrastruktur als politische Aufgabe*, 137.

31. François Ewald, *The Birth of Solidarity: The History of the French Welfare State*, trans. Melinda Cooper (Durham, NC: Duke University Press); Abram de Swaan, *Der sorgende Staat: Wohlfahrt, Gesundheit und Bildung in Europa und den USA der Neuzeit* (Frankfurt am Main: Campus, 1993).

32. Rankin, "Infrastructure and the International Governance of Economic Development."

33. Ragnar Nurske, *Problems of Capital Formation in Underdeveloped Countries* (Oxford: Oxford University Press, 1953), 154.

34. Reimut Jochimsen, *Theorie der Infrastruktur: Grundlagen der marktwirtschaftlichen Entwicklung* (Tübingen: Mohr, 1966).

35. Zef Segal, "Communication and State Construction: The Postal Service in German States, 1815–1866," *Journal of Interdisciplinary History* 44.4 (2014): 453–473.

36. Eugen Weber, *Peasants into Frenchmen: The Modernization of Rural France 1870–1914* (Stanford, CA: Stanford University Press, 1988); Germà Bel i Queralt, *Infrastructure and the Political Economy of Nation Building in Spain, 1720–2010* (Portland, UK: Sussex Academic Press, 2012).

37. Todd Diacon, *Stringing Together a Nation: Cândido Mariano da Silva Rondon and the Construction of a Modern Brazil 1906–1930* (Durham, NC: Duke University Press, 2004).

38. Michel Foucault, *Security, Territory, Population: Lectures at the Collège de France 1977—1978*, trans. Graham Burchell (New York: Palgrave Macmillan, 2009), 55–59.

39. Dirk van Laak, "Infrastruktur und Macht," in François Duceppe-Lamarre and Jens Ivo Engels (eds.), *Umwelt und Herrschaft in der Geschichte / Environnement et pouvoir: une approche historique* (Munich: Oldenbourg, 2008), 106–111.

40. Michael Mann, *The Sources of Social Power*, Vol. 2, *The Rise of Classes and Nation States 1760–1914* (Cambridge: Cambridge University Press, 1993), 59.

41. Michael Mann, "Infrastructural Power Revisited," *Studies in Comparative International Development* 43.3–4 (2008): 355–365.

42. Trutz von Trotha, "Was war Kolonialismus? Einige zusammenfassende Befunde zur Soziologie und Geschichte des Kolonialismus und der Kolonialherrschaft," *Saeculum* 55 (2004): 49–95, at 68.

43. Jens Ivo Engels and Gerrit Jasper Schenk, "Infrastrukturen der Macht—Macht der Infrastrukturen. Überlegungen zu einem Forschungsfeld," in Birte Förster and Martin Bauch (eds.), *Wasserinfrastrukturen und Macht von der Antike bis zur Gegenwart* (Munich: Oldenbourg, 2015), 22–58, at 24.

44. Engels and Schenk, "Infrastrukturen der Macht," 56.

45. Dirk van Laak, "Technokratie im Europa des 20. Jahrhunderts—eine einflussreiche 'Hintergrund-ideologie;'" in Lutz Raphael (ed.), *Theorien und Experimente der Moderne: Europas Gesellschaften im 20. Jahrhundert* (Cologne: Böhlau, 2012), 101–128. On the whole, however, sociological study of infrastructures is just beginning; see Eva Barlösius, "Infrastrukturen als gesellschaftliche Weichensteller," *SozBlog: Blog der Deutschen Gesellschaft für Soziologie (DGS)*, February 25, 2015.

46. Keller Easterling, *Extrastatecraft: The Power of Infrastructure Space* (London: Verso, 2014), 15.

47. Goran Ahrne and Nils Brunsson, *Meta-Organizations* (Cheltenham, UK: Edward Elgar, 2008), 169; Lawrence Busch, *Standards: Recipes for Reality* (Cambridge, MA: MIT Press, 2011).

48. Achille Mbembe, *On the Postcolony* (Berkeley: University of California Press, 2001).

49. Tess Lea and Paul Pholeros, "This Is Not a Pipe: The Treacheries of Indigenous Housing," *Public Culture* 22.1 (2010): 187–209; Brian Larkin, "The Politics and Poetics of Infrastructure," *Annual Review of Anthropology* 42 (2013): 327–343; Tanja Winther, *The Impact of Electricity: Development, Desires and Dilemmas* (New York: Berghahn, 2008).

50. Penelope Harvey and Hannah Knox, "The Enchantments of Infrastructure," *Mobilities* 7.4 (2012): 521–536.

51. Dimitris Dalakoglou, "The Road from Capitalism to Capitalism: Infrastructures of (Post)Socialism in Albania," *Mobilities* 7.4 (2012): 571–586; Penelope Harvey and Hannah Knox, *Roads: An Anthropology of Infrastructure and Expertise* (Ithaca, NY: Cornell University Press, 2015).

52. Penelope Harvey, "Cementing Relations: The Materiality of Roads and Public Spaces in Provincial Peru," *Social Analysis* 54.2 (2010): 28–46; Adeline Masquelier, "Road Mythographies: Space, Mobility, and the Historical Imagination in Postcolonial Niger," *American Ethnologist* 29.4 (2002): 829–856; Michael K. Bess, "Routes of Conflict: Building Roads and Shaping the Nation in Mexico, 1941–1952," *Journal of Transport History* 35.1 (2014): 78–96.

53. Ferdinand Fried, *Umsturz der Gesellschaft* (Stuttgart: Union, 1950), 336–337.

54. Anthony Giddens, *The Constitution of Society: Outline of the Theory of Structuration* (Cambridge, UK: Polity 1984); Lewis Mumford, *The Myth of the Machine* (New York: Harcourt Brace Jovanovich, 1970).

55. James R. Beniger, *The Control Revolution: Technological and Economic Origins of the Information Society* (Cambridge, MA: Harvard University Press, 1986).

56. Anna von Schnitzler, "Citizenship Prepaid: Water, Calculability, and Techno-Politics in South Africa," *Journal of Southern African Studies* 34.4 (2008): 899–917; Erik Swynegedouw, *Social Power and the Urbanization of Water: Flows of Power* (Oxford: Oxford University Press, 2004).

57. The classic study is Thomas P. Hughes, *Networks of Power: Electrification in Western Society, 1880–1930* (Baltimore: Johns Hopkins University Press, 1983).

58. Bernward Joerges, "Die Brücken des Robert Moses: Stille Post in der Stadt- und Techniksoziologie," *Leviathan* 27.1 (1999): 43–63, at 60.

59. Elisabeth Heidenreich, "Natur und Kultur heute: verwickelt in technische Fliessräume," in Susanne Frank and Matthew Gandy (eds.), *Hydropolis: Wasser und die Stadt der Moderne* (Frankfurt am Main: Campus, 2006), 57–73, at 71.

60. Hans Blumenberg, "Dogmatische und rationale Analyse von Motivationen technischen Fortschritts," *Zeitschrift für Kulturphilosophie* 7.2 (2013): 407–422.

61. Wolfgang Essbach, "Zur Anthropologie artifizieller Umwelt," in Kurt W. Alt and Natascha Rauschenberger (eds.), *Ökohistorische Reflexionen: Mensch und Umwelt zwischen Steinzeit und Silicon Valley* (Freiburg im Breisgau: Rombach, 2001), 171–195, at 19. Freud advanced similar claims in *Civilization and Its Discontents* (1930).

62. Radkau, *Das Zeitalter der Nervosität.*

63. Paul Virilio, *Speed and Politics*, trans. Mark Polizzotti (Los Angeles: Semiotext[e], 2006); Hartmut Rosa, *Beschleunigung: Die Veränderung der Zeitstrukturen in der Moderne* (Frankfurt am Main: Suhrkamp, 2005); Peter Borscheid, *Das Tempo-Virus. Eine Kulturgeschichte der Beschleunigung* (Frankfurt am Main: Campus, 2004).

64. Susan Leigh Star, "The Ethnography of Infrastructure," *American Behavioral Scientist* 43.3 (1999): 377–391.

65. Christa Kamleithner, "Neue Mischungsverhältnisse: Zum Gebrauch von Infrastrukturen," in Nathalie Bredella and Chris Dähne (eds.), *Infrastrukturen des Urbanen: Soundscapes, Landscapes, Netscapes* (Bielefeld: transcript, 2013), 253–274, at 253; Sebastian Giessmann, *Netze und Netzwerke: Archäologie einer Kulturtechnik, 1740–1840* (Bielefeld: transcript, 2006).

66. See, for example, Stefan Kaufmann, *Kommunikationstechnik und Kriegführung 1815–1945: Stufen telemedialer Rüstung* (Munich: Fink, 1996).

67. Stefan Höhne, *New York City Subway: Die Erfindung des urbanen Passagiers* (Cologne: Böhlau, 2017); Stefan Höhne, "Token Tactics: Artefaktpolitiken in der New Yorker City Subway," in Wiebke Porombka, Heinz Reif, and Erhard Schütz (eds.), *Versorgung und Entsorgung der Moderne: Logistiken und Infrastrukturen der 1920er und 1930er Jahre* (Frankfurt am Main: Peter Lang, 2011), 24–33 and 73–88. See also Benjamin Lee and

Edward LiPuma, "Cultures of Circulation: The Imaginations of Modernity," *Public Culture* 14.1 (2002): 191–213.

68. Joachim Fritz-Vannahme, "Der Anthropologe im Dschungel der Grossstadt: Ein Porträt von Marc Augé," *Die Zeit*, August 23, 1996.

69. Dennis Rogers and Bruce O'Neill, "Infrastructural Violence: Introduction to the Special Issue," *Ethnography* 13.4 (2012): 401–412.

70. Swati Chattopadhyay, *Unlearning the City: Infrastructure in a New Optical Field* (Minneapolis: University of Minnesota Press, 2012).

71. Ulrika Trovalla and Eric Trovalla, "Infrastructure Turned Suprastructure: Unpredictable Materialities and Visions of a Nigerian Nation," *Journal of Material Culture* 20.1 (2015): 43–57; Lisa Bjorkman, *Pipe Politics, Contested Waters: Embedded Infrastructures of Millennial Mumbai* (Durham, NC: Duke University Press, 2015).

72. Abdou Maliq Simone, "People as Infrastructure: Intersecting Fragments in Johannesburg," *Public Culture* 16.3 (2004): 407–429. A similar argument is made by Julia Elyachar, "Phatic Labor, Infrastructure, and the Question of Empowerment in Cairo," *American Ethnologist* 37.3 (2010): 452–464.

73. See, for example, Stephen Graham and Colin McFarlane (eds.), *Infrastructural Lives: Urban Infrastructure in Context* (New York: Routledge, 2015).

74. Ash Amin, "Lively Infrastructure," *Theory, Culture & Society* 31.7–8 (2014): 137–161.

CHAPTER 1

1. Dietmar Dath, "Das schreibende Flussgespenst," *Frankfurter Allgemeine Zeitung*, March 3, 2005.

2. Birte Förster and Martin Bauch (eds.), *Wasserinfrastrukturen und Macht von der Antike bis zur Gegenwart*, (Munich: Oldenbourg, 2015).

3. Lionel T. C. Rolt, *From Sea to Sea: The Canal du Midi* (Columbus: Ohio University Press, 1973); Chandra Mukerji, *Impossible Engineering: Technology and Territoriality on the Canal du Midi* (Princeton, NJ: Princeton University Press, 2009).

4. Jens Ivo Engels, "Kanalregionen im Frankreich der Sattelzeit: Elemente für die Erforschung der Raumwirkungen von Infrastrukturen," *Francia* 37 (2010): 149–165, at 149.

5. Werner Abelshauser, "Staat, Infrastruktur und regionaler Wohlstandsausgleich im Preussen der Hochindustrialisierung," in Fritz Blaich (ed.), *Staatliche Umverteilungspolitik in historischer Perspektive* (Berlin: Duncker & Humblot, 1980), 9–58.

6. Christopher F. Jones, "A Landscape of Energy Abundance: Anthracite Coal Canals and the Roots of American Fossil Fuel Dependence, 1820–1860," *Environmental History* 15 (2010): 449–484.

7. See Forest de Belidor, *Architectura Hydraulica*; Carl Christian Rommerdt, *Handbuch der Land- und Wasserbaukunst: Für angehende und ausübende Forstmänner, Kameralisten und Oekonomen* (Erfurt: Henning, 1828).

8. Jo Guldi, *Roads to Power: Britain Invents the Infrastructure State* (Cambridge, MA: Harvard University Press, 2012).

9. Bel i Queralt, *Infrastructure and the Political Economy of Nation Building*; Allan Mitchell, *The Great Train Race: Railways and the Franco-German Rivalry, 1815–1914* (New York: Berghahn, 2000).

10. Jürgen Osterhammel, *Die Verwandlung der Welt* (Munich: Beck, 2009), 1010ff.

11. Harry W. Lintsen (ed.), *Geschiedenis van de Techniek in Nederland: De wording van een moderne samenleving 1800–1890*, 6 vols. (Zutphen: Stichting Historie der Techniek, 1992–1995).

12. Nicolai Hannig, "Die Suche nach Prävention: Naturgefahren im 19. und 20. Jahrhundert," *Historische Zeitschrift* 300.1 (2015): 33–65; John R. Gold and George Revill, "Landscapes of Defence," *Landscape Research* 24 (1999): 229–239.

13. David Blackbourn, *The Conquest of Nature: Water, Landscape and the Making of Modern Germany* (London: Pimlico, 2007).

14. Marc Cioc, *The Rhine: An Eco-Biography, 1815–2000* (Seattle: University of Washington Press, 2002), 14; see Christoph Bernhardt, *Im Spiegel des Wassers: Eine transnationale Geschichte des Oberrheins (1800–2000)* (Cologne: Böhlau, 2016).

15. Uwe Lübken, "'Der grosse Brückentod': Überschwemmungen als infrastrukturelle Konflikte im 19. und 20. Jahrhundert," *Saeculum* 58.1 (2007): 89–114.

16. Blackbourn, *The Conquest of Nature*; Thomas Etzemüller, "Romantischer Rhein—eiserner Rhein: Ein Fluss als *imaginary landscape* der Moderne," *Historische Zeitschrift* 295.2 (2012): 390–424.

17. Guido Thiemeyer and Isabel Tölle, "Supranationalität im 19. Jahrhundert? Die Beispiele der Zentralkommission für die Rheinschifffahrt und des Oktroyvertrages 1804–1832," *Journal of European Integration History* 17 (2011): 177–196.

18. Albert Demangeon and Lucien Febvre, *Le Rhin: Problèmes d'histoire et d'économie* (Paris: Colin, 1935); Horst Johannes Tümmers, *Der Rhein: Ein europäischer Fluss und seine Geschichte* (Munich: Beck, 1994).

19. Andreas Teuscher, *Schweiz am Meer—Pläne für den "Central-Hafen" Europas inklusive Alpenüberquerung mit Schiffen im 20. Jahrhundert* (Zurich, Switzerland: Limmat, 2014).

20. Jiri Janác, *European Coasts of Bohemia: Negotiating the Danube-Oder-Elbe Canal in a Troubled Twentieth Century* (Amsterdam: Amsterdam University Press, 2012).

21. Lina Schröder, *Der Rhein-(Maas-)Schelde-Kanal als geplante Infrastrukturzelle von 1946 bis 1986: Eine Studie zur Infrastruktur- und Netzwerk-Geschichte* (Münster: Waxmann, 2017).

22. Eike-Christian Heine, *Die Geschichte des Nord-Ostsee-Kanals* (Berlin: Kadmos, 2015).

23. Marko Zejnelovic, *Der "Highway of Nations": Eine postkoloniale Studie zum Panamakanal als imperiale Infrastruktur* (Augsburg: Institut für Spanien, Portugal- und Lateinamerikastudien, 2014); Ashley Carse, *Beyond the Big Ditch: Politics, Ecology, and Infrastructure at the Panama Canal* (Cambridge, MA: MIT Press, 2014).

24. Cornelius Torp, "Revolution zur See: Maritime Frachtkosten und Weltmarkt im 'langen' 19. Jahrhundert," in Jürgen Elvert, Lutz Feldt, Ingo Löppenberg, and

Jens Ruppenthal (eds.), *Das maritime Europa. Werte—Wissen—Wirtschaft* (Stuttgart: Steiner, 2016), 93–108.

25. Angelika Jordans, "Ein Stück London in der Wüste," *Frankfurter Allgemeine Zeitung*, January 20, 1994.

26. Robert von Schlagintweit, *Die Santa Fe- und Südpacificbahn in Nordamerika* (Cologne: Mayer, 1884).

27. Frank Norris, *The Octopus: A Story of California* (New York: Penguin, 1994).

28. Manfred Ohlsen, *Der Eisenbahnkönig Bethel Henry Strousberg. Eine preussische Gründerkarriere* (Berlin: Verlag der Nation, 1987).

29. John Lauritz Larson, *Internal Improvement: National Public Works and the Promise of Popular Government in the Early United States* (Chapel Hill: University of North Carolina Press, 2001).

30. Mike Chrimes (ed.), *The Civil Engineering of Canals and Railways before 1850* (Aldershot, UK: Ashgate, 1997); Hans-Peter Bärtschi, "Durchmessene Räume—durchmessene Zeiten: Die Eisenbahn als Landschaftsgestalterin," in David Gugerli (ed.), *Vermessene Landschaften: Kulturgeschichte und technische Praxis im 19. und 20. Jahrhundert* (Zurich, Switzerland: Chronos, 1999), 79–87.

31. Volker Mehnert, "Bier und Bratwurst statt Ruhm und Ehre," *Frankfurter Allgemeine Zeitung*, November 20, 2014. For the United States, see Henry Petroski, *The Road Taken: The History and Future of America's Infrastructure* (New York: Bloomsbury, 2016).

32. Stephen Minicucci, "Internal Improvements and the Union, 1790–1860," *Studies in American Political Development* 18.2 (2004): 160–185.

33. Quoted in Nurske, *Problems of Capital Formation*, 153.

34. Stefan Kaufmann, "Landschaft beschriften: Zur Logik des American Grid Systems," in Stefan Kaufmann (ed.), *Ordnungen der Landschaft. Natur und Raum technisch und symbolisch entwerfen* (Würzburg: Ergon, 2002), 73–94; Jens Ivo Engels, "Machtfragen: Aktuelle Entwicklungen und Perspektiven der Infrastrukturgeschichte," *Neue Politische Literatur* 55.1 (2010): 51–70. On competing metaphors (*maze*, *net*, and *grid*), see Osterhammel, *Die Verwandlung der Welt*, 1011.

35. Richard R. John, "Private Enterprise, Public Good? Communications Deregulation as a National Political Issue, 1839–1851," in Jeffrey L. Pasley, Andrew W. Robertson, and David Waldstreicher (eds.), *Beyond the Founders. New Approaches to the Political History of the Early American Republic* (Chapel Hill: University of North Carolina Press, 2004), 328–354.

36. Larson, *Internal Improvement*.

37. Douglas J. Puffert, "The Standardization of the Track Gauge on North American Railways, 1830–1890," *Journal of Economic History* 60.4 (2000): 933–960.

38. Ferdinand von Richthofen, "Land Communication between Europe and China," *Geographical Magazine*, 1 (1874): 144–146 at 146; the original German, "read before the Berlin Geographical Society, 11th of April, 1874," mentions "overpopulation" and the benefits of "forcing [Chinese] entry into the cultural battle of the [world's] peoples."

39. Richard Mowbray Haywood, *The Beginning of Railway Development in Russia in the Reign of Nicholas I, 1835–1842* (Durham, NC: Duke University Press, 1969); Roland Cvetkovski, *Modernisierung durch Beschleunigung: Raum und Mobilität im Zarenreich* (Frankfurt am Main: Campus, 2006), 233–244.

40. Frithjof Benjamin Schenk, *Russlands Fahrt in die Moderne: Mobilität und sozialer Raum im Eisenbahnzeitalter* (Stuttgart: Steiner, 2014), 377–378; Karl Schlögel, *Das sowjetische Jahrhundert: Archäologie einer untergegangenen Welt* (Munich: Beck, 2017), 795–816.

41. Walter Sperling, *Der Aufbruch der Provinz. Die Eisenbahn und die Neuordnung der Räume im Zarenreich* (Frankfurt am Main: Campus, 2011), 10.

42. Schenk, *Russlands Fahrt in die Moderne*, 380; see also Alfred Gottwaldt, "Der Bahnhof," in Alexa Geisthövel and Habbo Knoch (eds.), *Orte der Moderne: Erfahrungswelten des 19. und 20. Jahrhunderts* (Frankfurt am Main: Campus, 2005), 17–26.

43. Schenk, *Russlands Fahrt in die Moderne*, 311ff.

44. Schenk, *Russlands Fahrt in die Moderne*, 385.

45. Reinhart Koselleck, *Zeitschichten: Studien zur Historik* (Frankfurt am Main: Suhrkamp, 2000), 157.

46. Johann Peter Eckermann, *Conversations with Goethe: In the Last Years of His Life*, Ritchie Robertson (trans.) (New York: Penguin, 2022), chap. 287 (October 23, 1828).

47. Markus Völkel, "Einigkeit und Freiheit. Die Eisenbahn ein Mittel nationaler Politik?" in *Zug der Zeit—Zeit der Züge: Deutsche Eisenbahn 1835–1985*, vol. 1 (Berlin: Siedler, 1985), 218–227.

48. Quoted in Marcus Junkelmann, "Die Eisenbahn im Krieg: Militärische Theorien und Kriegsgeschehen bis zum Ausbruch des Ersten Weltkriegs," in *Zug der Zeit—Zeit der Züge: Deutsche Eisenbahn 1835–1985*, vol. 1, ed. Harm-Hindreich Brandt (Munich: Orbis, 1989), 232–245, at 233.

49. Quoted in Koselleck, *Zeitschichten*, 193.

50. Giessmann, *Netze und Netzwerke*, and *Die Verbundenheit der Dinge*.

51. Hermann Lübbe, "Technik in Politik und Ideologie," in Hardi Fischer (ed.), *Technik wozu und wohin?* (Zurich, Switzerland: Artemis, 1981), 131–140.

52. Julia Obertreis, "Karrieren, Patronage und 'Infrastrukturpoesie': Dimensionen der Infrastrukturgeschichte am Beispiel des russländischen und sowjetischen Zentralasien," in Förster and Bauch, *Wasserinfrastrukturen und Macht*, 232–265.

53. Hans Jürgen Teuteberg and Cornelius Neutsch (eds.), *Vom Flügeltelegraphen zum Internet: Geschichte der modernen Telekommunikation* (Stuttgart: Steiner, 1998); Michael North (ed.), *Kommunikationsrevolutionen: Die neuen Medien des 16. Und 19. Jahrhunderts* (Cologne: Böhlau, 2001).

54. Torsten Kathke, *Wires That Bind: Nation, Region, and Technology in the Southwestern United States, 1854–1920* (Bielefeld: transcript, 2017).

55. Kurt Wiedenfeld, "Der volkswirtschaftliche Einfluss der modernen Verkehrsmittel und die deutsche Volkswirtschaftslehre des 19. Jahrhunderts," in *Die Entwicklung*

der deutschen Volkswirtschaftslehre im neunzehnten Jahrhundert. Festgabe für Gustav Schmoller, 2. Teil (Leipzig: Duncker & Humblot, 1908), 1–28, at 11–12.

56. Quoted in Koselleck, *Zeitschichten*, 159.

57. Patrice Flichy, *Une histoire de la communication moderne* (Paris: La Découverte, 2004).

58. Ralf Roth, *Städte im europäischen Raum: Verkehr, Kommunikation und Urbanität im 19. und 20. Jahrhundert* (Stuttgart: Steiner, 2009), 16.

59. Anton A. Huurdeman, *The Worldwide History of Telecommunications* (Hoboken, NJ: Wiley, 2003); Roland Wenzlhuemer, *Connecting the Nineteenth-Century World: The Telegraph and Globalization* (Cambridge: Cambridge University Press, 2013), 15.

60. Artur Dix, *Deutschland auf den Hochstrassen des Weltwirtschaftsverkehrs* (Jena: Fischer, 1901), 7.

61. Artur Dix, "Wirtschaftsstruktur und Geopolitik," *Volkswirtschaftliche Blätter* 26.7–9 (1927): 465–484, at 472.

62. Marshall McLuhan, *The Gutenberg Galaxy: The Making of Typographic Man* (Toronto: University of Toronto Press, 2017).

63. Michael Geistbeck, *Weltverkehr. Die Entwicklung von Schifffahrt, Eisenbahn, Post und Telegraphie bis zum Ende des 19. Jahrhunderts* (Hildesheim: Gerstenberg, 1986), 530.

64. Deep Kanta Lahiri-Choudhury, *Telegraphic Imperialism: Crisis and Panic in the Indian Empire, c. 1830–1920* (New York: Palgrave Macmillan, 2010), 122.

65. Daniel R. Headrick, *The Tools of Empire: Technology and European Imperialism in the Nineteenth Century* (Oxford: Oxford University Press, 1981); Maria Paula Diogo and Dirk van Laak, *Europeans Globalizing: Mapping, Exploiting, Exchanging* (New York: Palgrave Macmillan, 2016).

66. Wenzlhuemer, *Connecting the Nineteenth-Century World*, 156–161; Ruth Oldenziel, "Islands: The United States as a Networked Empire," in Gabrielle Hecht (ed.), *Entangled Geographies. Empire and Technopolitics in the Global Cold War* (Cambridge, MA: MIT Press, 2011), 13–41.

67. Simone M. Müller, *Wiring the World: The Social and Cultural Creation of Global Telegraph Networks* (New York: Columbia University Press, 2016).

68. Quoted in Michael Salewski, "Technik als Vision der Zukunft um die Jahrhundertwende," in Michael Salewski and Ilona Stölken-Fitschen (eds.), *Moderne Zeiten: Technik und Zeitgeist im 19. und 20. Jahrhundert* (Stuttgart: Steiner, 1994), 77–91, at 83.

69. Christian Holtorf, *Der erste Draht zur neuen Welt: Die Verlegung des transatlantischen Telegrafenkabels* (Göttingen: Wallstein, 2013), 276.

70. David Paul Nickles, "Telegraph Diplomats: The United States' Relations with France in 1848 and 1870," *Technology and Culture* 40.1 (1999): 1–25.

71. Karl Marx, *Selected Writings*, ed. David McLellan (Oxford: Oxford University Press, 2000), 363 (first published in the *New York Daily Tribune*, August 8, 1853). Indeed, the German electrical engineering company Siemens & Halske would lay the cable from London to Calcutta in 1869–1870.

72. Michael Radunski, "Schöne Grüsse aus dem Central Telegraph Office," *Frankfurter Allgemeine Zeitung*, July 13, 2013.

73. Paul Dehn, *Weltwirtschaftliche Neubildungen* (Berlin: Allgemeiner Verein für Deutsche Literatur, 1904), 354.

74. Ronen Steinke, "Stopp," *Süddeutsche Zeitung*, July 20/21, 2013.

75. Iwan Rhys Morus, "'The Nervous System of Britain': Space, Time and the Electric Telegraph in the Victorian Age," *British Journal for the History of Science* 33.4 (2000): 455–475, at 475.

76. Ute Daniel, "Der Krimkrieg 1853–1856 und die Entstehungskontexte medialer Kriegsberichterstattung," in Ute Daniel (ed.), *Augenzeugen: Kriegsberichterstattung vom 18. zum 21. Jahrhundert* (Göttingen: Vandenhoeck & Ruprecht, 2006), 40–67.

77. David Paul Nickles, *Under the Wire: How the Telegraph Changed Diplomacy* (Cambridge, MA: Harvard University Press, 2003).

78. Steinke, "Stopp." On the Ems Dispatch, see Tobias Nanz, *Grenzverkehr: Eine Mediengeschichte der Diplomatie* (Berlin: Diaphanes, 2010), 169–190.

79. Vanessa Ogle, "Whose Time Is It? The Pluralization of Time and the Global Condition, 1870s to 1940s," *American Historical Review* 120.5 (2013): 1376–1402.

80. Tobias Nanz, "Das 'Rote Telefon': Ein hybrides Objekt des Kalten Krieges," *Zeitschrift für Literaturwissenschaft und Linguistik* 43.173 (2014): 135–148.

81. Maria Kaika, "City of Flows: Der Wandel der symbolischen Bedeutung technischer Infrastrukturen in der Moderne," in Timothy Moss, Matthias Naumann, and Markus Wissen (eds.), *Infrastrukturnetze und Raumentwicklung* (Munich: Oekom, 2008), 87–111, at 108.

82. Christoph Asendorf, *Ströme und Strahlen: Das langsame Verschwinden der Materie um 1900* (Giessen: Anabas, 1989), 60; Franziska Bollerey and Axel Föhl, "Unter dem Pflaster der Strasse: Facetten subterraner Netze," in Manfred Rasch and Dietmar Bleidick (eds.), *Technikgeschichte im Ruhrgebiet, Technikgeschichte für das Ruhrgebiet* (Essen: Klartext, 2004), 438–473.

83. Dieter Büker, *Mensch—Kultur—Abwasser. Von der Annehmlichkeit für wenige zur Existenzfrage der Gesellschaft: Der Umgang des Menschen mit Abwässern. Ein kulturhistorischer Längsschnitt von den Anfängen bis zum Beginn des 20. Jahrhunderts* (Essen: Die blaue Eule, 2000).

84. Marion Hombach, "Kaiserreich in den Tropen," *GeoEpoche* 71 (2015): 84–97, at 93.

85. Schott, *Europäische Urbanisierung (1000–2000)*; hence Manchester's nickname of Shock City.

86. Donald Worster, *Under Western Skies: Nature and History in the American West* (Oxford: Oxford University Press, 1992); Jones, "A Landscape of Energy Abundance."

87. "Electricity, communications, traffic, and information all learned to flow from water." Heidenreich, "Natur und Kultur heute," 57.

88. Martin V. Melosi, *The Sanitary City: Environmental Services in Urban America from the Colonial Times to the Present* (Pittsburgh, PA: University of Pittsburgh Press, 2008);

Frank and Gandy, *Hydropolis*; Matthew Gandy, *The Fabric of Space: Water, Modernity, and Urban Imagination* (Cambridge, MA: MIT Press, 2014).

89. Ragnhild Münch, *Gesundheitswesen im 18. und 19. Jahrhundert: Das Berliner Beispiel* (Berlin: Akademie, 1995).

90. Schott, *Europäische Urbanisierung*, 226–230; Dieter Hoffmann-Axthelm, "Stadttechnik und Stadtkultur," *Stadtbauwelt* 103 (1989): 1666–1673, at 1668.

91. Friedrich Lenger, *Metropolen der Moderne: Eine europäische Stadtgeschichte* (Munich: Beck, 2013).

92. Lincoln Steffens, *The Shame of the Cities* (New York: McClure, Phillips, 1904).

93. Daniel Ziblatt, "Why Some Cities Provide More Public Goods than Others: A Subnational Comparison of the Provision of Public Goods in German Cities in 1912," *Studies in Comparative International Development* 43 (2008): 273–289.

94. This is the thesis advanced in de Swaan, *Der sorgende Staat.*

95. Johannes Depenbrock, *Die Stellung der Kommunen in der Versorgungswirtschaft: Ein Beitrag zum Wirtschaftsverfassungsrecht* (Munich: Beck, 1981), 21.

96. Theodore M. Porter, "Objectivity and Authority: How French Engineers Reduced Public Utility to Numbers," *Poetics Today* 12.2 (1991): 245–265, at 253. See also David Rodogno, Bernhard Struck, and Jakob Vogel (eds.), *Shaping the Transnational Sphere: Experts, Networks, and Issues from the 1840s to the* 1930s (New York: Berghahn, 2014).

97. Wolfgang Krabbe, "Die Entfaltung der kommunalen Leistungsverwaltung in deutschen Städten des späten 19. Jahrhunderts," in Hans-Jürgen Teuteberg (ed.), *Urbanisierung im 19. und 20. Jahrhundert* (Cologne: Böhlau, 1983), 373–391; Ekke Feldmann, *Bauordnungen und Baupolizei: Zur Entwicklungsgeschichte zwischen 1850 und 1950* (Frankfurt am Main: Lang, 2011).

98. Van Laak, "Technokratie im Europa des 20. Jahrhunderts.

99. Porter, "Objectivity and Authority," 256; and Theodore M. Porter, *Trust in Numbers: The Pursuit of Objectivity in Science and Public Life* (Princeton, NJ: Princeton University Press, 1995).

100. Constantin Goschler, "Infrastruktur-Liberalismus: Rudolf Virchow als Berliner Kommunalpolitiker," *Jahrbuch zur Liberalismus-Forschung* 18 (2006): 45–58, at 55.

101. Goschler, "Infrastruktur-Liberalismus," 57.

102. Uwe Kühl, *Der Munizipalsozialismus in Europa* (Munich: Oldenbourg, 2001).

103. Ziblatt, "Why Some Cities Provide More Public Goods than Others," 283.

104. Ausubel and Herman, *Cities and Their Vital Systems*, 3.

105. Peter Payer, *Der Gestank von Wien: Über Kanalgase, Totendünste und andere üble Geruchskulissen* (Vienna: Döcker, 1997).

106. Stephen Halliday, *The Great Stink of London: Sir Joseph Bazalgette and the Cleansing of the Victorian Metropolis* (Stroud: Sutton, 2001). On a similar event in Paris in 1880, see David S. Barnes, *The Great Stink of Paris and the Nineteenth-Century Struggle against Filth and Germs* (Baltimore: Johns Hopkins University Press, 2006).

107. Andreas Bernard, *Die Geschichte des Fahrstuhls: Über einen beweglichen Ort der Moderne* (Frankfurt am Main: Fischer, 2006), 64.

108. Christian Kassung noted this in a lecture at the Justus-Liebig University, Giessen, November 30, 2015; and see Helmut Lackner, "Ein 'blutiges Geschäft.' Kommunale Vieh- und Schlachthöfe im Urbanisierungsprozess des 19. Jahrhunderts: Ein Beitrag zur Geschichte der städtischen Infrastruktur," *Technikgeschichte* 71.2 (2004): 89–138; Paula Young Lee (ed.), *Meat, Modernity, and the Rise of the Slaughterhouse* (Durham: University of New Hampshire Press, 2008); and Dorothee Brantz, "Slaughter in the City: The Establishment of Public Abattoirs in Paris and Berlin, 1780–1914," PhD diss., University of Chicago, 2003.

109. Benson Bobrick, *Labyrinths of Iron: Subways in History, Myth, Art, Technology, and War* (New York: Henry Holt, 1994).

110. Peter Galison, *Einstein's Clocks, Poincaré's Maps: Empires of Time* (New York: Norton, 2003).

111. Peter R. Gleichmann, "Die Verhäuslichung körperlicher Verrichtungen," in Peter R. Gleichmann, Johan Goudsblom, and Hermann Korte (eds.), *Materialien zu Norbert Elias' Zivilisationstheorie* (Frankfurt am Main: Suhrkamp, 1979), 254–278.

112. Peter Payer, *Unentbehrliche Requisiten der Grossstadt: Eine Kulturgeschichte der öffentlichen Bedürfnisanstalten von Wien* (Vienna: Löcker, 2000).

113. Payer, *Unentbehrliche Requisiten der Grossstadt*, 17; Alain Corbin, *The Foul and the Fragrant: Odor and the Social Imagination*, trans. Miriam L. Kochan (Cambridge, MA: Harvard University Press, 1986).

114. Manuel Frey, *Der reinliche Bürger: Entstehung und Verbreitung bürgerlicher Tugenden in Deutschland, 1780–1860* (Göttingen: Vandenhoeck & Ruprecht, 1997); Richard Sennett, *The Fall of Public Man* (New York: Norton, 2017).

115. Bernard, *Die Geschichte des Fahrstuhls.*

116. Dietmar Arnold, Ingmar Arnold, and Frieder Salm, *Dunkle Welten: Bunker, Tunnel und Gewölbe unter Berlin* (Berlin: Links, 2013).

117. Max Winter, *Im unterirdischen Wien* (Berlin: Seemann, 1905).

CHAPTER 2

1. Karl Marx and Friedrich Engels, *Collected Works*, vol. 46 (New York: International Publishers, 1992), 449; Engels is referring to the French engineer Marcel Depréz (1843–1918), who pioneered long-distance electricity distribution.

2. Vaclav Smil, *Creating the Twentieth Century: Technical Innovations of 1867–1914 and Their Lasting Impact* (Oxford: Oxford University Press, 2005).

3. Smil, *Creating the Twentieth Century*, 279.

4. August Bebel, *Woman and Socialism*, trans. Meta L. Stern (New York: Socialist Literature, 1910), 233–238, "The Revolution in Domestic Life."

5. Wolfgang Schivelbusch, *Disenchanted Night: The Industrialization of Light in the Nineteenth Century*, trans. Angela Davies (Berkeley: University of California Press, 1995); Nicolas Kenny, "City Glow: Streetlights, Emotions, and Nocturnal Life,

1880s–1910s," *Journal of Urban History* 43.1 (2017): 91–114; Alice Barnaby, *Light Touches: Cultural Practices of Illumination, 1780–1900* (London: Routledge, 2016).

6. Ernst Rebske, *Lampen, Laternen, Leuchten: Eine Historie der Beleuchtung* (Stuttgart: Franckh, 1962).

7. Hellmut von Laer, "Gleicht Berlin bei Nacht bald Bielefeld?" *Frankfurter Allgemeine Zeitung*, June 25, 2014.

8. Depenbrock, *Die Stellung der Kommunen in der Versorgungswirtschaft*.

9. Charles Bazerman, *The Language of Edison's Light* (Cambridge, MA: MIT Press, 1999).

10. Hans-Liudger Dienel and Barbara Schmucki (eds.), *Mobilität für alle: Die Geschichte des öffentlichen Personennahverkehrs in der Stadt zwischen technischem Fortschritt und sozialer Pflicht* (Stuttgart: Steiner, 1997).

11. Takahito Mori, *Elektrifizierung als Urbanisierungsprozess: Frankfurt am Main, 1886–1933* (Darmstadt: Hessisches Wirtschaftsarchiv, 2014), 8; Dieter Schott, *Die Vernetzung der Stadt: Kommunale Energiepolitik, öffentlicher Nahverkehr und die "Produktion" der modernen Stadt: Darmstadt—Mannheim—Mainz 1880–1918* (Darmstadt: Wissenschaftliche Buchgesellschaft, 1999), 8; John P. McKay, *Tramways and Trolleys: The Rise of Urban Mass Transport in Europe* (Princeton, NJ: Princeton University Press, 1976).

12. Barbara Schmucki, "The Machine in the City: Public Appropriation of the Tramway in Britain and Germany, 1870–1915," *Journal of Urban History* 38.6 (2012): 1060–1093, at 1064.

13. Wiebke Porombka, *Medialität urbaner Infrastrukturen: Der öffentliche Nahverkehr, 1870–1933* (Bielefeld: transcript, 2013), 169.

14. David Gugerli, *Redeströme: Zur Elektrifizierung der Schweiz 1880–1914* (Zurich, Switzerland: Chronos, 1996).

15. Graeme Gooday, *Domesticating Electricity: Technology, Uncertainty and Gender, 1880–1914* (Pittsburgh, PA: University of Pittsburgh Press, 2008); for Zanzibar's adoption, see Tanja Winther, *The Impact of Electricity: Development, Desires and Dilemmas* (New York: Berghahn, 2008).

16. Gerhard Paul, *BilderMACHT: Studien zur Visual History des 20. und 21. Jahrhunderts* (Göttingen: Wallstein, 2013), 17–44.

17. Giessmann, *Die Verbundenheit der Dinge*.

18. Beate Binder, *Elektrifizierung als Vision: Zur Symbolgeschichte einer Technik im Alltag* (Tübingen: Tubinger Vereinigug fur Volkskunde, 1999), 11.

19. Mori, *Elektrifizierung als Urbanisierungsprozess*, 7.

20. Nat Brandt, *Chicago Death Trap: The Iroquois Theatre Fire of 1903* (Carbondale: Southern Illinois University Press, 2006).

21. Stefan Zweig, "Blick über die elektrische Stadt in die Zukunft hinein," *Neue Rundschau* 1 (1992): 109–114, at 113.

22. Marc Pitzke, "Blackout von 1977: New Yorks dunkelste Nacht," *Der Spiegel*, July 13, 2007. On similar cases, see Stephen Graham (ed.), *Disrupted Cities: When Infrastructure Fails* (New York: Routledge, 2010).

23. Daniel Wilhelm, *Die Kommunikation infrastruktureller Grossprojekte: Die Elektrifizierung Oberschwabens durch die OEW in der ersten Hälfte des 20. Jahrhunderts* (Stuttgart: Steiner, 2014), 36.

24. Wilhelm, *Die Kommunikation infrastruktureller Grossprojekte*. Christoph Hoeft, Sören Messinger-Zimmer, and Julia Zilles (eds.), *Bürgerprotest in Zeiten der Energiewende: Lokale Konflikte um Windkraft, Stromtrassen und Fracking* (Bielefeld: transcript, 2017).

25. Quoted in Wilhelm, *Die Kommunikation infrastruktureller Grossprojekte*, 247.

26. Hanno Trurnit, *Geschichte(n) hinterm Zähler: Die Beziehungen zwischen Energieversorgern und ihren Kunden* (Munich: Trurnit, 1996), 23.

27. Alice Tisdale Hobart, *Oil for the Lamps of China* (Indianapolis, IN: Bobbs-Merrill, 1933); Xijujie Wu, *Ein Jahrhundert Licht: Eine technikethnologische Studie zur Beleuchtung im chinesischen ländlichen Alltag* (Wiesbaden: Harrassowitz, 2009).

28. Ursula Schneider and Heide David, *"Das Paradies kommt wieder . . ." Zur Kulturgeschichte und Ökologie von Herd, Kühlschrank und Waschmaschine: Katalogbuch* (Hamburg: Museum der Arbeit, 1993); Lutz Pape and Hans-Jürgen Weinert, *Bottichwaschmaschine und Haustelegraph: Anfänge der Elektrotechnik im Haushalt* (Braunschweig: Westermann, 1993); Martina Hessler, "Die Einführung elektrischer Haushaltsgeräte in der Zwischenkriegszeit: Der Angebotspush der Produzenten und die Reaktion der Konsumentinnen," *Technikgeschichte* 65 (1998): 297–311.

29. Trurnit, *Geschichte(n) hinterm Zähler*, 36–47.

30. Ruth Schwartz Cowan, *More Work for Mother: The Ironies of Household Technology from the Open Hearth to the Microwave* (New York: Basic Books, 1983).

31. Julius Maier and William Henry Preece, *Das Telephon und dessen praktische Verwendung* (Stuttgart: Enke, 1889), 1, quoted in Kirsten von Hagen, *Telefonfiktionen* (Munich: Fink, 2015), 23.

32. Eckart Menzler-Trott, "Vom Telefon zum Call Center: Eine kleine Geschichte des Telefonierens," *Telepolis*, May 21, 1998, www.heise.de/tp/artikel/2/2335/1.html.

33. Von Hagen, *Telefonfiktionen*, 12.

34. Claude S. Fisher, *America Calling: A Social History of the Telephone to 1940* (Berkeley: University of California Press, 1992).

35. Edwards, "Infrastructure and Modernity," 201.

36. Von Hagen, *Telefonfiktionen*, 25.

37. Irina Lazarova, *"Hier spricht Lenin": Das Telefon in der russischen Literatur der 1920er und 30er Jahre* (Cologne: Böhlau, 2010).

38. Karl Veidt, "Als Divisionspfarrer im Argonnerwald, 1914–1917," *Hessische Quellen zum Ersten Weltkrieg*, https://www.lagis-hessen.de/de/purl/resolve/subject/qhg/id/35-4; retrieved December 17, 2017.

39. Paul Neubaur, "Die Usambarabahn in Deutsch-Ostafrika," *Velhagen und Klasings Monatshefte* 17.11 (1903): 575–582, at 575.

40. Rudolf Schmidt, "Die Nachrichtenmittel (1921)," quoted in Thomas Jander and Veit Didczuneit, *Netze des Krieges: Kommunikation 1914/1918* (Königswinter: Lempertz, 2014), 32.

41. Martin van Creveld, *Supplying War: Logistics from Wallenstein to Patton* (Cambridge: Cambridge University Press, 2004), 1.

42. Mark Whitmore, "Transport and Supply during the First World War," https://www.iwm.org.uk/history/transport-and-supply-during-the-first-world-war; retrieved December 20, 2017.

43. Kaufmann, *Kommunikationstechnik und Kriegführung 1815–1945*, 170–261.

44. Jander and Didczuneit, *Netze des Krieges*, 10.

45. Ernst Jünger, *In Stahlgewittern* (1930), quoted in Jander and Didczuneit, *Netze des Krieges*, 24.

46. Christoph Nübel, *Durchhalten und Überleben an der Westfront: Raum und Körper im Ersten Weltkrieg* (Paderborn: Schöningh, 2014).

47. Quoted in Raul Hilberg, *Sonderzüge nach Auschwitz* (Mainz: Dumjahn, 1981), 159.

48. Regine Zennss-Reimann, *Unterm Flügelrad: Frauen bei der Eisenbahn* (Berlin: Museum für Verkehr & Technik, 1993), 9–12, 17–20.

49. Jochen Oltmer, "Unentbehrliche Arbeitskräfte: Kriegsgefangene in Deutschland 1914–1918," in Jochen Oltmer (ed.), *Kriegsgefangene im Europa des Ersten Weltkriegs* (Paderborn: Schöningh, 2006), 67–96, at 71.

50. Dirk Böndel, Anna Döpfner, and Alfred Gottwaldt (eds.), *Ich diente nur der Technik: Sieben Karrieren zwischen 1940 und 1950* (Berlin: Nicolai, 1995).

51. Richard Löwenthal. "Bonn und Weimar: Zwei deutsche Demokratien," in Heinrich August Winkler *(ed.), Politische Weichenstellungen im Nachkriegsdeutschland 1945–1953* (Göttingen: Vandenhoeck & Ruprecht, 1979), 10–11.

52. "Technical Relief Grows in Germany: Hardships of Industrial Conflicts Mitigated by 1,500 Local Groups," *New York Times*, December 3, 1922.

53. Barbara Schmucki, *Der Traum vom Verkehrsfluss* (Frankfurt am Main: Campus, 2001).

54. Werner Durth and Niels Gutschow, *Träume in Trümmern: Planungen zum Wiederaufbau zerstörter Städte im Westen Deutschlands 1940–1950* (Wiesbaden: Vieweg, 1988); Jeffry M. Diefendorf, *In the Wake of War: The Reconstruction of German Cities after World War II* (Oxford: Oxford University Press, 1993).

55. Ernst Randzio, *Unterirdischer Städtebau, besonders mit Beispielen aus Gross-Berlin* (Bremen: Dorn, 1951), 8.

56. Albert Osswald, *Eine Zeit vergeht* (Giessen: Ferber, 1993), 34.

57. Dirk van Laak, "Mythos 'Hessenplan': Aufstieg und Wandel einer Landesplanung nach dem Zweiten Weltkrieg," in Wendelin Strubelt and Detlef Briesen (eds.), *Raumplanung nach 1945: Kontinuitäten und Neuanfänge in der Bundesrepublik Deutschland* (Frankfurt am Main: Campus, 2015), 127–149.

58. Axel Dossmann, Jan Wenzel, and Kai Wenzel, *Architektur auf Zeit: Baracken, Pavillons, Container* (Berlin: b-books, 2006).

59. Whitmore, "Transport and Supply during the First World War."

60. Richard Vahrenkamp, "Logistik als Metasystem der Infrastruktur: Die Grenzen der Eisenbahn-Logistik und der Aufstieg der Lastwagen-Logistik 1900 bis 1938," in Wiebke Porombka, Heinz Reif, and Erhard Schütz (eds.), *Versorgung und Entsorgung der Moderne: Logistiken und Infrastrukturen der 1920er und 1930er Jahre* (Frankfurt am Main: Lang, 2011), 123–150.

61. Irmela Schneider, "Radiophone Praktiken des (Selbst-)Regierens in der Weimarer Republik," in Irmela Schneider and Isabell Otto (eds.), *Formationen der Mediennutzung II: Strategien der Verdatung* (Bielefeld: transcript, 2007), 37–53.

62. Christian Schreiber, "Der letzte Liftboy Europas," *Next floor: Das Magazin für die Kunden der Schindler-Aufzüge AG* 1 (2013): 8–10.

63. Habbo Knoch, "Das Grandhotel," in Alexa Geisthövel and Habbo Knoch (eds.), *Orte der Moderne: Erfahrungswelten des 19. und 20. Jahrhunderts* (Frankfurt am Main: Campus, 2005), 131–140; Habbo Knoch, *Grandhotels: Luxusräume und Gesellschaftswandel in New York, London und Berlin um 1900* (Göttingen: Wallstein, 2016).

64. Alexandre Tessier, "La grande hôtellerie parisienne face à l'innovation: entre devoir d'adoption et facteur de distinction (1855–1914)," in Marc Gigase, Cédric Humair, and Laurent Tissot (eds.), *Le tourisme comme facteur de transformations économiques, techniques et sociales (XIXe–XXe siècles). Tourism as a factor of economic, technical and social transformations (XIXth—XXth centuries)* (Neuchâtel: Alphil, 2014), 179–195; Alexandre Tessier, *Le Grand Hôtel, L'invention du luxe hôtelier, 1862–1972* (Rennes, France: Presses universitaires de Rennes, 2012).

65. Dagmar Bellmann, *Von Höllengefährten zu schwimmenden Palästen: Die Passagierschifffahrt auf dem Atlantik (1840–1930)* (Frankfurt am Main: Campus, 2015).

66. Eckart Gruber, *Fünfuhr-Tee im Adlon: Menschen und Hotels* (Berlin: Fannei & Walz, 1994), 148.

67. Knoch, "Das Grandhotel," 139; Habbo Knoch, "Life on Stage: Grand Hotels as Urban Interzones around 1900," in Martina Hessller and Clemens Zimmermann (eds.), *Creative Urban Milieus: Historical Perspectives on Culture, Economy, and the City* (Frankfurt am Main: Campus, 2008), 137–158.

68. Rüdiger Hachtmann, "Tourismus und Tourismusgeschichte," *Docupedia-Zeitgeschichte*, December 22, 2010, https://www.docupedia.de/zg/Tourismus_und_Tourismusgeschichte; Ueli Gyr, "Geschichte des Tourismus: Strukturen auf dem Weg zur Moderne," *Europäische Geschichte Online* (EGO), December 3, 2010, http://ieg-ego.eu/de/threads/europa-unterwegs/tourismus/ueli-gyr-geschichte-des-tourismus.

69. "Wie Strassen die Schwächen Berns offenlegen: Gespräch mit Daniel Flückiger," *Berner Zeitung*, October 20, 2011; Daniel Flückiger, *Strassen für alle: Infrastrukturpolitik im Kanton Bern 1790–1850* (Baden: Hier & Jetzt, 2011).

70. Ariane Leendertz, *Ordnung schaffen: Deutsche Raumplanung im 20. Jahrhundert* (Göttingen: Wallstein, 2008).

71. Richard W. Butler, "The Concept of a Tourist Area Cycle of Evolution: Implications for Management of Resources," *Canadian Geographer* 24.1 (1980): 5–12.

72. Ekkehard Schönherr, "Infrastrukturen des Glücks: Eine Bild-, Raum- und Infrastrukturgeschichte Mallorcas im 19. und 20. Jahrhundert unter Berücksichtigung des Tourismus," PhD diss., Erfurt, 2014.

73. Christoph Hennig, *Reiselust: Touristen, Tourismus und Urlaubskultur* (Frankfurt am Main: Suhrkamp, 1999), 58.

74. Kurt Tucholsky, *Von der Kunst, richtig zu reisen* (Münster: Coppenrath 2015); John Urry, *The Tourist Gaze: Leisure and Travel in Contemporary Societies* (London: Sage, 1990).

75. Richard Mullen and James Munson, *The Smell of the Continent: The British Discover Europe* (London: Pan Books, 2010); Laurent Tissot, *Naissance d'une industrie touristique: Les Anglais et la Suisse au XIXe siècle* (Lausanne, Switzerland: Payot, 2000).

76. Robert Walser, "Etwas über die Eisenbahn" (1907), quoted in Steffen Richter, "Tunnelblicke: Zur literarischen Repräsentation von Machtverhältnissen anhand einer schweizerischen Infrastruktureinrichtung," in Gonçalo Vilas-Boas and Teresa Oliviera (eds.), *Macht in der Deutschschweizer Literatur* (Berlin: Frank & Timme, 2012), 267–286, at 267.

77. Judith Schueler, *Materialising Identity: The Co-construction of the Gotthard Railway and Swiss National Identity* (Amsterdam: Aksant, 2008). On a similarly ambitious project, see Teuscher, *Schweiz am Meer*.

78. Laurent Tissot, "Tourismus," *Historisches Lexikon der Schweiz* (www.hls-dhsdss.ch/textes/d/D14070.php; retrieved December 17, 2017).

79. Ariane Devanthéry, "Schweizerreisen," *Historisches Lexikon der Schweiz*, May 1, 2015 (https://hls-dhs-dss.ch/de/articles/024575/2015-01-05/.

80. Wolfgang König, *Bahnen und Berge: Verkehrstechnik, Tourismus und Naturschutz in den Schweizer Alpen 1870–1939* (Frankfurt am Main: Campus, 2000).

81. Hasso Spode, "Romantische Zeitreise: Tourismus als Chronotopie," in Brigitta Schmidt-Lauber (ed.), *Sommerfrische: Bilder, Orte, Praktiken* (Vienna: Institut für Europäische Ethnologie, 2014), 33–45.

82. Höhne, "Token Tactics," 73–88.

83. Mullen and Munson, *The Smell of the Continent*, xiii.

84. Mullen and Munson, *The Smell of the Continent*, 243.

85. Wolfgang Sachs, *Die Liebe zum Automobil: Ein Rückblick in die Geschichte unserer Wünsche* (Reinbek: Rowohlt, 1984).

86. Paul Virilio, *Polar Inertia*, trans. Patrick Camiller (London: Sage, 2000).

87. Alexa Geisthövel, "Das Auto," in Geisthövel and Knoch, *Orte der Moderne*, 37–46, at 43.

88. Thomas Kühne, "Massenmotorisierung und Verkehrspolitik im 20. Jahrhundert: Technikgeschichte als politische Sozial- und Kulturgeschichte," *Neue Politische Literatur* 41.2 (1996): 196–229, at 199.

89. Sachs, *Die Liebe zum Automobil*, 120; Heike Weber, *Das Versprechen mobiler Freiheit: Zur Kultur- und Technikgeschichte von Kofferradio, Walkman und Handy* (Bielefeld: transcript, 2008).

90. Gijs Mom, *Atlantic Automobilism: Emergence and Persistence of the Car, 1895–1940* (New York: Berghahn, 2015).

91. Liane Löwe, "'A Nation Built on Transport': Das Auto und die US-amerikanische Gesellschaft," in Anke Köth, Anna Minta, and Andreas Schwarting (eds.), *Building America: Die Erschaffung einer neuen Welt* (Dresden: Thelem, 2005), 247–270.

92. Engels, "Machtfragen," 52; Volker Mehnert, "Schnell sind nur die Schmuggler," *Frankfurter Allgemeine Zeitung*, May 20, 2009.

93. Timothy Davis, *National Park Roads: A Legacy in the American Landscape* (Charlottesville: University of Virginia Press, 2016).

94. Christoph Mauch and Thomas Zeller (eds.), *The World Beyond the Windshield: Roads and Landscapes in the United States and Europe* (Athens: Ohio University Press, 2008).

95. Piotr Heller, "Besonders schön ist's in der Rhön," *Frankfurter Allgemeine Sonntagszeitung*, January 6, 2013.

96. Uwe Fraunholz, *Motorphobia: Anti-automobiler Protest in Kaiserreich und Weimarer Republik* (Göttingen: Vandenhoeck & Ruprecht, 2002).

97. Christoph Maria Merki, "Vom 'Herrenfahrer' zum 'Balkanraser': Zur Geschichte des Automobilismus in der Schweiz," *Schweizerische Zeitschrift für Geschichte* 56.1 (2006): 46–56, at 53.

98. Kühne, "Massenmotorisierung und Verkehrspolitik," 204; Kurt Möser, "Zwischen Systemopposition und Systemteilnahme: Sicherheit und Risiko im motorisierten Strassenverkehr 1890–1930," in Armin Hermann and Harry Niemann (eds.), *Geschichte der Strassenverkehrssicherheit im Wechselspiel zwischen Fahrzeug, Fahrbahn und Mensch* (Bielefeld: transcript, 2006), 159–167, at 164.

99. Harald Welzer, *Mentale Infrastrukturen—wie das Wachstum in die Welt und in die Seelen kam* (Berlin: Heinrich-Böll-Stiftung, 2011), 35 and 39.

100. Johannes Hörl and Dietmar Schöndorfer (eds.), *Die Grossglockner Hochalpenstrasse: Erbe und Auftrag* (Vienna: Böhlau, 2015).

101. Axel Dossmann, "Wie wir die Autobahnen lieben lernten," *Sowi: Sozialwissenschaftliche Informationen* 4 (1996): 235–242, at 236.

102. Werner Jochmann (ed.), *Adolf Hitler: Monologe im Führerhauptquartier 1941–1944. Die Aufzeichnungen Heinrich Heims* (Hamburg: Knaus, 1980), 39, 192.

103. Franz Freiherr Karaisl von Karais and Helmut Stabina, *Deutsche Strassenfibel* (Leipzig: Staackmann, 1937).

104. Wolfgang Benz, *Geschichte des Dritten Reiches* (Bonn: Bundeszentrale für Politische Bildung 2008), 120–121.

105. Karl Friedrich Schinkel, *Anweisung zum Bau und zur Unterhaltung der Kunststrassen* (Berlin: Zentral- & Landesbibliothek, 1834).

106. Kühne, "Massenmotorisierung und Verkehrspolitik," 204.

107. Eugen Diesel, *Völkerschicksal und Technik* (Stuttgart: Cotta, 1930), 97.

108. Diesel, *Völkerschicksal und Technik*, 105.

109. Ingrid Heckmann-Strohkark, "Der Traum von einer europäischen Gemeinschaft: Die Internationalen Autobahnkongresse 1931 und 1932," in Martin Heller and Andreas Volk (eds.), *Die Schweizer Autobahn* (Zurich, Switzerland: Museum für Gestaltung, 1999), 32–45.

110. Frank Schipper, *Driving Europe: Building Europe on Roads in the Twentieth Century* (Amsterdam: Aksant, 2008), 98.

111. Vincent Lagendijk, *Electrifying Europe: The Power of Europe in the Construction of Electricity Networks* (Amsterdam: Amsterdam University Press, 2008); Alexander Badenoch and Andreas Fickers (eds.), *Materializing Europe: Transnational Infrastructures and the Project of Europe* (Basingstoke, UK: Palgrave Macmillan, 2010); Irene Anastasiadou, *Constructing Iron Europe: Transnationalism and Railways in the Interbellum* (Amsterdam: Amsterdam University Press, 2011); Suzanne Lommers, *Europe—on Air: Interwar Projects for Radio Broadcasting* (Amsterdam: Amsterdam University Press, 2012). For a more general viewpoint, see Högselius, Kaijser, and van der Vleuten, *Europe's Infrastructure Transition*; and Gerold Ambrosius and Christian Henrich-Franke, *Integration von Infrastrukturen in Europa im historischen Vergleich* (Baden-Baden: Nomos, 2013).

112. Oskar Kahrstedt, *Internationale Bekämpfung der Arbeitslosigkeit durch Erschliessung überseeischer Gebiete: Zugleich ein Beitrag zum Problem der Vergrösserung des Welthandelsvolumens* (Berlin: Hobbing, 1931); Peo Hansen and Stefan Jonsson, *Eurafrica: The Untold History of European Integration and Colonialism* (London: Bloomsbury, 2014).

113. Bernhard Kellermann, *Der Tunnel* (Frankfurt am Main: Suhrkamp, 1952), 153; Thomas Hahn, "Tunnel und Damm als Medien des Weltverkehrs: Populäre Kommunikation in der modernen Raumrevolution," in Hartmut Böhme (ed.), *Topographien der Literatur: Deutsche Literatur im transnationalen Kontext* (Stuttgart: Metzler, 2005), 479–500.

114. Johan Schot and Vincent Lagendijk, "Technocratic Internationalism in the Interwar Years: Building Europe on Motorways and Electricity Networks," *Journal of Modern European History* 6.2 (2008): 196–216.

115. David Arnold, *Everyday Technology: Machines and the Making of India's Modernity* (Chicago: University of Chicago Press, 2013).

116. David J. Garrow (ed.), *The Montgomery Bus Boycott and the Women Who Started It: The Memoir of Jo Ann Gibson Robinson* (Knoxville: University of Tennessee Press, 1987).

117. Quoted in Sudhir Rao, "A Rail Passenger Called Gandhi . . . Railways Played a Role in the Making of the Mahatma," *Governance Now*, January 30, 2012, https://www.governancenow.com/views/columns/rail-passenger-called-gandhi; for the context, see Diogo and van Laak, *Europeans Globalizing*, chap. 5.

118. Eduard A. Pfeiffer, *Technik der Stadt: Vom Schaffen der technischen Betriebe im lebendigen Kreislauf der grossen Siedlungen* (Stuttgart: Franckh, 1937), 1.

119. John M. Jordan, "'Society Improved the Way You Can Improve a Dynamo': Charles P. Steinmetz and the Politics of Efficiency," *Technology and Culture* 30.1 (1989): 57–82.

120. Heinrich Zeiss, "Hygiene und Technik," *Technikgeschichte* 25 (1936): 66–73, at 70; Tony D. Sampson, *Virality: Contagion Theory in the Age of Networks* (Minneapolis: University of Minnesota Press, 2012).

121. Gustavus H. Robinson, "The Public Utility Concept in American Law," *Harvard Law Review* 41.3 (1928): 277–308.

122. J. Friedrich Zwicky, *Public Utilities* (Jena: Osterwalg, 1937), 9.

123. Richard R. John, "Telecommunications," *Enterprise and Society* 9 (2008): 507–520, at 516.

124. Jack London, *War of the Classes* (New York: Regent Press, 1905), preface.

125. Joseph Henry Beale, *Cases on Public Service Companies: Public Carriers, Public Works, and Other Public Utilities* (Cambridge, MA: Harvard University Press, 1902); Oscar Lewis Pond, *Municipal Control of Public Utilities: A Study of the Attitude of Our Courts toward an Increase of the Sphere of Municipal Activity* (New York: Columbia University Press, 1906).

126. Robert Beck, *Der Plan Freycinet und die Provinzen: Aspekte der infrastrukturellen Entwicklung der französischen Provinzen durch die Dritte Republik* (Frankfurt am Main: Lang, 1986).

127. Lorenz Jellinghaus, *Zwischen Daseinsvorsorge und Infrastruktur: Zum Funktionswandel von Verwaltungswissenschaften und Verwaltungsrecht in der zweiten Hälfte des 19. Jahrhunderts* (Frankfurt am Main: Klostermann, 2006), 13ff.

128. Wolfgang Rüfner, "§ 80 Daseinsvorsorge und soziale Sicherheit," in Josef Isensee and Paul Kirchhof (eds.), *Handbuch des Staatsrechts der Bundesrepublik Deutschland, Bd. III: Das Handeln des Staates* (Heidelberg: Müller, 1996), 1038–1084.

129. Ernst Forsthoff, *Die Verwaltung als Leistungsträger* (Stuttgart: Kohlhammer, 1938).

130. Archiv des Deutschen Museums München, estate of Anton Zischka.

131. Wolfgang Schivelbusch, *Entfernte Verwandtschaft: Faschismus, Nationalsozialismus, New Deal, 1933–1939* (Frankfurt am Main: Fischer, 2008); Thomas P. Hughes, *American Genesis: A Century of Invention and Technological Enthusiasm, 1870–1970* (New York: Viking, 1989).

132. Michael Wildt, *Volk, Volksgemeinschaft, AfD* (Hamburg: Hamburger, 2017).

133. John Kenneth Galbraith, *The Economic Effects of the Federal Public Works Expenditures 1933–1938* (Washington, DC: Government Printing Office, 1940), 2.

134. Galbraith, *The Economic Effects*, 107; Jason Scott Smith, *Building New Deal Liberalism: The Political Economy of Public Works, 1933–1956* (Cambridge: Cambridge University Press, 2006).

135. Benjamin Brendel, "Konvergente Konstruktionen: Eine Kulturgeschichte des Staudammbaus in den USA, Spanien und Ägypten 1933–1971," PhD diss., Giessen, 2017.

136. Herward Sieberg, *Colonial Development: Die Grundlegung moderner Entwicklungspolitik durch Grossbritannien, 1919–1949* (Stuttgart: Steiner, 1985); Michael Mann, *Wiring the Nation: Telecommunication, Newspaper-Reportage and Nation-Building in British India, 1850–1930* (Oxford: Oxford University Press, 2017).

137. Robert Millward, *Private and Public Enterprise in Europe: Energy, Telecommunication and Transport, 1830–1990* (Cambridge: Cambridge University Press, 2005).

138. Quoted in Wolfgang Lotz and Gerd R. Ueberschär, *Die Deutsche Reichspost 1933–1945: Eine politische Verwaltungsgeschichte*, 2 vols. (Berlin: Nicolai, 1999), 78.

139. Bernhard Stier, *Staat und Strom: Die politische Steuerung des Elektrizitätssystems in Deutschland 1890–1950* (Ubstadt-Weiher: Verlag Regionalkultur, 1999), 456.

140. "Angezettelt," exhibition at the Frankfurter Museum für Kommunikation, 2014; see, also, Claudia Michels, "Worte, die ausgrenzen," *Frankfurter Rundschau*, June 6, 2014.

141. Michael Wildt, "'Der muss hinaus! Der muss hinaus!' Antisemitismus in deutschen Nord- und Ostseebädern 1920–1935," *Mittelweg* 36.4 (2001): 2–25; Frank Bajohr, *"Unser Hotel ist judenfrei." Bäder-Antisemitismus im 19. und 20. Jahrhundert* (Frankfurt am Main: Fischer, 2003).

142. Stephen A. Berrey, *The Jim Crow Routine: Everyday Performances of Race, Civil Rights, and Segregation in Mississippi* (Chapel Hill: University of North Carolina Press, 2015).

143. Pape and Weinert, *Bottichwaschmaschine und Haustelegraph*, 27–28.

144. Michaela Wiegel, "Ins Wesen von Paris fahren," *Frankfurter Allgemeine Zeitung*, July 19, 2000, 13.

145. Anne Frank, *The Diary of a Young Girl* (New York: Bantam, 1995), entry for May 26, 1944.

146. Michels, "Worte, die ausgrenzen."

147. Hilberg, *Sonderzüge nach Auschwitz*; Zygmunt Bauman, *Modernity and the Holocaust* (Ithaca, NY: Cornell University Press, 2002).

148. Dirk van Laak and Dirk Rose (eds.), *Schreibtischtäter: Begriff, Geschichte, Typologie* (Göttingen: Wallstein, 2018).

149. Michael Makropoulos, "Die infrastrukturelle Konstruktion der 'Volksgemeinschaft': Aspekte des Autobahnbaus im nationalsozialistischen Deutschland," in Ulrich Bröckling, Axel T. Paul, and Stefan Kaufmann (eds.), *Vernunft—Entwicklung—Leben: Schlüsselbegriffe der Moderne. Festschrift für Wolfgang Essbach* (Munich: Fink, 2004), 185–203, at 190–192; see also Erhard Schütz and Eckhard Gruber, *Mythos Reichsautobahn: Bau und Inszenierung der "Strassen des Führers" 1933–1941* (Berlin: Links, 1996); and Benjamin Steininger, *Raum-Maschine Reichsautobahn: Zur Dynamik eines bekannt / unbekannten Bauwerks* (Berlin: Kadmos, 2005).

CHAPTER 3

1. Lutz Niethammer, "'Normalisierung' im Westen: Erinnerungsspuren in die 50er Jahre," in Dan Diner (ed.), *Ist der Nationalsozialismus Geschichte? Zu Historisierung und Historikerstreit* (Frankfurt am Main: Fischer, 1987), 153–184, at 177.

2. Christian Pfister, "Das '1950er Syndrom': Die umweltgeschichtliche Epochenschwelle zwischen Industriegesellschaft und Konsumgesellschaft," in Christian Pfister (ed.), *Das 1950er Syndrom: Der Weg in die Konsumgesellschaft* (Bern: Haupt, 1995), 51–95.

3. Charles Dickens, *Hard Times* (New York: Penguin, 1995), 27.

4. John R. McNeill, *Something New Under the Sun: An Environmental History of the Twentieth-Century World* (New York: Norton, 2001), 296–297.

5. See Andrea Westermann, *Plastik und politische Kultur in Westdeutschland* (Zurich, Switzerland: Chronos, 2007); Eli Rubin, *Synthetic Socialism: Plastic and Dictatorship in the German Democratic Republic* (Chapel Hill: University of North Carolina Press, 2008).

6. Arne Andersen, *Der Traum vom guten Leben: Alltags- und Konsumgeschichte vom Wirtschaftswunder bis heute* (Frankfurt am Main: Campus, 1997).

7. Niethammer, "'Normalisierung' im Westen," 161.

8. Sandra Evans, *Sowjetisch wohnen: Eine Literatur- und Kulturgeschichte der Kommunalka* (Bielefeld: transcript, 2011); Schlögel, *Das sowjetische Jahrhundert*, 324–345.

9. Arthur Miller, *Death of a Salesman: Certain Private Conversations in Two Acts and a Requiem* (New York: Penguin, 1998), 54.

10. Markus Krajewski, "Fehler-Planungen: Zur Geschichte und Theorie der industriellen Obsoleszenz," *Technikgeschichte* 81.1 (2014): 91–114.

11. Peter Quester, "Höchstens haltbar bis . . . Wie das 20. Jahrhundert das eingebaute Kaputtgehen erfand," *TrenntMagazin*, https://trenntmagazin.de/obsolescence-hochstens-haltbar-bis/#.VijmprxFvFa; retrieved December 17, 2017; see Christian Kreiss, *Geplanter Verschleiss: Wie die Industrie uns zu immer mehr und immer schnellerem Konsum antreibt—und wie wir uns dagegen wehren können* (Berlin: Europa, 2014).

12. Frederick W. Cottrell, *Energy and Society: The Relationship between Energy, Social Change, and Economic Development* (New York: McGraw-Hill, 1955); Vaclav Smil, *Energy in World History* (London: Routledge, 2020).

13. Sophie Gerber, *Küche, Kühlschrank, Kilowatt: Zur Geschichte des privaten Energiekonsums in Deutschland, 1945–1990* (Bielefeld: transcript, 2014).

14. Miriam A. Bader-Gassner, *Pipelineboom: Internationale Ölkonzerne im westdeutschen Wirtschaftswunder* (Baden-Baden: Nomos, 2014); Benjamin Steininger, "Pipeline: Am Puls der fossilen Moderne," in Kijan Espahangizi and Barbara Orland (eds.), *Stoffe in Bewegung* (Zurich, Switzerland: Diaphanes, 2014), 231–244.

15. Ulrich Hellmann, *Künstliche Kälte: Die Geschichte der Kühlung im Haushalt* (Giessen: Anabas, 1990).

16. Jean-Claude Kaufmann, *Kochende Leidenschaft: Soziologie vom Kochen und Essen* (Konstanz: UVK, 2006), 56–57.

17. Alexander Klose, *Das Container-Prinzip: Wie eine Box unser Leben verändert* (Hamburg: Mare, 2009); Michael B. Miller, *Europe and the Maritime World: A Twentieth-Century History* (Cambridge: Cambridge University Press, 2012).

18. Martina Hessler, *"Mrs. Modern Woman": Zur Sozial- und Kulturgeschichte der Haushaltstechnisierung* (Frankfurt am Main: Campus, 2011); Martina Hessler, *Kulturgeschichte der Technik* (Frankfurt am Main: Campus, 2012), 72–89.

19. Welzer, *Mentale Infrastrukturen.*

20. Nils Gilman, *Mandarins of the Future: Modernization Theory in Cold War America* (Baltimore: Johns Hopkins University Press, 2004).

21. Rankin, "Infrastructure and the International Governance of Economic Development, 1950–1965," 70.

22. See Jochimsen, *Theorie der Infrastruktur.*

23. Hans-Dietrich Genscher, "Infrastruktur als öffentliche Aufgabe: Vortrag am 8. Juni 1970," *BMI-Dokumentation* 30 (1970): 8.

24. Sebastian Strube, *Euer Dorf soll schöner werden: Ländlicher Wandel, staatliche Planung und Demokratisierung in der Bundesrepublik Deutschland* (Göttingen: Vandenhoek & Ruprecht, 2013), 184–185.

25. Frank Uekötter, *Die Wahrheit ist auf dem Feld: Eine Wissensgeschichte der deutschen Landwirtschaft* (Göttingen: Vandenhoek & Ruprecht, 2010); Eva Barlösius and Michèle Spohr, "Rückzug 'vom Lande': Die sozialräumliche Neuordnung durch Infrastrukturen," in Peter A. Berger, Carsten Keller, Andreas Klärner, and Rainer Neef (eds.), *Urbane Ungleichheiten: Neue Entwicklungen zwischen Zentrum und Peripherie* (Wiesbaden: Springer, 2014), 233–251.

26. Wolfgang Schmidt, *Integration und Wandel: Die Infrastruktur der Streitkräfte als Faktor sozioökonomischer Modernisierung in der Bundesrepublik 1955 bis 1975* (Munich: Oldenbourg, 2007).

27. Dirk Rose, "'Im Vorort': Zu einem Romanfragment Detlev von Liliencrons," *Zeitschrift für Germanistik* 16.3 (2006): 638–639.

28. Graeme Davison, *Car Wars: How the Car Won Our Hearts and Conquered Our Cities* (Crows Nest, Sydney, Australia: Allen & Unwin, 2004).

29. Joachim Scheiner, "Stellt den McDrive doch unter Denkmalschutz!" *Frankfurter Allgemeine Zeitung*, May 28, 2013.

30. Dirk van Laak, "From the Conservative Revolution to Technocratic Conservatism," Jan-Werner Müller (ed.), *German Ideologies since 1945: Studies in the Political Thought and Culture of the Bonn Republic* (New York: Palgrave Macmillan, 2003), 147–160.

31. Robert A. Caro, *The Power Broker: Robert Moses and the Fall of New York* (New York: Knopf, 1974).

32. Anthony Flint, *Wrestling with Moses: How Jane Jacobs Took On New York's Master Builder and Transformed the American City* (New York: Random House, 2011).

33. Joerges, "Die Brücken des Robert Moses," 59.

34. Sebastian Haumann, *"Schade, dass Beton nicht brennt . . .": Planung, Partizipation und Protest in Philadelphia und Köln 1940–1990* (Stuttgart: Steiner, 2011).

35. Pat Choate and Susan Walter, *America in Ruins: Beyond the Public Works Pork Barrel* (Washington, DC: Council of State Planning Agencies, 1981).

36. "Die Bundesrepublik—ein unterentwickeltes Land," *Der Spiegel*, August 30, 1961.

37. Dietmar Klenke, *Bundesdeutsche Verkehrspolitik und Motorisierung: Konfliktträchtige Weichenstellungen in den Jahren des Wiederaufstiegs* (Stuttgart: Steiner, 1993), 350.

38. "Tempo 20," *Der Spiegel*, August 20, 1963.

39. Quoted in Reinhard Loske, "Die politische Ökologie der Infrastrukturen," in Reinhard Loske and Roland Schaeffer (eds.), *Die Zukunft der Infrastrukturen: Intelligente Netzwerke für eine nachhaltige Entwicklung* (Marburg: Metropolis, 2005), 23–44, at 24. See also Gilles Duranton and Matthew A. Turner, "The Fundamental Law of Road Congestion: Evidence from US Cities," *American Economic Review* 101 (2011): 2616–2652.

40. Alain Monnier, *Givrée* (Paris:Flammarion, 2009).

41. Jon Agar, *The Government Machine: A Revolutionary History of the Computer* (Cambridge, MA: MIT Press, 2003).

42. Slava Gerovitch, *From Newspeak to Cyberspeak: A History of Soviet Cybernetics* (Cambridge, MA: MIT Press, 2002); Ronald R. Kline, *The Cybernetics Moment: Or Why We Call Our Age the Information Age* (Baltimore: Johns Hopkins University Press, 2015).

43. Philippe Rivière, "Der Staat als Maschine: Das Kybernetik-Experiment in Allendes Chile," *Le Monde Diplomatique*, November 12, 2010; Sascha Reh, *Gegen die Zeit* (Frankfurt am Main: Schöffling, 2015).

44. Paul N. Edwards, "Meteorology as Infrastructural Globalism," *Osiris* 21 (2006): 229–250.

45. Frank Dittmann, "Technik versus Konflikt: Wie Datennetze den Eisernen Vorhang durchdrangen," *Osteuropa* 59.10 (2009): 101–119; for the broader context, see Martin Schmitt, *Internet im Kalten Krieg: Eine Vorgeschichte des globalen Kommunikationsnetzes* (Bielefeld: transcript, 2016); Benjamin Peters, *How Not to Network a Nation: The Uneasy History of the Soviet Internet* (Cambridge, MA: MIT Press, 2016).

46. Caroline Humphrey, "Ideology in Infrastructure: Architecture and Soviet Imagination," *Journal of the Royal Anthropological Institute* 11.1 (2005): 39–58.

47. Ralph Kaschka, *Auf dem falschen Gleis: Infrastrukturpolitik und -entwicklung der DDR am Beispiel der Deutschen Reichsbahn 1949–1989* (Frankfurt am Main: Campus, 2011).

48. Schlögel, *Das sowjetische Jahrhundert*, 716–724.

49. Ilko-Sascha Kowalczuk and Arno Polzin (eds.), *Fasse Dich kurz! Der grenzüberschreitende Telefonverkehr der Opposition in den 1980er Jahren und das Ministerium für Staatssicherheit* (Göttingen: Vandenhoeck & Ruprecht, 2015).

50. *Fasse Dich kurz . . . Telefonieren in der DDR*, directed by Jörg Mischke (MDR documentary, 2007).

51. Klaus Gestwa and Johannes Grützenmacher, "Infrastrukturen," in Stefan Plaggenborg (ed.), *Handbuch der Geschichte Russlands, Bd. 5.2: Vom Ende des Zweiten Weltkriegs bis zum Zusammenbruch der Sowjetunion* (Stuttgart: Hiersemann, 2003): 1089–1152, at 1147–1148.

52. *Fasse Dich kurz . . . Telefonieren in der DDR.*

53. Alec Nove 1988 statement, quoted in Gestwa and Grützenmacher, "Infrastrukturen," 1127.

54. Lyubomir Pozharliev, "Collectivity vs. Connectivity: Highway Peripheralization in Former Yugoslavia (1940s–1980s)," *Journal of Transport History* 37.2 (2016): 194–213.

55. Dalakoglou, "The Road from Capitalism to Capitalism."

56. "E5: Terror von Blech und Blut," *Der Spiegel*, August 25, 1975.

57. Merki, "Vom 'Herrenfahrer' zum 'Balkanraser.'"

58. Luminita Gatejel, *Warten, Hoffen und endlich Fahren: Auto und Sozialismus in der Sowjetunion, in Rumänien und der DDR (1956–1989/91)* (Frankfurt am Main: Campus, 2014); Lewis H. Siegelbaum (ed.), *The Socialist Car: Automobility in the Eastern Bloc* (Ithaca, NY: Cornell University Press, 2011).

59. Gestwa and Grützenmacher, "Infrastrukturen," 1129.

60. Axel Dossmann, *Begrenzte Mobilität: Eine Kulturgeschichte der Autobahnen in der DDR* (Essen: Klartext, 2003), 173.

61. Christopher J. Ward, *Brezhnev's Folly: The Building of BAM and Late Soviet Socialism* (Pittsburgh, PA: University of Pittsburgh Press, 2009); Johannes Grützmacher, *Die Baikal-Amur-Magistrale: Vom stalinistischen Lager zum Mobilisierungsprojekt unter Breznev* (Munich: Oldenbourg, 2012); Andreas Röhr, *Die Bajkal-Amur-Magistrale: Geschichte eines sibirischen Raumerschliessungsprojektes* (Berlin: Logos, 2012).

62. Timothy Moss, "Divided City, Divided Infrastructures: Securing Energy and Water Services in Postwar Berlin," *Journal of Urban History* 35 (2009): 923–942.

63. Axel Klausmeier, *Hinter der Mauer: Zur militärischen und baulichen Infrastruktur des Grenzkommandos Mitte* (Berlin: Links, 2012).

64. Peter E. Fässler, *Durch den "Eisernen Vorhang": Die deutsch-deutschen Wirtschaftsbeziehungen 1949–1969* (Cologne: Böhlau, 2006).

65. Astrid M. Eckert, "Geteilt, aber nicht unverbunden: Grenzgewässer als deutsch-deutsches Umweltproblem," *Vierteljahrshefte für Zeitgeschichte* 62.1 (2014): 69–99; Jinhee Park, "Von der Müllkippe zur Abfallwirtschaft: Die Entwicklung der Hausmüllentsorgung in Berlin (West) von 1945 bis 1990," PhD diss., TU Berlin, 2004.

66. Kowalczuk and Polzin, *Fasse Dich kurz!*.

67. Josef Foschepoth, *Überwachtes Deutschland: Post- und Telefonüberwachung in der alten Bundesrepublik* (Göttingen: Vandenhoeck & Ruprecht, 2012).

68. Gestwa and Grützenmacher, "Infrastrukturen," 1120.

69. Helmut Koziolek, "Umfassende Intensivierung der Infrastruktur," *Sitzungsberichte der Akademie der Wissenschaften der DDR* 2 (1986): 12.

70. Felix Robin Schulz, *Death in East Germany 1945–1990* (Oxford: Berghahn, 2013), 69.

71. Susanne Bauer, "Lenin spannt aus: Alexander Sinowjews 'Homo sovieticus'—eine mehrfache Provokation," *Zeithistorische Forschungen* 11 (2014): 485–492, at 489.

72. Robert Momberg, *Theorie und Politik der Infrastruktur unter Berücksichtigung institutionen- und politökonomischer Einflussfaktoren* (Frankfurt am Main: Lang, 2000), 23. Ten years after reunification, the figure reached 75 percent.

73. See, for example, Genscher, "Infrastruktur als öffentliche Aufgabe"; Helmut Kohl, "Infrastrukturreform als Innenpolitik–Möglichkeiten, Grenzen, Prioritäten," *Bergedorfer Gesprächskreis zu Fragen der freien industriellen Gesellschaft* 38 (1971).

74. Wolfgang Kil, *Luxus der Leere: Vom schwierigen Rückzug aus der Wachstumswelt: Eine Streitschrift* (Wuppertal: Müller & Busmann, 2004); Hans-Christian Schink, *Verkehrsprojekte Deutsche Einheit: Begleitbuch zur Ausstellung* (Ostfildern-Ruit: Cantz, 2004).

75. See Pankaj Mishra, *From the Ruins of Empire: The Revolt against the West and the Remaking of Asia* (New York: Penguin, 2013).

76. Jalāl Āl-Ahmad, *Occidentosis: A Plague from the West*, trans. R. Campbell (Berkeley, CA: Mizan Press, 1984).

77. Headrick, *The Tools of Empire*; Daniel R. Headrick, *The Tentacles of Progress: Technology Transfer in the Age of Imperialism, 1850–1940* (Oxford: Oxford University Press, 1988).

78. See, for example, Zejnelovic, *Der "Highway of Nations."*

79. Sieberg, *Colonial Development*; Dirk van Laak, *Imperiale Infrastruktur: Deutsche Planungen für eine Erschliessung Afrikas, 1880–1960* (Paderborn: Schöningh, 2004).

80. James C. Scott, *Seeing like a State: How Certain Schemes to Improve the Human Condition Have Failed* (New Haven, CT: Yale University Press, 1998), 253.

81. World Bank, *World Development Report 1994: Infrastructure for Development* (Oxford: Oxford University Press, 1994), https://openknowledge.worldbank.org/handle/10986/5977; Peter Lanjouw, "Infrastructure: A Ladder for the Poor," *Finance & Development* 32.1 (1995): 33–35.

82. Franco Moretti and Dominique Prestre, "Bankspeak: The Language of World Bank Reports," *New Left Review* 92 (2015): 75–99.

83. "Immer mehr Städter erzwingen hohe Investitionen in die Infrastruktur," *Frankfurter Allgemeine Zeitung*, November 28, 2015.

84. Irfan Habib, "Science, Technological Education and Industrialization: Contours of a Bhadralok Debate," in Robin MacLeod and David Kumar (eds.), *Technology and the Raj: Western Technology and Technical Transfers to India, 1700–1947* (New Delhi: Sage, 1995), 235–249.

85. Dan Morse, "Still Called by Faith to the Booth: As Pay Phones Vanish, Amish and Mennonites Build Their Own," *Washington Post*, September 3, 2006.

86. Fred A. Bleakley, "Rural County Balks at Joining Global Village," *Wall Street Journal*, January 4, 1996.

87. Chandrika Kaul, *Communications, Media and the Imperial Experience: Britain and India in the Twentieth Century* (New York: Palgrave Macmillan, 2014).

88. Andreas Eckert, "Exportschlager Wohlfahrtsstaat? Europäische Sozialstaatlichkeit und Kolonialismus in Afrika nach dem Zweiten Weltkrieg," *Geschichte und Gesellschaft* 32.4 (2006): 467–488.

89. Vijay Prashad, *The Darker Nations: A People's History of the Third World* (New York: New Press, 2007).

90. Ali Mazuri, "Nkrumah: The Leninist Czar," *Transition* 26 (1966): 8–17.

91. Robert H. Jackson and Carl G. Rosberg, *Personal Rule in Black Africa: Prince, Autocrat, Prophet, Tyrant* (Berkeley: University of California Press, 1982).

92. Allen Isaacman, "Displaced People, Displaced Energy, and Displaced Memories: The Case of Cahora Bassa, 1970–2004," *International Journal of African Historical Studies* 38.2 (2005): 201–238.

93. Julia Tischler, *Light and Power in a Multiracial Nation: The Kariba Dam Scheme in the Central African Federation* (Basingstoke, UK: Palgrave Macmillan, 2013), 225.

94. David Edgerton, "Creole Technologies and Global Histories: Rethinking How Things Travel in Space and Time," *Journal of History of Science and Technology* 1.1 (2007): 3–31.

95. Brian Larkin, *Signal and Noise: Media, Infrastructure, and Urban Culture in Nigeria* (Durham, NC: Duke University Press, 2008), 6.

96. Ravi Ahuja, *Pathways of Empire: Circulation, "Public Works" and Social Space in Colonial Orissa (c. 1780–1914)* (Hyderabad, India: Orient BlackSwan, 2009), 9.

97. Nicholas W. Balabkins, "Barriers to Technological Transfer: Infrastructural Difficulties in Nigeria," in Karin Peschel (ed.), *Infrastructure and the Space-Economy: Essays in Honor of Rolf Funck* (Berlin: Springer, 1990), 101–115; Trovalla and Trovalla, "Infrastructure Turned Suprastructure."

98. Hans Seligo, *Afrika im Umbau* (Jena: Urania-Freidenker, 1931); Gustav-Adolf Gedat, *Was wird aus diesem Afrika? Erlebter Kampf um einen Erdteil* (Stuttgart: Steinkopf, 1938); Wilhelm R. Schmidt, *Als Telegrafenbauer in Deutsch-Südwest* (Erfurt: Sutton, 2006).

99. Hubertus Büschel, *Hilfe zur Selbsthilfe: Deutsche Entwicklungsarbeit in Afrika 1960–1975* (Frankfurt am Main: Campus, 2014), 85ff.

100. Jonas van der Straeten, "Measuring Progress in Megawatts? Development Narratives and the Shaping of the Tanzanian Power Infrastructure 1906–1980," PhD diss., Technische Universität Berlin, 2017.

101. Georg Haneke and Claudia Stahl, *Infrastruktur und sozialer Wandel: Sozialanthropologische Beobachtungen entlang einer Strasse im Norden Kenias* (Saarbrücken: Breitenbach, 1994); Masquelier, "Road Mythographies"; Harvey, "Cementing Relations."

102. Winther, *The Impact of Electricity*.

103. Han van Dijk and Roche Mongbo, "Report of the workshop 'Urban Public Services,'" *Bulletin de l'APAD*, 23–24 (2002). https://journals.openedition.org/apad/155.

104. Cottrell, *Energy and Society*, 288.

105. Andy Hahnemann, *Texturen des Globalen: Geopolitik und populäre Literatur in der Zwischenkriegszeit 1918–1939* (Heidelberg: Winter, 2010).

106. Jens Hohensee, *Der erste Ölpreisschock 1973/74: Die politischen und gesellschaftlichen Auswirkungen der arabischen Erdölpolitik auf die Bundesrepublik Deutschland und Westeuropa* (Stuttgart: Steiner, 1996).

107. Clarence B. Davis and Kenneth E. Wilburn (eds.), *Railway Imperialism* (New York: Greenwood, 1991).

108. Christoph Hein, "Eine Billion Dollar für Brücken, Strassen, Flughäfen," *Frankfurter Allgemeine Sonntagszeitung*, July 2, 2006; Petra Kolonko, "Seide und Zement," *Frankfurter Allgemeine Zeitung*, May 13, 2017.

109. Matthias Rüb, "Kontinentale Durchbrüche," *Frankfurter Allgemeine Zeitung*, May 20, 2015.

110. John F. Jungclaussen, "Hilflos im Wasser: Regengüsse legen offen, wie marode die britische Infrastruktur ist. Geflickt wird mit Geld aus China," *Die Zeit*, February 20, 2014.

111. Björn Blass, "Räume der Entsorgung: Werte, Praktiken und Topographien von Müll (1880–1930)," PhD diss., Freie Universität Berlin, 2020.

112. Heike Weber, "'Entschaffen' Reste und das Ausrangieren, Zerlegen und Beseitigen des Gemachten," *Technikgeschichte* 81.1 (2014): 1–32.

113. Gundula Madeleine Tegtmeyer, "Die Müllspezialisten von Kairo," *Natürlich: Das Magazin für ganzheitliches Leben* 3 (2006): 52–59.

114. Raymond G. Stokes, Roman Köster, and Stephen C. Sambrook, *The Business of Waste: The United Kingdom and Germany, 1945 to the Present* (Cambridge: Cambridge University Press, 2013).

115. Susan Strasser, *Waste and Want: A Social History of Trash* (New York: Henry Holt, 1999).

116. Ludolf Kuchenbuch, "Abfall: Eine Stichwortgeschichte," in Hans-Georg Soeffner (ed.), *Kultur und Alltag* (Göttingen: Schwartz, 1988), 155–170, at 155.

117. Sonja Windmüller, *Die Kehrseite der Dinge: Müll, Abfall, Wegwerfen als kulturwissenschaftliches Problem* (Münster: LIT, 2004).

118. Frey, *Der reinliche Bürger*.

119. Friedrich Wilhelm Büsing, *Die Städtereinigung* (Stuttgart: Bergstraesser, 1897), 3.

120. Kaika, "City of Flows," 107.

121. Elisabeth Heidenreich, *Fliessräume: Die Vernetzung von Natur, Raum und Gesellschaft seit dem 19. Jahrhundert* (Frankfurt am Main: Campus, 2004), 65.

122. Ivan Illich, *H_2O und die Wasser des Vergessens* (Reinbek: Rowohlt, 1994); Karlheinz Cless, *Menschen am Brunnen: Ethnologische Perspektiven zum Umgang mit Wasser* (Bielefeld: transcript, 2014).

123. Reiner Burger, "In der Mitte entsteht ein Fluss," *Frankfurter Allgemeine Zeitung*, August 12, 2015.

124. Alessandra Ponte, "Müllschlucker: Die Domestizierung des Abfalls," *Arch+* 191/192 (2009): 78–83.

125. Goschler, "Infrastruktur-Liberalismus," 55.

126. Heike Weber and Ruth Oldenziel, "Recycling and Re-Use in the Twentieth Century," *Contemporary European History*, 22.3 (2013): 347–370; Stefan Krebs, Gabriele Schabacher, and Heike Weber (eds.), *Kulturen des Reparierens* (Bielefeld: transcript, 2018).

127. "Die Charlottenburger Drei," *TrenntMagazin*, https://trenntmagazin.de/recycling-die-charlottenburger-drei/#.VAR4Kih4bII; retrieved December 17, 2017.

128. Franz-Josef Brüggemeier, *Das unendliche Meer der Lüfte: Luftverschmutzung, Industrialisierung und Risikobewusstsein im 19. Jahrhundert* (Essen: Klartext, 1996).

129. David Edgerton, *The Shock of the Old: Technology and Global History since 1900* (London: Profile, 2019).

130. Eckert, "Geteilt, aber nicht unverbunden," 69–99.

131. Brüggemeier, *Das unendliche Meer der Lüfte*.

132. Joachim Radkau, *Die Ära der Ökologie: Eine Weltgeschichte* (Munich: Beck, 2011); Joachim Radkau, *Geschichte der Zukunft: Prognosen, Visionen, Irrungen in Deutschland von 1945 bis heute* (Munich: Hanser, 2017).

133. Roman Köster, "Abschied von der 'verlorenen Verpackung': Das Recycling von Hausmüll in Westdeutschland 1945–1990," *Technikgeschichte* 81 (2014): 33–60.

134. Werner Schenkel, "Entstehung, Entsorgung und Wiederverwertung von Müll—ein globales Problem," in Werner Nachtigall and Charlotte Schönbeck (eds.), *Technik und Natur* (Düsseldorf: VDI, 1994), 483–520, at 484.

135. Bernhard Schäfers, "Zur Genesis und zum Stellenwert von Partizipationsforderungen im Infrastrukturbereich," *Raumforschung und Raumordnung* 32.1 (1974): 1–6.

136. Loske, "Die politische Ökologie der Infrastrukturen," 23.

137. Dietrich Henkel, "Soziale Infrastruktur: Anpassung oder Rückbau?" *Bauwelt* 76.24 (1985): 124–128, at 126.

138. Matthew Gandy, *Recycling and the Politics of Urban Waste* (New York: Routledge, 1994).

139. Tegtmeyer, "Die Müllspezialisten von Kairo," 53.

CHAPTER 4

1. Traugott Haefeli-Meylan, *Die Entstehung der Briefmarke und ihre weltweite Verbreitung* (Heidelberg: Decker, 1985); Bernhard Siegert, *Relais: Geschicke der Literatur als Epoche der Post, 1751–1913* (Berlin: Brinkmann & Bose, 1993), 110–118.

2. Garrett Hardin, "The Tragedy of the Commons," *Science* 162 (1968): 1243–1248; for an opposing view, see Elinor Ostrom, *Institutional Incentives and Sustainable Development: Infrastructure Policies in Perspective* (Boulder, CO: Westview, 1993); Elinor Ostrom and Silke Hilfreich (eds.), *Was mehr wird, wenn wir teilen: Vom gesellschaftlichen Wert der Gemeingüter* (Munich: Oekom, 2011).

3. Nil Disco and Eda Kranakis (eds.), *Cosmopolitan Commons: Sharing Resources and Risks across Borders* (Cambridge, MA: MIT Press, 2013).

4. Bernd Müllender, "Ode an die Pappendecke," *Die Tageszeitung (taz Magazin)*, July 7, 2001, www.taz.de/1/archiv/?dig=2001/07/07/a0206.

5. Wolfgang Blümel, "Empirische Erfassung der Präferenzen für öffentliche Güter: Die Debatte über den 'Freifahrer,'" *Zeitschrift für Wirtschaftspolitik* 37 (1988): 65–71.

6. In social terms, disconnection has come to count as a particularly severe stigma; Laura Cwiertnia, "Wer lebt im Dunkeln?" *Die Zeit*, November 30, 2017.

7. Nikhil Anand, "Consuming Citizenship: Prepaid Meters and the Politics of Technology in Mumbai," *Occasional Papers of the School of Social Science* 53 (2014): 47–70;

Anne Clendinning, *Demons of Domesticity: Women and the English Gas Industry, 1889–1939* (Aldershot, UK: Ashgate, 2004).

8. In South African townships, for example, metered devices helped break rent strikes, which had forced political concessions from the apartheid government during the 1980s. As one Soweto protester complained, "Everything has to be paid for in advance today. We pay in advance for electricity, for the telephone, and now we pay in advance for water. Who knows, maybe soon we'll have to pay in advance for the sun to be turned on and off." Quoted in Schnitzler, "Citizenship Prepaid," 900.

9. Susanne Frank and Matthew Gandy, introduction to *Hydropolis: Wasser und die Stadt der Moderne* (Frankfurt am Main: Campus, 2006), 30. For a similar dispute that arose in Thessaloniki, Greece, see Christiane Schlözer, "Rein ins kalte Wasser," *Süddeutsche Zeitung*, May 9, 2014.

10. Wilhelm Heinrich Riehl, *Die Naturgeschichte des Volkes, Erster Band: Land und Leute* (Stuttgart: Cotta, 1925), 70.

11. Gerd Habermann, *Der Wohlfahrtsstaat: Die Geschichte eines Irrwegs* (Frankfurt am Main: Ullstein, 1994).

12. Christian Siebenbiedel, "Leuchttürme machen den Seehandel sicher," *Frankfurter Allgemeine Sonntagszeitung*, October 9, 2011.

13. Dimitri Ladischensky, "Geliebt, gehasst, gequält: Mythos und Realität des blinden Passagiers," *Mare* 60 (2007): 66–71; Angela Köckritz, "Der Stromdieb," *Die Zeit*, April 12, 2017.

14. Gerald Hubmayr, *Schwarzfahren: Die Kunst des tariffreien Netzgleitens* (Cologne: Böhlau, 2002); Marianne Moesle, "Der Schein und das Bewusstsein," *Süddeutsche Zeitung Magazin* 4 (2014): 8–13.

15. Maren Lorenz, *Vandalismus als Alltagsphänomen* (Hamburg: Hamburger, 2009).

16. Gabriella Bassu, "Schwarzfahren kostet jetzt 60 Euro," *Frankfurter Allgemeine Zeitung*, July 1, 2015.

17. Alfred C. Mierzejewski, *The Most Valuable Asset of the Reich: A History of the German National Railway. Vol. 1: 1920–1932* (Chapel Hill: University of North Carolina Press, 1999).

18. Sabine Dworog, "Der Wandel von Staatsaufgaben: Die Konflikte um den Ausbau des Frankfurter Flughafens im Kontext gesellschaftlicher Umbrüche (1960er bis 1980er)," PhD diss., Giessen, 2007.

19. Evelyn Hagenah, *Staatsaufgaben* (Frankfurt am Main: Suhrkamp, 1996).

20. Louis T. Wells and Eric S. Gleason, "Is Foreign Infrastructure Investment Still Risky?" *Harvard Business Review* 9–10 (1995): 44–55, at 47.

21. Ralf Roth, "Wie wurden die Eisenbahnen der Welt finanziert? Einige Vergleiche," *Zeitschrift für Weltgeschichte* 10.1 (2009): 55–80, at 79.

22. Wells and Gleason, "Is Foreign Infrastructure Investment Still Risky?" 55.

23. Volker Schneider and Marc Tenbücken (eds.), *Der Staat auf dem Rückzug: Die Privatisierung öffentlicher Infrastrukturen* (Frankfurt am Main: Campus, 2004).

24. Tim Engartner, "11 000 Jahre Verspätung," *Die Zeit*, March 11, 2009.

25. Nicola Liebert, "Die Lektion British Rail," *Edition Le Monde diplomatique: Ausverkauft. Wie das Gemeinwohl zur Privatsache wird* 6 (2009): 4.

26. Georg Hermes, *Staatliche Infrastrukturverantwortung: Rechtliche Grundstrukturen netzgebundener Transport- und Übertragungssysteme zwischen Daseinsvorsorge und Wettbewerbsregulierung am Beispiel der leitungsgebundenen Energieversorgung in Europa* (Tübingen: Mohr, 1998).

27. Florian C. Haus, *Zugang zu Netzen und Infrastruktureinrichtungen* (Cologne: Heymann, 2022).

28. Eva Barlösius, "Gleichwertig ist nicht gleich," *Politik und Zeitgeschichte* 37 (2006): 16–23.

29. Jochen Monstadt, "Der räumliche Wandel der Stromversorgung und die Auswirkungen auf die Raum- und Infrastrukturplanung," in Timothy Moss, Matthias Naumann, and Markus Wissen (eds.), *Infrastrukturnetze und Raumentwicklung: Zwischen Universalisierung und Differenzierung* (Munich: Oekom, 2008), 187–224, at 223.

30. Graham and Marvin, *Splintering Urbanism*; Timothy Moss, "'Cold Spots' of Urban Infrastructure: Shrinking Processes in Eastern Germany and the Modern Infrastructural Ideal," *International Journal of Urban and Regional Research* 32 (2008): 436–451.

31. Michael Hascher, "Maut als Mittel der Strassenfinanzierung zwischen öffentlichen und privaten Interessen, 1908–2008," in Hans-Liudger Dienel and Hans-Ulrich Schiedt (eds.), *Die moderne Strasse: Planung, Bau und Verkehr vom 18. bis zum 20. Jahrhundert* (Frankfurt am Main: Campus, 2010), 147–169.

32. Schmidt, "Alles muss raus."

33. Werner Rügemer, *Cross Border Leasing: Ein Lehrstück zur globalen Enteignung der Städte* (Münster: Westfälisches Dampfboot, 2005).

34. Nadine Oberhuber, "Auf Sand gebaut," *Die Zeit*, October 4, 2007.

35. Hein, "Eine Billion Dollar für Brücken, Strassen, Flughäfen."

36. Alex von Tunzelmann, "The Railway Man," *Guardian*, June 24, 2015.

37. Ian Buruma, "The Worst Railroad Job," *New York Review of Books* 61.18 (2014): 19–21.

38. Adam Hochschild, *Schatten über dem Kongo: Die Geschichte eines großen, fast vergessenen Menschheitsverbrechens* (Reinbek: Rowohlt, 2000), 245.

39. Franz Sonnenberger, "'Kolonisieren heisst transportieren'—Europa und der Beginn des Eisenbahnzeitalters in Afrika," in *Zug der Zeit—Zeit der Züge: Deutsche Eisenbahn 1835–1985* (Berlin: Siedler, 1985), 1:228–231, at 228.

40. Susanne Kill, Christopher Kopper, and Jan-Henrik Peters, *Die Reichsbahn und der Strafvollzug in der DDR: Häftlingszwangsarbeit und Gefangenentransport in der SED-Diktatur* (Essen: Klartext, 2016).

41. Syed Hussein Alatas, *The Myth of the Lazy Native: A Study of the Image of the Malays, Filipinos and Javanese from the 16th to the 20th Century and Its Function in the Ideology of Colonial Capitalism* (London: Routledge, 1977); Anton Markmiller, *Die Erziehung des*

Negers zur Arbeit: Wie die koloniale Pädagogik afrikanische Gesellschaften in die Abhängigkeit führte (Berlin: Reimer, 1995).

42. Nikole K. Longen, "Fronarbeiten zur Finanzierung von Infrastruktur: Der Ausbau des Strassennetzes im Kurtrierer Raum 1716–1841," in Dienel and Schiedt, *Die moderne Strasse*, 23–48.

43. Ursula Krechel, *Landgericht* (Salzburg: Jung & Jung, 2012), 85.

44. An example is the Czech Aktion Z. See Wikipedia, s.v. "Aktion Z," last modified March 26, 2021, https://de.wikipedia.org/wiki/Aktion_Z.

45. Birgit Wolf, *Sprache in der DDR: Ein Wörterbuch* (Berlin: De Gruyter, 2000), 151.

46. Nikola Baković, "'No One Here Is Afraid of Blisters or Work!' Social Integration, Mobilization and Cooperation in Yugoslav Youth Brigades: The Example of Čačak Region Brigades (1946–1952)," *Hungarian Historical Review* 4.1 (2015): 29–55.

47. Corinna Kuhr-Korolev, *Gezähmte Helden: die Formierung der Sowjetjugend 1917–1932* (Essen: Klartext, 2005).

48. Kiran Klaus Patel, *Soldaten der Arbeit: Arbeitsdienste in Deutschland und den USA, 1933–1939/1942* (Göttingen: Vandenhoeck & Ruprecht, 2003).

49. Robert D. Leighninger Jr., "Cultural Infrastructure: The Legacy of New Deal Public Space," *Journal of Architectural Education* 49.4 (1996): 226–236.

50. José A. Gómez-Ibáñez, *Regulating Infrastructure: Monopoly, Contracts, and Discretion* (Cambridge, MA: Harvard University Press, 2003).

51. Gerold Ambrosius, *Der Staat als Unternehmer: Öffentliche Wirtschaft und Kapitalismus seit dem 19. Jahrhundert* (Göttingen: Vandenhoeck & Ruprecht, 1984); Gerold Ambrosius, *Hybride Eigentums- und Verfügungsrechte: Öffentlich-private Kooperationen in systematisch-theoretischer und historisch-empirischer Perspektive* (Berlin: BWV, 2012).

52. George Monbiot, *Captive State: The Corporate Takeover of Britain* (London: Pan, 2001), 18–58; Werner Rügemer, *"Heuschrecken" im öffentlichen Raum: Public Private Partnership—Anatomie eines globalen Finanzinstruments* (Bielefeld: transcript, 2012).

53. See the mission statement of Rijkswaterstaat, https://www.rijkswaterstaat.nl/en; retrieved May 30, 2022.

54. "About the Bundesnetzagentur," https://www.bundesnetzagentur.de/EN/General/Bundesnetzagentur/About/start.html; retrieved May 30, 2022.

55. Rüdiger Soldt, "Wir fahr'n, fahr'n, fahr'n auf der Autobahn," *Frankfurter Allgemeine Zeitung*, November 29, 2014.

56. Catalina Schröder, "Wenn Anlieger zu Anlegern werden," *Die Zeit*, March 11, 2013.

57. Christopher Schrader, "Hauptgewinn: eine Latrine," *Süddeutsche Zeitung*, April 17, 2015.

58. Sally Denton, *The Profiteers: Bechtel and the Men Who Built the World* (New York: Simon & Schuster, 2016).

59. Hartmut Berghoff and Jakob Vogel (eds.), *Wirtschaftsgeschichte als Kulturgeschichte: Dimensionen eines Perspektivenwechsels* (Frankfurt am Main: Campus, 2004).

60. Hans Nordmann, *Die Frühgeschichte der Eisenbahn* (Berlin: Akademie, 1947), 11–15.

61. Thomas Hahn, "Teures Pflaster," *Süddeutsche Zeitung*, August 8/9, 2015.

62. See the homepage of *Heimatverein Niederzimmern*, https://www.heimatverein-niederzimmern.de/; retrieved May 30, 2022.

CHAPTER 5

1. Brian Hayes, *Infrastructure: A Guide to the Industrial Landscape* (New York: Norton, 2014), 1.

2. Hayes, *Infrastructure*, 5.

3. Larkin, "The Politics and Poetics of Infrastructure."

4. Carl D. Goerdeler, "Fetisch einer Minderheit," *Die Zeit*, May 27, 1994.

5. Alain Corbin, *Les Cloches de la terre : paysage sonore et culture sensible dans les campagnes au XIXe siècle* (Paris : Flammarion, 2013).

6. "CST: Nun heulen sie wieder," *Die Zeit*, November 27, 2014.

7. In 1827 the poem was set to music by Franz Schubert for the *Winterreise* song cycle.

8. Wilhelm Prausnitz (ed.), *Atlas und Lehrbuch der Hygiene mit besonderer Berücksichtigung der Städte-Hygiene* (Munich: Lehmann, 1909), 24; Pfeiffer, *Technik der Stadt*, 43.

9. Sabine Dworog, "Luftverkehrsinfrastruktur: Zur Rolle des Staates bei der Integration eines Flughafens in seine Umgebung," *Saeculum* 58.1 (2007): 115–149.

10. Rudolf Vierhaus (ed.), *Das Tagebuch der Baronin von Spitzemberg, geb. Freiin von Varnbüler: Aufzeichnungen aus der Hofgesellschaft des Hohenzollernreiches* (Göttingen: Vandenhoeck & Ruprecht, 1989), 381.

11. Theodor Lessing, *Der Lärm: Eine Kampfschrift gegen die Geräusche unseres Lebens* (Wiesbaden: Bergmann, 1909); Theodor Lessing, *Der Antirüpel: Monatsblätter zum Kampf gegen Lärm, Rohheit und Unkultur im deutschen Wirtschafts-, Handels- und Verkehrsleben* (1908/09–1911); Klaus Saul, "'Kein Zeitalter seit Erschaffung der Welt hat so viel und so ungeheuerlichen Lärm gemacht': Lärmquellen, Lärmbekämpfung und Antilärmbewegung im Deutschen Kaiserreich," in Günter Bayerl, Norman Fuchsloch, and Torsten Mayer (eds.), *Umweltgeschichte. Methoden, Themen, Potentiale* (Münster: Waxmann, 1996), 187–218.

12. Volker Ullrich, "Quietschen, Hupen, Fauchen," *ZeitGeschichte* 2 (2013): 24–26.

13. Daniel Morat, "'Automobile gehen über mich hin': Urbane Dispositive akustischer Innervation um 1900," in Sylvia Mieszkowski and Sigrid Nieberle (eds.), *Unlaute: Noise / Geräusch in Kultur, Medien und Wissenschaften seit 1900* (Bielefeld: transcript, 2017), 127–148.

14. Karin Bijsterveld, Eefje Cleophas, Stefan Krebs, and Gijs Mom (eds.), *Sound and Safe: A History of Listening Behind the Wheel* (Oxford: Oxford University Press, 2014).

15. Hansjörg Küster, "Das Nichts am Ende des Tunnels," *Frankfurter Allgemeine Zeitung*, March 8, 1999.

16. See Karin Bijsterveld (ed.): *Soundscapes of the Urban Past: Staged Sound as Mediated Cultural Heritage* (Bielefeld: transcript, 2013); Daniel Morat (ed.), *Sounds of Modern History: Auditory Cultures in 19th- and 20th-Century Europe* (New York: Berghahn, 2014).

17. Kaika, "City of Flows."

18. Kate Ascher, *The Works: Anatomy of a City* (London: Penguin, 2007), 152.

19. *Baumeister: Zeitschrift für Architektur* 3 (1998): 43.

20. Kaika, "City of Flows."

21. Gottfried Knapp, "Orte der Verdrängung: Was nicht geliebt wird, muss auch nicht schön sein," *Baumeister: Zeitschrift für Architektur* 3 (1998): 38–41.

22. Engels and Schenk, "Macht der Infrastrukturen," 52.

23. Star, "The Ethnography of Infrastructure," 379.

24. Lewis Mumford, *The City in History: Its Origins, Its Transformations, and Its Prospects* (New York: Harvest, 1989), 481–482.

25. Edwards, "Infrastructure and Modernity."

26. Star, "The Ethnography of Infrastructure," 381.

27. Porombka, *Medialität urbaner Infrastrukturen*, 409.

28. Robert Musil, "Denkmale," in *Gesammelte Werke*, vol. 7 (Reinbek: Rowohlt, 1978), 506–509, at 506.

29. Mimi Melnick, *Manhole Covers* (Cambridge, MA: MIT Press, 1994).

30. Axel Dossmann, "Erbschaft unterm Hakenkreuz," *Die Tageszeitung* (*taz*), December 19, 2006.

31. Dirk van Laak, "Vom Lebensraum zum Leitungsweg: Die Stadtstraße als soziale Arena," in Michael Flitner, Julia Lossau, and Anna-Lisa Müller (eds.), *Infrastrukturen der Stadt* (Wiesbaden: Springer, 2017), 145–162.

32. Scott, *Seeing like a State* (especially chap. 2); see also Anton Tantner, *Die Hausnummer: Eine Geschichte von Ordnung und Unordnung* (Marburg: Jonas, 2007).

33. Christine Domke, *Die Betextung des öffentlichen Raumes: Eine Studie zur Spezifik von Meso-Kommunikation am Beispiel von Bahnhöfen, Innenstädten und Flughäfen* (Heidelberg: Winter, 2014).

34. *ADAC Motorwelt* no. 9 (2012): 66–67.

35. Clay McShane, "The Origins and Globalization of Traffic Control Signals," *Journal of Urban History* 25.3 (1999): 379–404; Frank Schipper, "Unravelling Hieroglyphs: Urban Traffic Signs and the League of Nations," *Métropoles* 6 (2009).

36. Craig Murphy and JoAnn Yates, *ISO, the International Organization for Standardization: Global Governance through Voluntary Consensus* (London: Routledge, 2008), 107.

37. Sebastian Giessmann, "Synchronisation im Diagramm: Henry C. Beck und die Londoner Tube Map von 1933," in Christian Kassung and Thomas Macho (eds.), *Kulturtechniken der Synchronisation* (Munich: Fink, 2013), 339–366.

38. Roland Wenzlhuemer, "Verbrechen, Verbrechensbekämpfung und Telegrafie: Kriminalhistorische Perspektiven auf die Entkoppelung von Transport und Kommunikation im langen 19. Jahrhundert," *Historische Zeitschrift* 301.2 (2015): 347–373; Roland Wenzlhuemer, *Globalgeschichte schreiben: Eine Einführung in 6 Episoden* (Munich: UVK/Lucius, 2017), 221–255.

39. Leon Hempel, Susanne Krasmann, and Ulrich Bröckling (eds.), *Sichtbarkeitsregime: Überwachung, Sicherheit und Privatheit im 21. Jahrhundert* (Wiesbaden: VS Verlag für Sozialwissenschaften, 2011); Byung-Chul Han, *The Transparency Society*, trans. Erik Butler (Stanford, CA: Stanford University Press, 2015).

40. Natalia G. Tomilina (ed.), *Пять колец под кремлевскими звездами* (Moscow: MFD, 2011), 31–32.

41. Dirk van Laak, *Weisse Elefanten: Anspruch und Scheitern technischer Grossprojekte im 20. Jahrhundert* (Stuttgart: Deutsche Verlagsanstalt, 1999).

42. Tobias Birkenhof, *Die Auswirkungen von Großveranstaltungen auf die langfristige ökonomische Entwicklung von Schwellenländern: Eine wachstumstheoretische und empirische Untersuchung am Beispiel der Olympischen Sommerspiele in Beijing* (Berlin: Logos, 2009).

43. Nick Malkoutzis, "How the 2004 Olympics Triggered Greece's Decline," *Bloomberg Business*, August 2, 2012.

44. John R. Gold and Margaret M. Gold (eds.), *Olympic Cities: City Agendas, Planning, and the World's Games, 1896–2016* (London: Routledge, 2011).

45. Christine Kalb, *Weltausstellungen im Wandel der Zeit und ihre infrastrukturellen Auswirkungen auf Stadt und Region* (Frankfurt am Main: Lang, 1994).

46. Alexander C. T. Geppert, *Fleeting Cities: Imperial Expositions in Fin-de-Siècle Europe* (New York: Palgrave Macmillan, 2010).

47. Chandra Mukerji, "Intelligent Uses of Engineering and the Legitimacy of State Power," *Technology and Culture* 44.4 (2003): 655–676, at 659.

48. On this topos, see David Feest and Lutz Häfner (eds.), *Die Zukunft der Rückständigkeit: Chancen—Formen—Mehrwert* (Cologne: Böhlau, 2016).

49. Frank Westerman, *Ingenieure der Seele: Schriftsteller unter Stalin—eine Erkundungsreise* (Berlin: BVT, 2003).

50. Obertreis, "Karrieren, Patronage und 'Infrastrukturpoesie.'"

51. Van Laak, "Technokratie im Europa des 20. Jahrhunderts—eine einflussreiche 'Hintergrundideologie.'"

52. Hugh R. Trevor-Roper (ed.), *Hitler's Table Talk 1941–1944: Secret Conversations* (London: Enigma, 2000), 64.

53. Larkin, *Signal and Noise*, 19.

54. Bent Flyvbjerg, Nils Bruzelius, and Werner Rothengatter, *Megaprojects and Risk: An Anatomy of Ambition* (Cambridge: Cambridge University Press, 2003).

55. Christian Fuchs, Stephan Lebert, and Daniel Müller, "Die unheimliche Firma," *Die Zeit*, July 16, 2015; Bettina Malter, "Hier ruhen unsere Milliarden," *Die Zeit*, July 3, 2014.

56. Dieter Bartetzko, "Unten bleiben!" *Frankfurter Allgemeine Sonntagszeitung*, February 10, 2013.

57. Thomas P. Hughes, *Rescuing Prometheus: Four Monumental Projects That Changed the Modern World* (New York: Vintage, 1998).

58. Merlind Theile, "Die Achse des Guten," *Die Zeit*, September 4, 2015; Dirk van Laak, "Zukunft konkret: Zeithistorische Anmerkungen zum Handeln der praktisch Planenden," in Markus Bernhardt, Stefan Brakensiek, and Benjamin Scheller (eds.), *Ermöglichen und Verhindern: Vom Umgang mit Kontingenz* (Frankfurt am Main: Campus, 2016), 191–208.

59. Deckers, "Unter den Wolken"; Timo Kotowski, "Teure Luftnummern in der Provinz," *Frankfurter Allgemeine Zeitung*, July 30, 2014.

60. Hanna Decker, "Grossprojekte werden zu optimistisch geplant," *Frankfurter Allgemeine Zeitung*, October 8, 2015.

61. Susanne Schattenberg, *Stalins Ingenieure: Lebenswelten zwischen Technik und Terror in den 1930er Jahren* (Munich: Oldenbourg, 2002).

62. Oliver Cyriax, "Planning Officers," in *The Penguin Encyclopedia of Crime* (London: Penguin, 1996), 447–448.

CHAPTER 6

1. Andreas Platthaus, "Schriftsteller, ans Telefon!" *Frankfurter Allgemeine Zeitung*, January 31, 2014.

2. Jonathan Waldman, *Rust: The Longest War* (New York: Simon & Schuster, 2015).

3. This is the view presented in the educational film *Singing Wires* (1951), about the electrification of the rural United States.

4. Max Scharnigg, "Mein halbes digitales Leben," *Süddeutsche Zeitung*, August 8/9, 2015.

5. Joachim Scheiner and Christian Holz-Rau (eds.), *Räumliche Mobilität und Lebenslauf: Studien zu Mobilitätsbiografien und Mobilitätssozialisation* (Wiesbaden: Springer, 2015).

6. Nikolaus Piper, "Im Zeichen des Windhunds," *Süddeutsche Zeitung*, October 18/19, 2014.

7. Andreas Fickers, "Visibly Audible: The Radio Dial as Mediating Interface," in Karin Bijsterveld and Trevor Pinch (eds.), *Oxford Handbook of Sound Studies* (Oxford: Oxford University Press, 2012), 411–439.

8. Susan J. Douglas, *Listening In: Radio and the American Imagination* (Minneapolis: University of Minnesota Press, 2004).

9. Weber, *Das Versprechen mobiler Freiheit.*

10. Wolfgang Riepl, *Das Nachrichtenwesen des Altertums mit besonderer Rücksicht auf die Römer* (Leipzig: Teubner, 1913), 5–6.

11. Julia Zons, *Casellis Pantelegraph: Geschichte eines vergessenen Mediums* (Bielefeld: transcript, 2015).

12. Burkhard Strassmann, "Bewahren Sie Ruhe," *Die Zeit*, November 26, 2015.

13. Ingmar Arnold, *Luft-Züge: Die Geschichte der Rohrpost* (Berlin: Links, 2016); see also Laura Meneghello, "Die Rohrpost als Metainfrastruktur," PhD diss., Siegen, 2017.

14. Andrew Emmerson, *Discovering Subterranean London* (Oxford, UK: Shire, 2009), 38.

15. Georg Schmid, *Die Spur und die Trasse: (Post)moderne Wegmarken in der Geschichtswissenschaft* (Cologne: Böhlau, 1988).

16. Gregg Marland and Alvin M. Weinberg, "Longevity of Infrastructure," in Ausubel and Herman, *Cities and Their Vital Systems*, 312–332, at 331.

17. Tom Pleiner, *Überplanung von Infrastruktur* (Tübingen: Mohr Siebeck, 2016).

18. Edgerton, *The Shock of the Old.*

19. Steinke, "Stopp."

20. Monika Goetsch, "Wie im Traum," *Süddeutsche Zeitung*, October 18/19, 2015.

21. See the webpage, Gaslicht-Kultur eV, 2021, http://www.gaslicht-kultur.de.

22. Edo Reents, "Freudig begrüssen wir den zuckenden Schein," *Frankfurter Allgemeine Zeitung*, March 1, 2005.

23. Examples include Siegfried Lenz, *The Lightship*, trans. Michael Bullock (London: Faber & Faber, 2010); and the 2004 film directed by Philippe Lioret *The Light.*

24. Jürgen Osterhammel, "Grenzen und Brücken," in *Die Flughöhe der Adler: Historische Essays zur globalen Gegenwart* (Munich: Beck, 2017), 82–100.

25. Ascher, *The Works*, 35.

26. Marc Buggeln and Inge Marszolek (eds.), *Bunker: Kriegsort, Zuflucht, Erinnerungsraum* (Frankfurt am Main: Campus, 2008).

27. Edward Wong, "The Air-Raid-Shelter Apartments Under Beijing," *New York Times Magazine*, April 22, 2011.

28. Konrad Putzier, "Warum bauen die Schweizer so viele Bunker?" *Die Welt*, July 12, 2011; Silvia Berger Ziauddin, "Überlebensinsel und Bordell: Zur Ambivalenz des Bunkers im atomaren Zeitalter," in David Eugster and Sibylle Marti (eds.), *Das Imaginäre des Kalten Krieges: Beiträge zu einer Kulturgeschichte des Ost-West-Konfliktes in Europa* (Essen: Klartext, 2015), 69–93.

29. Elian Stefa and Gyler Mydyti, *Concrete Mushrooms: Reusing Albania's 750,000 Bunkers* (Milan: ACTAND, 2013).

30. Corey Kilgannon, "The Truth about Alligators in the Sewers of New York," *New York Times*, February 26, 2020.

31. Robert Daley, *The World Beneath the City* (Philadelphia: Lippincott, 1959), 187ff.

32. Maurice Yacowar, "The Bug in the Rug: Notes on the Disaster Genre," in Barry Keith Grant (ed.) *Film Genre Reader III* (Austin: University of Texas Press, 2003), 277–295; Stephen Keane, *Disaster Movies: The Cinema of Catastrophe* (London: Wallflower, 2006).

33. David E. Nye, *Technology Matters: Questions to Live With* (Cambridge, MA: MIT Press, 2007), 162.

34. Hans Blumenberg, *Shipwreck with Spectator: Paradigm of a Metaphor for Existence*, trans. Steven Rendall (Cambridge, MA: MIT Press, 1997), 67.

35. Matthias Bickenbach and Michael Stolzke, *Die Geschwindigkeitsfabrik: Eine fragmentarische Kulturgeschichte des Autounfalls* (Berlin: Kadmos, 2014).

36. Henry Petroski, *To Engineer Is Human: The Role of Failure in Successful Design* (New York: St. Martin's, 1992); Henry Petroski, *Success through Failure: The Paradox of Design* (Princeton, NJ: Princeton University Press, 2006).

37. Pfeiffer, *Technik der Stadt*, 36.

38. Ellis L. Armstrong, Michael C. Robinson, and Suellen M. Hoy, *History of Public Works in the United States 1776–1976* (Chicago: American Public Works Association, 1976), xi.

39. Julius Rodenberg, "Sonntag vor dem Landberger Tor (1880)," *Bilder aus dem Berliner Leben*, vol. 1 (Berlin: Paetel, 1891), 23.

40. "Das mobile Chaos," *Der Spiegel*, July 12, 1998.

41. Marie Antoinette Glaser (ed.), *Baustelle: Metamorphosen in der Stadt* (Baden: Müller, 2008).

42. Rodenberg, *Bilder aus dem Berliner Leben*, 2:165.

43. Kristen L. Sanford, Joel A. Tarr, and Sue McNeil, "Crisis Perception and Policy Outcomes: Comparison between Environmental and Infrastructure Crises," *Journal of Infrastructure Systems* 1.4 (1995): 195–203, at 199.

44. Sanford, Tarr, and McNeil, "Crisis Perception and Policy Outcomes," 199.

45. David E. Nye, *When the Lights Went Out: A History of Blackouts in America* (Cambridge, MA: MIT Press, 2010).

46. Henry Petroski, *To Forgive Design: Understanding Failure* (Cambridge, MA: Harvard University Press, 2012).

47. Bernard J. Feldman, "What to Say about the Tacoma Narrows Bridge to Your Introductory Class," *Physics Teacher* 41 (2003): 92–96.

48. Chris LeRose, "The Collapse of the Silver Bridge," *West Virginia Historical Society Quarterly*, 15.4 (2001).

49. Hence the initiative, founded in 2013, Damit Deutschland vorne bleibt (So Germany Stays Ahead).

50. Krajewski, "Fehler-Planungen."

51. Homepage, COG Deutschland eV.

52. Christian Lotz, "Nachhaltigkeit neuskalieren: Internationale forstwissenschaftliche Kongresse und Debatten um die Ressourcenversorgung der Zukunft im Nord- und Ostseeraum (1870–1914)," PhD diss., Giessen, 2015, 65.

53. George Rainer, *Understanding Infrastructure: A Guide for Architects and Planners* (New York: Wiley, 1990), xxiii.

54. Albrecht Göschel, Heidrun Kunert-Schroth, and Klaus Mittag, *Revision der sozialen Infrastruktur: Ein Rückblick auf die 80er Jahre* (Berlin: Institut für Urbanistik, 1992), 12.

55. "Netzanschluss ist Menschenrecht," *Frankfurter Allgemeine Zeitung*, October 31, 2011.

56. Georg Küffner, "Bröckelnde Brücken," *Frankfurter Allgemeine Zeitung*, March 5, 2013.

57. Nikolaus Doll, S. Fründt, E. A. Ginten, T. Heuzeroth, B. Nicolai, A. Tauber, and D. Wetzel, "Das unfassbare deutsche Infrastruktur-Desaster," *Die Welt*, May 12, 2013; Wolf Dombrowsky, "4 Menschen und Katastrophen," in *BBK: Notfall- und Katastrophenpharmazie I: Bevölkerungsschutz und Medizinische Notfallversorgung* (Bonn: Bundesamt für Bevölkerungsschutz und Katastrophenhilfe, 2009), 249–288.

58. Doll et al., "Das unfassbare deutsche Infrastruktur-Desaster."

59. Sebastian Schoepp, "Letzter Zug nach Nirgendwo," *Süddeutsche Zeitung*, May 17, 2010.

60. Schoepp, "Letzter Zug nach Nirgendwo."

CHAPTER 7

1. Franz Vogt, *Anstandsbüchlein für das Volk* (Donauwörth: Auer, 1894), 141.

2. Vogt, *Anstandsbüchlein für das Volk*, 142.

3. Vogt, *Anstandsbüchlein für das Volk*, 145.

4. Dietmar Fack, *Automobil, Verkehr und Erziehung: Motorisierung und Sozialisation zwischen Beschleunigung und Anpassung 1885–1945* (Opladen: VS Verlag für Sozialwissenschaften, 2000), 197–198.

5. Robert Schmidt, "Technik, Risiko und das Zusammenspiel von Habitat und Habitus," in Gunter Gebauer et al. (eds.), *Kalkuliertes Risiko: Technik, Spiel und Sport an der Grenze* (Frankfurt am Main: Campus, 2006), 78–95; Lorenz Engell, Bernhard Siegert, and Joseph Vogl (eds.), *Gefahrensinn* (Munich: Fink, 2009).

6. Jeremy Packer, "Disciplining Mobility," in Jack Z. Bratich, Jeremy Packer, and Cameron McCarthy (eds.), *Foucault, Cultural Studies, and Governmentality* (Albany: State University of New York Press, 2003), 135–164.

7. McShane, "The Origins and Globalization of Traffic Control Signals," 385.

8. Thomas Fischermann et al., "Fussgänger—Angst bis in die Zehenspitzen," *Die Zeit*, February 11, 2016; Fisher, *America Calling*.

9. Van Laak, "Vom Lebensraum zum Leitungsweg."

10. Manuel Castells, "Space of Flows—Raum der Ströme. Eine Theorie des Raums in der Informationsgesellschaft," in Peter Noller, Walter Prigge, and Klaus Ronneberger (eds.), *Stadt-Welt: Über die Globalisierung städtischer Milieus* (Frankfurt am Main: Campus, 1994), 120–133.

11. Bruce Schneier, *Beyond Fear: Thinking Sensibly about Security in an Uncertain World* (New York: Copernicus, 2006).

12. Achim Saupe, "'Human Security' and the Challenge of Automobile and Road Traffic Safety: A Cultural Historical Perspective," *Historical Social Research* 35.4 (2010): 102–121.

13. Wolfgang Schivelbusch, *The Railway Journey: The Industrialization of Time and Space in the Nineteenth Century* (Berkeley: University of California Press, 2014), 54.

14. Alexander Friedrich Marfeld, *Am Anfang und am Ende—der Passagier* (Berlin: Safari, 1967).

15. Ariane Bentner, "Fliegende Mädchen—Muster an Takt und Charme: Aus den Anfängen eines weiblichen Dienstleistungsberufes," *Feministische Studien* 8.2 (1990): 105–115; Jörg Reichle, "Wo wir sind, ist oben," *Süddeutsche Zeitung*, January 1/2, 2016.

16. Ralf Wiegand, "Gedenkenlos," *Süddeutsche Zeitung*, April 18/19, 2015.

17. Kim Förster, "Sicherheitsschleuse," *Arch+* 191/192 (2009): 102; Annette Vowinckel, *Flugzeugentführungen: Eine Kulturgeschichte* (Göttingen: Wallstein, 2011); Jörg Potthast, "Politische Soziologie der Zugänge: Das Beispiel der Flughafensicherheit," in Leon Hempel, Susanne Krasmann, and Ulrich Bröckling (eds.), *Sichtbarkeitsregime: Überwachung, Sicherheit und Privatheit im 21. Jahrhundert* (Wiesbaden: VS Verlag für Sozialwissenschaften, 2011), 223–241; Anke Hagemann, "Filter, Ventile und Schleusen: Die Architektur der Zugangsregulierung," in Volker Eick, Jens Sambale, and Eric Töpfer (eds.), *Kontrollierte Urbanität: Zur Neoliberalisierung städtischer Sicherheitspolitik* (Bielefeld: transcript, 2007), 301–328.

18. Bruce Schneier, "Beyond Security Theater," *New Internationalist* 427 (2009): 10–13.

19. Anke Ortlepp, "Airport Security and the Limits of Mobility: The Case of the United States," *Traverse: Zeitschrift für Geschichte / Révue Historique* 16.1 (2009): 75–88.

20. Thomas Lindenberger, "'Streikexzesse': Zur öffentlichen Un-Ordnung im Kaiserreich," *Deutsches Historisches Museum*, https://www.dhm.de/ausstellungen/streik/html/kaiserzeit1.html; retrieved December 18, 2017.

21. Klaus Rainer Röhl, *Nähe zum Gegner: Kommunisten und Nationalsozialisten im Berliner BVG-Streik von 1932* (Frankfurt am Main: Campus, 1994).

22. Kwabena Akurang-Parry, "Anticolonial Movements, Africa," in Thomas Benjamin (ed.): *Encyclopedia of Western Colonialism since 1450*, vol. 1 (Detroit, MI: Macmillan Reference, 2007), 74–81, at 79.

23. Walter Gordon Merritt, *Strikes and Public Utilities: A Remedy* (New York: American Anti-Boycott Association, 1917).

24. Inge Scherer, *Grenzen des Streikrechts in den Arbeitsbereichen der Daseinsvorsorge: Schutz der Bürger vor gezielter Schädigung durch Streiks* (Berlin: Duncker & Humblot, 2000); Lena Rudkowski, *Der Streik in der Daseinsvorsorge* (Munich: Beck, 2010); Gregor Thüsing, *Arbeitskampf in der Daseinsvorsorge* (Tübingen: Mohr Siebeck, 2013).

25. *Frankfurter Allgemeine Zeitung*, September 8, 2014.

26. Yvonne Staat, "Gute Fahrt," *Frankfurter Allgemeine Sonntagszeitung*, January 4, 2015.

27. Martin Randelhoff, "Resiliente Infrastrukturen und Städte: Kritikalität und Interdependenzen," *Zukunft Mobilität*, December 17, 2013, https://www.zukunftmobilitaet.net/40882/analyse/resilienz-infrastruktur-stadt-wirtschaft-zukunft-resiliente-infrastrukturen.

28. Marc Widmann, "Menschens Kinder," *Süddeutsche Zeitung*, May 4/5, 2014.

29. Tillmann Schulze, *Bedingt abwehrbereit: Schutz kritischer Informations-Infrastrukturen in Deutschland und den USA* (Wiesbaden: Verlag für Sozialwissenschaften, 2006); Richard G. Little, "Managing the Risk of Cascading Failure in Complex Urban Infrastructures,"

in Stephen Graham (ed.), *Disrupted Cities: When Infrastructure Fails* (London: Routledge, 2010), 27–40.

30. Jörg Potthast, *Die Bodenhaftung der Netzwerkgesellschaft: Eine Ethnografie von Pannen an Grossflughäfen* (Bielefeld: transcript, 2007), 13.

31. Frank Patalong, "Untersee-Kabel: Die fragilen Lebensadern des Internets," *Der Spiegel*, February 2, 2015, https://www.spiegel.de/netzwelt/web/untersee-kabel-die-fragilen-lebensadern-des-internets-a-1015809.html.

32. Edwards, "Infrastructure and Modernity," 195.

33. Nye, *Technology Matters*, 164.

34. See Karl Leopold von Lichtenfels, *Lexikon des Überlebens: Handbuch für Krisenzeiten* (Cologne: Anaconda, 2006).

35. Thomas Petermann, Harald Bradke, Arne Lüllmann, Maik Poetzsch, and Ulrich Riehm, *Gefährdung und Verletzbarkeit moderner Gesellschaften, am Beispiel eines grossräumigen Ausfalls der Stromversorgung* (Berlin: Büro für Technikfolgen-Abschätzung beim Deutschen Bundestag, 2010), 193.

36. Wolf R. Dombrowsky, "Schutz kritischer Infrastrukturen" (unpublished manuscript).

37. Dombrowsky, "4 Menschen und Katastrophen," 257.

38. Marc Elsberg, *Blackout: Morgen ist es zu spät* (Munich: Blanvalet, 2012); Per Högselius, Arne Kaijser, and Erik van der Vleuten, *Europe's Infrastructure Transition: Economy, War, Nature* (New York: Palgrave Macmillan, 2015).

39. Christian Meier, *Das Gebot zu vergessen und die Unabweisbarkeit des Erinnerns: Vom öffentlichen Umgang mit schlimmer Vergangenheit* (Berlin: Siedler, 2010).

40. Giessmann, *Die Verbundenheit der Dinge*, 381ff.

41. *The Jewish Peril: Protocols of the Learned Elders of Zion* (London: Eyre & Spottiswoode, 1920), 29–30; Johannes Roskothen, *Verkehr: Zu einer poetischen Theorie der Moderne* (Munich: Fink, 2003), 188.

42. Jeffrey L. Sammons (ed.), *Die Protokolle der Weisen von Zion : die Grundlage des modernen Antisemitismus—eine Fälschung : Text und Kommentar* (Göttingen: Wallstein, 2018), 64–65.

43. Haruki Murakami, *Underground: The Tokyo Gas Attack and the Japanese Psyche*, trans. Alfred Birnbaum and Philip Gabriel (New York: Vintage, 2000).

44. Quoted in James Fenelon, *Four Hours of Fury: The Untold Story of World War II's Largest Airborne Invasion and the Final Push into Nazi Germany* (New York: Scribner, 2019), 148.

45. Junkelmann, "Die Eisenbahn im Krieg," 1:236.

46. Friedrich Ratzel, *Erdenmacht und Völkerschicksal: Eine Auswahl aus seinen Werken*, ed. Karl Haushofer (Stuttgart: Kroner, 1940), 137–148, at 147.

47. Willibald Stavenhagen, *Verkehrs-, Beobachtungs- und Nachrichten-Mittel in militärischer Beleuchtung: Für Offiziere aller Waffen des Heeres und der Marine* (Göttingen: Peters, 1905), 1.

48. Thomas Speckmann, "Von unschuldigen Zivilisten," *Frankfurter Allgemeine Zeitung*, December 2, 2009.

49. Joachim Schickel, *Guerrilleros, Partisanen: Theorie und Praxis* (Munich: Hanser, 1970).

50. Robert H. Ferrell, "The Mukden Incident: September 18–19, 1931," *Journal of Modern History* 27.1 (1955): 66–72.

51. United States Strategic Bombing Survey, *The Effects of Strategic Bombing on German Morale* (Washington, DC: Government Printing Office, 1947), 2 vols. See Martina Metzger, *Bewältigung, Auswirkungen und Nachwirkungen des Bombenkrieges in Berlin und London 1940–1955: Zerstörung und Wiederaufbau zweier europäischer Hauptstädte* (Stuttgart: Steiner, 2013).

52. Erich Hampe, *Der Zivile Luftschutz im Zweiten Weltkrieg: Dokumentation und Erfahrungsberichte über Aufbau und Einsatz* (Frankfurt am Main: Berhard & Graefe, 1963).

53. Klaus Grot, *So war's, damals: Dienstchronik eines Pionieroffiziers im Kalten Krieg 1954–1991* (Berlin: Miles, 2014).

54. *Freundschaftsgesellschaft BRD-Kuba eV*, https://fgbrdkuba.de/

55. Carola Dietze, *Die Erfindung des Terrorismus: Die Erfindung des Terrorismus in Europa, Russland und den USA 1858–1866* (Hamburg: Hamburger, 2016).

56. Jens Ivo Engels and Alfred Nordmann (eds.), *Was heisst Kritikalität? Zu einem Schlüsselbegriff der Debatte um Kritische Infrastrukturen* (Bielefeld: transcript, 2018).

57. *Schutz kritischer Infrastrukturen—Basisschutzkonzept Empfehlungen für Unternehmen* (Berlin: Bundesministerium des Innern, 2005), 29.

58. Schulze, *Bedingt abwehrbereit*, 96.

59. Gero von Randow, "Der Angriff auf die Moderne," *Frankfurter Allgemeine Zeitung*, September 19, 2001.

60. Emmerson, *Discovering Subterranean London*, 35.

61. Bernd Kremer, *Die Kunst zu überleben: Zivilverteidigung in der Bundesrepublik* (Munich: Odang, 1966), 93.

62. William Rankin, "Infrastructure and the International Governance," 72.

63. Gerard Alberts and Ruth Oldenziel (eds.), *Hacking Europe: From Computer Cultures to Demoscenes* (London: Springer, 2014).

64. Edwards, "Infrastructure and Modernity," 219.

65. Malte Grehsin, "Deutschland hat angefangen," *Frankfurter Allgemeine Zeitung*, January 4, 2014.

66. Stephen Graham (ed.), *Cities, War and Terrorism: Towards an Urban Geopolitics* (Oxford: Blackwell, 2004); Martin Coward, "Network-Centric Violence: Critical Infrastructure and the Urbanization of Security," *Security Dialogue* 40.4–5 (2009): 399–418.

67. Quoted in Schulze, *Bedingt abwehrbereit*, 133.

68. *Civilization Forum*, July 26/27, 2002, https://www.civforum.de/showthread.php?6208-Zerst%F6rung-der-Infrastruktur!!!.

CHAPTER 8

1. Michael Fisch, "Tokyo's Commuter Train Suicides and the Society of Emergence," *Cultural Anthropology* 28.2 (2013): 320–343.

2. Fisch, "Tokyo's Commuter Train Suicides."

3. George P. Fletcher, *A Crime of Self-defense: Bernhard Goetz and the Law on Trial* (Chicago: University of Chicago Press, 1990).

4. Otto Ullrich, "Technologie," in Wolfgang Sachs (ed.), *Wie im Westen so auf Erden: Ein polemisches Handbuch zur Entwicklungspolitik* (Reinbek: Rowohlt, 1993), 390–408, at 405.

5. Lars Wilhelmer, *Transit-Orte in der Literatur: Eisenbahn—Hotel—Hafen—Flughafen* (Bielefeld: transcript, 2015).

6. See Clelia Caruso, "Zur Einübung in die Kulturtechnik des Telefonierens," PhD diss., Universität des Saarlandes, 2010.

7. Kurt Tucholsky, "Berlin! Berlin!" *Gesammelte Werke in zehn Bänden*, vol. 2 (Reinbek: Rowohlt, 1960), 130.

8. Willy Hellpach, *Nervosität und Kultur* (Berlin 1902), 28–29.

9. Robert Musil, "Geschwindigkeit ist eine Hexerei," *Gesammelte Werke*, vol. 2 (Reinbek: Rowohlt 1978), 685.

10. Georg Simmel, "The Metropolis and Mental Life," in David Frisby and Mike Featherstone (eds.), *Simmel on Culture: Selected Writings* (London: Sage, 1997), 174–186.

11. Walther Kiaulehn, *Berlin: Schicksal einer Weltstadt* (Munich: Beck, 1997), 20–21.

12. Urs Stäheli, "Infrastrukturen des Kollektiven: alte Medien—neue Kollektive?" *Zeitschrift für Medien- und Kulturforschung* 12.2 (2012): 99–116, at 107.

13. Walt Whitman, *The Portable Walt Whitman*, ed. Michael Warner (New York: Penguin, 2004), 475.

14. Stäheli, "Infrastrukturen des Kollektiven," 99–116.

15. Roskothen, *Verkehr*.

16. Höhne, *New York City Subway*, 330.

17. Helmut Lethen, *Cool Conduct: The Culture of Distance in Weimar Germany*, trans. Don Reneau (Berkeley: University of California Press, 2002).

18. Trevor Pinch, "The Invisible Technologies of Goffman's Sociology from the Merry-Go-Round to the Internet," *Technology and Culture* 51.2 (2010): 409–424.

19. Gottfried Korff, "Mentalität und Kommunikation in der Grossstadt: Berliner Notizen zur 'inneren' Urbanisierung," in Theodor Kohlmann and Hermann Bausinger (eds.), *Grossstadt: Aspekte empirischer Kulturforschung* (Berlin 1985), 343–361.

20. Juliane Löffler, "So gross kann Dein ******* nicht sein!" *Der Freitag*, December 19, 2014, https://www.freitag.de/autoren/juloeffl/so-gross-kann-dein-nicht-sein.

21. Michael Polanyi, *Implizites Wissen* (Frankfurt am Main: Suhrkamp, 1985).

22. Hannelore Bublitz, Irina Kaldrack, Theo Röhle, and Mirna Zeman (eds.), *Automatismen—Selbst-Technologien* (Munich: Fink, 2013); Kai Nowak, "Teaching Self-Control: Road Safety and Traffic Education in Postwar Germany," *Historical Social Research / Historische Sozialforschung* 41.1 (2016): 135–153.

23. Erhard Schüttpelz, "Körpertechniken," *Zeitschrift für Medien- und Kulturforschung* 2.1 (2010): 101–120.

24. Quoted in Dirk Rose, "Peripherie und Perspektive: Infrastrukturgeschichtliche Überlegungen zu Liliencron und Bölsche," *Zeitschrift für Germanistik* 15.2 (2005): 311–326, at 313.

25. Petra Löffler, *Verteilte Aufmerksamkeit: Eine Mediengeschichte der Zerstreuung* (Zurich, Switzerland: Diaphanes, 2014); Hartmut Winkler, *Switching / Zapping: Ein Text zum Thema und ein parallel laufendes Unterhaltungsprogramm* (Darmstadt: Häusser, 1991).

26. Porombka, *Medialität urbaner Infrastrukturen*, 248.

27. Ulrich Bröckling, *Gute Hirten führen sanft: Über Menschenregierungskünste* (Berlin: Suhrkamp, 2017), 175–196.

28. Schivelbusch, *The Railway Journey*.

29. See Martha Lampland and Susan Leigh Star, *Standards and Their Stories: How Quantifying, Classifying and Formalizing Practices Shape Everyday Life* (Ithaca, NY: Cornell University Press, 2009).

30. Alan Latham and Peter R. H. Wood, "Inhabiting Infrastructure: Exploring the Interactional Spaces of Urban Cycling," *Environment and Planning* 47 (2015): 300–319.

31. "Nicht neben einer Frau," *Frankfurter Allgemeine Zeitung*, March 2, 2016.

32. Paul Dost, *Einmal Dritter Pasewalk? Wer reist in welcher Wagenklasse? Psychologische Soziologie der Wagenklassen* (Krefeld: Röhr, 1977).

33. Jennifer Toth, *The Mole People: Life in the Tunnels Beneath New York City* (Chicago: Chicago Review, 1993); see also Margaret Morton, *The Tunnel: The Underground Homeless of New York City* (New Haven, CT: Yale University Press, 1995).

34. Teun Voeten, *Tunnel People* (Oakland, CA: PM, 2010).

35. Jürgen von Rutenberg, "Zurück in die Gegenwart," *Zeit-Magazin* 29 (2015): 12–20.

36. Mark Siemons, "Die Wurzel der Menschlichkeit," *Frankfurter Allgemeine Zeitung*, September 26, 2014.

37. Philipp Glitz, "Generation Wisch und Klick Studie zur Smartphone-Sucht," *Tagesschau.de*, September 30, 2015, www.tagesschau.de/inland/smartphone-sucht-101.html (retrieved December 30, 2015); see also Martin Spiewak, "Bloss nicht offline sein," *Die Zeit*, November 9, 2017.

38. Markus Krajewski, *The Server: A Media History from the Present to the Baroque*, trans. Ilinca Iurascu (New Haven, CT: Yale University Press, 2018).

39. Schott, *Europäische Urbanisierung (1000–2000)*, 266–267.

40. Sarah Jost and Gabriela Wachter (eds.), *Die verschwundene Arbeit in Fotografien aus Berliner Sammlungen und Archiven* (Berlin: Parthas, 2008).

41. Ascher, *The Works*, 140.

42. K. Röhling, "Die Fahrgeldhinterziehung auf Strassenbahnen und ihre Bekämpfung," *Zeitschrift für Kleinbahnen* 23 (1916): 415.

43. Kurt Tucholsky, "Der Kontrollierte," *Gesammelte Werke: 1907–1918* (Reinbek: Rowohlt, 1975), 2:114.

44. Helmut Heiber (ed.), *Joseph Goebbels: Reden, Bd. 2: 1939–1945* (Düsseldorf: Droste, 1972), 305–313.

45. Marianne Erath, "Hallo, ist da das Fräulein vom Amt?" *Emma*, July/August 2009, https://www.emma.de/artikel/frauenberuf-hallo-ist-da-das-fraeulein-vom-amt-264018.

46. Rudi Palla, *Verschwundene Arbeit: Das Buch der untergegangenen Berufe* (Vienna: Brandstätter, 2014).

47. Peter Payer (ed.), *Sauberes Wien: Stadtreinigung und Abfallbeseitigung seit 1945* (Vienna: Holzhausen, 2006).

48. Muhammad Masood Rafi, Sarosh Hashmat Lodi, and Nadeem Manzoor Hasan, "Corruption in Public Infrastructure Service and Delivery: The Karachi Case Study," *Public Works Management & Policy* 17.4 (2012): 370–387; Werner Rügemer, *Colonia Corrupta: Globalisierung, Privatisierung und Korruption im Schatten des Kölner Klüngels* (Münster: Westfälisches Dampfboot, 2010).

49. Dost, *Einmal Dritter Pasewalk?*

50. Stefan Höhne, "Vereinzelungsanlagen: Die Genese des Drehkreuzes aus dem Geist automatischer Kontrolle," *Technikgeschichte* 83.2 (2016): 103–124.

51. Older people remember how much work it took just to get personal computers up and running—all the external components and hours spent wading through manuals full of jargon (whose authors presumed that readers were as up-to-date and tech-savvy as they were). See Joseph Corn, *User Unfriendly: Consumer Struggles with Personal Technologies, from Clocks and Sewing Machines to Cars and Computers* (Baltimore: Johns Hopkins University Press, 2011), 2.

52. Edward Tenner, *Why Things Bite Back: Technology and the Revenge of Unintended Consequences* (New York: Vintage, 1996).

53. Rüdiger Safranski, *Zeit: Was sie mit uns macht und was wir aus ihr machen* (Munich: Hanser, 2015).

54. Corn, *User Unfriendly*, 217.

55. Sabine Zelger, *Das Pferd frisst keinen Gurkensalat: Kulturgeschichte des Telefonierens* (Cologne: Böhlau, 1997); Margret Baumann and Helmut Gold (eds.), *Mensch Telefon: Aspekte telefonischer Kommunikation* (Heidelberg: Braus, 2000); Stefan Münkler and

Alexander Roesler (eds.), *Telefonbuch: Beiträge zu einer Kulturgeschichte des Telefons* (Frankfurt am Main: Suhrkamp, 2000).

56. Edward M. Hugh-Jones (ed.), *The Push-Button World: Automation Today* (Norman: University of Oklahoma Press, 1956); Heike Weber, "Stecken, drehen, drücken: Interfaces von Alltagstechniken und ihre Bediengesten," *Technikgeschichte* 76 (2009): 233–254.

57. Stefan Schmitt, "Anrufen wird bald das neue Rauchen," *Die Zeit*, December 10, 2015.

58. Sadie Plant, *On the Mobile: The Effects of Mobile Telephones on Social and Individual Life* (Schaumburg, IL: Motorola, 2001), 26; Rich Ling, *New Tech, New Ties: How Mobile Communication Is Reshaping Social Cohesion* (Cambridge, MA: MIT Press, 2008).

59. Corinna Peil, *Mobilkommunikation in Japan: Zur kulturellen Infrastruktur der Handy-Aneignung* (Bielefeld: transcript, 2011).

CONCLUSION

1. Jörg Schieb, "Kommunikation, Unterhaltung und Konsum künftig mit einem Gerät: Grossbaustelle Datenautobahn," *Frankfurter Allgemeine Zeitung*, January 9, 1995.

2. Edgar Wolfrum, *Welt im Zwiespalt: Eine andere Geschichte des 20. Jahrhunderts* (Stuttgart: Klett-Cotta, 2017), 348.

3. Holtorf, *Der erste Draht zur neuen Welt*, 278.

4. Andrew Blum, *Kabelsalat: Wie ich einem kaputten Kabel folgte und das Innere des Internets entdeckte* (Munich: Knaus, 2012), 15.

5. Neal Stephenson, "Mother Earth Mother Board," *Wired*, December 1, 1996, https://www.wired.com/1996/12/ffglass; see Nicole Starosielski, *The Undersea Network* (Durham, NC: Duke University Press, 2015).

6. Tim Wu, *The Master Switch: The Rise and Fall of Information Empires* (New York: Vintage, 2013).

7. Peter Thomas, "Hundert Jahre Einstellungssache," *Frankfurter Allgemeine Zeitung*, August 20, 2013.

8. Barbara Orland, "Milchpropaganda vor und nach dem Ersten Weltkrieg: Konvergenzen zwischen Wissenschaft, Wirtschaft und Ernährungsreform," in Manfred Rasch and Dietmar Bleidick (eds.), *Technikgeschichte im Ruhrgebiet—Technikgeschichte für das Ruhrgebiet* (Essen: Klartext, 2004), 909–933.

9. Kendra Smith-Howard, *Pure and Modern Milk: An Environmental History since 1900* (Oxford: Oxford University Press, 2013).

10. Christoph Bartmann, *Die Rückkehr der Diener: Das neue Bürgertum und sein Personal* (Munich: Hanser, 2016).

11. Hence the research project of the working group led by Monika Dommann at the University of Zurich, "Materialfluss: Warentransport, Güterdistribution und der Aufstieg der globalen Logistik, 1850–2000."

12. Patrick Illinger, "Koloss der Meere," *Süddeutsche Zeitung*, February 7/8, 2015.

13. Claas Tatje, "Die Pendlerrepublik: Millionen Deutsche macht der Weg zur Arbeit krank und einsam. Sie und ihre Arbeitgeber sollten etwas dagegen unternehmen," *Die Zeit*, May 22, 2014; Claas Tatje, *Fahrtenbuch des Wahnsinns: Unterwegs in der Pendlerrepublik* (Munich: Kösel, 2014).

14. Max Stadler, "Der User," in Alban Frei and Hannes Mangold (eds.), *Das Personal der Postmoderne: Inventur einer Epoche* (Bielefeld: transcript, 2015), 75–90.

15. Alexander Stirn, "Kehraus im Weltraum," *Süddeutsche Zeitung online*, June 4, 2017, https://www.sueddeutsche.de/wissen/weltraumschrott-kehraus-im-weltraum-1.3530728.

16. W. Boyd Rayward, *The Universe of Information: The Work of Paul Otlet* (Moskow: International Federation for Documentation, 1975); Martin Burckhardt, "Der heisse Krieger," *Frankfurter Allgemeine Zeitung*, June 20, 2016.

17. Julia Fleischhack, *Eine Welt im Datenrausch: Computeranlagen und Datenmengen als gesellschaftliche Herausforderung in der Bundesrepublik Deutschland (1965–1975)* (Zurich, Switzerland: Chronos, 2016), 78.

18. Monika Dommann, *Autoren und Apparate: Die Geschichte des Copyrights im Medienwandel* (Frankfurt am Main: Fischer, 2014).

19. Marina Weisband, *Wir nennen es Politik: Ideen für eine zeitgemässe Demokratie* (Stuttgart: Klett-Cotta, 2013).

20. Christopher Schrader, "Rettender Verzicht auf Neues," *Süddeutsche Zeitung online*, September 10, 2010, https://www.sueddeutsche.de/wissen/klimawandel-rettender-verzicht-auf-neues-1.998145.

21. Franz Mauelshagen, "'Anthropozän': Plädoyer für eine Klimageschichte des 19. und 20. Jahrhunderts," *Zeithistorische Forschungen / Studies in Contemporary History* 9 (2012): 131–137; Andreas Malm, *Fossil Capital: The Rise of Steam-Power and the Roots of Global Warming* (London: Verso, 2016).

22. Claus Leggewie, "Wenn unsere Infrastruktur unter die Räder kommt," *NDR-Kultur: Gedanken zur Zeit*, December 15, 2013, https://www.yumpu.com/de/document/view/24819951/wenn-unsere-infrastruktur-unter-dierader-kommt-von-ndr.

23. Parag Khannah, *Connectography: Mapping the Future of Global Civilization* (New York: Random House, 2016).

24. Elizabeth Drew, "A Country Breaking Down," *New York Review of Books*, February 25, 2016, https://www.nybooks.com/articles/2016/02/25/infrastructure-country-breaking-down.

25. Hans-Heinrich Bass, "Verkehrspolitik unter dem Druck der Strasse: Die Dortmunder Fahrpreisunruhen von 1971," *WerkstattGeschichte* 61 (2013): 49–64.

26. Leonie Feuerbach, "Die Toilette der Zukunft," *Frankfurter Allgemeine Zeitung*, November 4, 2014.

27. See the home page of the World Toilet Organization, https://www.worldtoilet.org.

28. Jürgen Osterhammel and Niels P. Petersson, *Geschichte der Globalisierung: Dimensionen, Prozesse, Epochen* (Munich: Beck, 2007); Olaf Bach, *Die Erfindung der Globalisierung: Entstehung und Wandel eines zeitgeschichtlichen Grundbegriffs* (Frankfurt am Main: Campus, 2013).

29. Saskia Sassen, *Global Networks, Linked Cities* (London: Routledge, 2002).

30. Günther Lachmann, *Von Not nach Elend: Eine Reise durch deutsche Landschaften und Geisterstädte von morgen* (Munich: Piper, 2008).

31. Peter Glaser, "Der blaue Planet," *Süddeutsche Zeitung*, January 30/31, 2016.

32. Adrian Lobe, "Sie setzen die Massstäbe der digitalen Welt," *Frankfurter Allgemeine Zeitung*, August 12, 2015.

33. Wissen and Naumann, *Raumdimensionen des Wandels technischer Infrastruktursysteme*.

34. Olivier Coutard and Jonathan Rutherford, "Beyond the Networked City: An Introduction," in Olivier Coutard and Jonathan Rutherford (eds.), *Beyond the Networked City: Infrastructure Reconfigurations and Urban Change in the North and South* (London: Routledge, 2016), 1–25.

35. Georg Seesslen, *Dietmar Schönherr: Von den phantastischen Abenteuern des Raumschiffes* Orion *zu* Wünsch Dir was. *Die Biografie* (Mannheim: Reichelt & Brockmann, 2012).

36. Harald Welzer, "Wasser aus der Leitung," *Frankfurter Allgemeine Zeitung*, August 6, 2013.

37. Roland Schaeffer, "Produktive Nachhaltigkeit: Infrastrukturinnovationen als politische Strategie," in Reinhard Loske and Roland Schaeffer (eds.), *Die Zukunft der Infrastrukturen: Intelligente Netzwerke für eine nachhaltige Entwicklung* (Marburg: Metropolis, 2005), 45–69.

38. Thomas Brock, "Archäologie an der Pipeline," *Süddeutsche Zeitung*, August 23, 2013. See, also, the materials presented in *Archäologie in Deutschland*, no. 6 (2015).

39. European Commission, "Ecosystem Services and Green Infrastructure," https://ec.europa.eu/environment/nature/ecosystems/index_en.htm; retrieved December 18, 2017.

40. Malcolm Wells, *Infra Structures: Life Support for the Nation's Circulatory Systems* (Brewster, MA: Underground Art Gallery, 1994).

41. Zygmunt Bauman, *Liquid Modernity* (Cambridge, UK: Polity, 2000).

42. Roger Willemsen, *Wer wir waren: Zukunftsrede* (Frankfurt am Main: Fischer, 2017), 35.

43. Max Scharnigg, "Mein halbes digitales Leben," *Süddeutsche Zeitung*, August 8/9, 2015. See, also, Peter Vorderer and Christoph Klimmt, "Das neue Normal," *Die Zeit*, January 28, 2016; Spiewak, "Bloss nicht offline sein."

44. Matthias Müller von Blumencron, "Die Emotionsmaschine," *Frankfurter Allgemeine Zeitung*, February 5, 2016.

45. Florian Gässgen, "Mr WLAN," *Der Stern* 2 (2016): 112.

46. Welzer, *Mentale Infrastrukturen*.

47. Ernst Forsthoff, *Die Daseinsvorsorge und die Kommunen: Vortrag anlässlich der Jahrestagung des Verbands Kommunaler Unternehmen e. V. am 16. 12. 1957 in Köln* (Cologne: Sigillum, 1958), 11–12. See, also, Claudia Neu (ed.), *Daseinsvorsorge: Eine gesellschaftswissenschaftliche Annäherung* (Wiesbaden: Verlag für Sozialwissenschaften, 2009).

48. Frank and Gandy, *Hydropolis*, 32.

49. World Wildlife Fund, "Earth Hour," March 26, 2022, https://www.worldwildlife.org/pages/earth-hour.

50. Headrick, *The Tentacles of Progress*.

51. See, for example, the exemplary essay by Amina Nolte, "Political Infrastructure and the Politics of Infrastructure: The Jerusalem Light Rail," *City* 20.3 (2016): 441–454.

52. Holger Dalkmann (ed.) *Wege von der nachholenden zur nachhaltigen Entwicklung: Infrastrukturen und deren Transfer im Zeitalter der Globalisierung* (Wuppertal: Wuppertal Institut für Klima, Umwelt, Energie, 2004).

BIBLIOGRAPHY

Abelshauser, Werner. "Staat, Infrastruktur und regionaler Wohlstandsausgleich im Preussen der Hochindustrialisierung." In *Staatliche Umverteilungspolitik in historischer Perspektive*, edited by Fritz Blaich, 9–58. Berlin: Duncker & Humblot, 1980.

Abram, Simone, Brit Ross Winthereik, and Thomas Yarrow, eds. *Electrifying Anthropology: Exploring Electrical Practices and Infrastructures*. London: Routledge, 2020.

ADAC Motorwelt, no. 9 (2012).

Agar, Jon. *The Government Machine: A Revolutionary History of the Computer*. Cambridge, MA: MIT Press, 2003.

Ahrne, Goran, and Nils Brunsson. *Meta-Organizations*. Cheltenham, UK: Edward Elgar, 2008.

Ahuja, Ravi. *Pathways of Empire: Circulation, "Public Works" and Social Space in Colonial Orissa (c. 1780–1914)*. Hyderabad, India: Orient BlackSwan, 2009.

Akurang-Parry, Kwabena. "Anticolonial Movements, Africa." In *Encyclopedia of Western Colonialism since 1450*, vol. 1, edited by Thomas Benjamin, 74–81. Detroit, MI: Macmillan Reference, 2007.

Āl-Ahmad, Jalāl. *Occidentosis: A Plague from the West*. Translated by Robert Campbell. Berkeley, CA: Mizan Press, 1984.

Alatas, Syed Hussein. *The Myth of the Lazy Native: A Study of the Image of the Malays, Filipinos and Javanese from the 16th to the 20th Century and Its Function in the Ideology of Colonial Capitalism*. London: Routledge, 1977.

Alberts, Gerard, and Ruth Oldenziel, eds. *Hacking Europe: From Computer Cultures to Demoscenes*. London: Springer, 2014.

Alcock, Susan E., John Bodel, and Richard J. A. Talbert, eds. *Highways, Byways, and Road Systems in the Pre-Modern World*. Chichester, UK: Wiley-Blackwell, 2012.

Allemeyer, Marie L. *"Kein Land ohne Deich . . . !" Lebenswelten einer Küstengesellschaft in der Frühen Neuzeit*. Göttingen: Vandenhoeck & Ruprecht, 2006.

"Alligator Found in Uptown Sewer." *New York Times*, February 10, 1935.

Ambrosius, Gerold. *Der Staat als Unternehmer: Öffentliche Wirtschaft und Kapitalismus seit dem 19. Jahrhundert*. Göttingen: Vandenhoeck & Ruprecht, 1984.

Ambrosius, Gerold. *Hybride Eigentums- und Verfügungsrechte: Öffentlich-private Kooperationen in systematisch-theoretischer und historisch-empirischer Perspektive*. Berlin: BWV, 2012.

Ambrosius, Gerold, and Christian Henrich-Franke. *Integration von Infrastrukturen in Europa im historischen Vergleich*. Baden-Baden: Nomos, 2013.

Amin, Ash. "Lively Infrastructure," *Theory, Culture & Society* 31.7–8 (2014): 137–161.

Anand, Nikhil. "Consuming Citizenship: Prepaid Meters and the Politics of Technology in Mumbai," *Occasional Papers of the School of Social Science* 53 (2014): 47–70.

Anastasiadou, Irene. *Constructing Iron Europe: Transnationalism and Railways in the Interbellum*. Amsterdam: Amsterdam University Press, 2011.

Andersen, Arne. *Der Traum vom guten Leben: Alltags- und Konsumgeschichte vom Wirtschaftswunder bis heute*. Frankfurt am Main: Campus, 1997.

Archäologie in Deutschland, no. 6 (2015).

Armstrong, Ellis L., Michael C. Robinson, and Suellen M. Hoy. *History of Public Works in the United States, 1776–1976*. Chicago: American Public Works Association, 1976.

Arnold, David. *Everyday Technology: Machines and the Making of India's Modernity*. Chicago: University of Chicago Press, 2013.

Arnold, Dietmar, Ingmar Arnold, and Frieder Salm. *Dunkle Welten: Bunker, Tunnel und Gewölbe unter Berlin*. Berlin: Links, 2013.

Arnold, Ingmar. *Luft-Züge. Die Geschichte der Rohrpost*. Berlin: Links, 2016.

Ascher, Kate. *The Works: Anatomy of a City*. London: Penguin, 2007.

Aselmeyer, Norman. "Capturing the Zeitgeist of Infrastructure." *TRAFO—Blog for Transregional Research*, January 31, 2019. https://trafo.hypotheses.org/17723.

Asendorf, Christoph. *Batterien der Lebenskraft: Zur Geschichte der Dinge und ihrer Wahrnehmung im 19. Jahrhundert*. Weimar: Verlag und Datenbank für Geisteswissenschaften, 2002.

Asendorf, Christoph. *Ströme und Strahlen: Das langsame Verschwinden der Materie um 1900*. Giessen: Anabas, 1989.

Ausubel, Jesse H., and Robert Herman, eds. *Cities and Their Vital Systems: Infrastructure—Past, Present, and Future*. Washington, DC: National Academy Press, 1988.

Bach, Olaf. *Die Erfindung der Globalisierung: Entstehung und Wandel eines zeitgeschichtlichen Grundbegriffs*. Frankfurt am Main: Campus, 2013.

Badenoch, Alexander, and Andreas Fickers, eds. *Materializing Europe: Transnational Infrastructures and the Project of Europe*. Basingstoke, UK: Palgrave Macmillan, 2010.

Bader-Gassner, Miriam A. *Pipelineboom: Internationale Ölkonzerne im westdeutschen Wirtschaftswunder*. Baden-Baden: Nomos, 2014.

Badger, Emily, and Darla Cameron. "How Railroads, Highways and Other Manmade Lines Racially Divide America's Cities." *Washington Post*, July 16, 2015.

Bajohr, Frank. *"Unser Hotel ist judenfrei." Bäder-Antisemitismus im 19. und 20. Jahrhundert*. Frankfurt am Main: Fischer, 2003.

Baković, Nikola. "'No One Here Is Afraid of Blisters or Work!' Social Integration, Mobilization and Cooperation in Yugoslav Youth Brigades: The Example of Čačak Region Brigades (1946–1952)." *Hungarian Historical Review* 4.1 (2015): 29–55.

Balabkins, Nicholas W. "Barriers to Technological Transfer: Infrastructural Difficulties in Nigeria." In *Infrastructure and the Space-Economy: Essays in Honor of Rolf Funck*, edited by Karin Peschel, 101–115. Berlin: Springer, 1990.

Balbi, Gabriele, and Andreas Fickers, eds. *History of the International Telecommunication Union (ITU): Transnational Techno-Diplomacy from the Telegraph to the Internet*. Berlin: De Gruyter Oldenbourg, 2020.

Barlösius, Eva. "Gleichwertig ist nicht gleich." *Politik und Zeitgeschichte* 37 (2006): 16–23.

Barlösius, Eva. "Infrastrukturen als gesellschaftliche Weichensteller." *SozBlog: Blog der Deutschen Gesellschaft für Soziologie (DGS)*, February 25, 2015.

Barlösius, Eva, and Michèle Spohr. "Rückzug 'vom Lande': Die sozialräumliche Neuordnung durch Infrastrukturen." In *Urbane Ungleichheiten: Neue Entwicklungen zwischen Zentrum und Peripherie*, edited by Peter A. Berger et al., 233–251. Wiesbaden: Springer, 2014.

Barnaby, Alice. *Light Touches: Cultural Practices of Illumination, 1780–1900*. London: Routledge, 2016.

Barnes, David S. *The Great Stink of Paris and the Nineteenth-Century Struggle against Filth and Germs*. Baltimore: Johns Hopkins University Press, 2006.

Bartetzko, Dieter. "Unten bleiben!" *Frankfurter Allgemeine Sonntagszeitung*, February 10, 2013.

Bartmann, Christoph. *Die Rückkehr der Diener: Das neue Bürgertum und sein Personal*. Munich: Hanser, 2016.

Bärtschi, Hans-Peter. "Durchmessene Räume—durchmessene Zeiten: Die Eisenbahn als Landschaftsgestalterin." In *Vermessene Landschaften: Kulturgeschichte und technische Praxis im 19. und 20. Jahrhundert*, edited by David Gugerli, 79–87. Zurich, Switzerland: Chronos, 1999.

Bass, Hans-Heinrich. "Verkehrspolitik unter dem Druck der Strasse: Die Dortmunder Fahrpreisunruhen von 1971." *WerkstattGeschichte* 61 (2013): 49–64.

Bassu, Gabriella. "Schwarzfahren kostet jetzt 60 Euro." *Frankfurter Allgemeine Zeitung*, July 1, 2015.

Bauer, Susanne. "Lenin spannt aus: Alexander Sinowjews 'Homo sovieticus'—eine mehrfache Provokation." *Zeithistorische Forschungen* 11 (2014): 485–492.

Bauman, Zygmunt. *Liquid Modernity*. Cambridge, UK: Polity, 2000.

Bauman, Zygmunt. *Modernity and the Holocaust*. Ithaca, NY: Cornell University Press, 2002.

Baumann, Margret, and Helmut Gold, eds. *Mensch Telefon: Aspekte telefonischer Kommunikation*. Heidelberg: Braus, 2000.

Baumeister. Zeitschrift für Architektur, no. 3 (1998).

Bazerman, Charles. *The Language of Edison's Light*. Cambridge, MA: MIT Press, 1999.

Beale, Joseph Henry. *Cases on Public Service Companies: Public Carriers, Public Works, and Other Public Utilities*. Cambridge, MA: Harvard University Press, 1902.

Bebel, August. *Woman and Socialism*. Translated by Meta L. Stern. New York: Socialist Literature, 1910.

Beck, Robert. *Der Plan Freycinet und die Provinzen: Aspekte der infrastrukturellen Entwicklung der französischen Provinzen durch die Dritte Republik*. Frankfurt am Main: Lang, 1986.

Bel i Queralt, Germà. *Infrastructure and the Political Economy of Nation Building in Spain: 1720–2010*. Portland, UK: Sussex Academic Press, 2012.

Bellmann, Dagmar. *Von Höllengefährten zu schwimmenden Palästen: Die Passagierschifffahrt auf dem Atlantik (1840–1930)*. Frankfurt am Main: Campus, 2015.

Beniger, James R. *The Control Revolution: Technological and Economic Origins of the Information Society*. Cambridge, MA: Harvard University Press, 1986.

Bentner, Ariane. "Fliegende Mädchen—Muster an Takt und Charme: Aus den Anfängen eines weiblichen Dienstleistungsberufes." *Feministische Studien* 8.2 (1990): 105–115.

Benz, Wolfgang. *Geschichte des Dritten Reiches*. Bonn: Bundeszentrale für Politische Bildung, 2008.

Berger Ziauddin, Silvia. "Überlebensinsel und Bordell: Zur Ambivalenz des Bunkers im atomaren Zeitalter." In *Das Imaginäre des Kalten Krieges: Beiträge zu einer*

Kulturgeschichte des Ost-West-Konfliktes in Europa, edited by David Eugster and Sibylle Marti, 69–93. Essen: Klartext, 2015.

Berghoff, Hartmut, and Jakob Vogel, eds. *Wirtschaftsgeschichte als Kulturgeschichte: Dimensionen eines Perspektivenwechsels*. Frankfurt am Main: Campus, 2004.

Bernard, Andreas. *Die Geschichte des Fahrstuhls: Über einen beweglichen Ort der Moderne*. Frankfurt am Main: Fischer, 2006.

Bernhardt, Christoph. *Im Spiegel des Wassers: Eine transnationale Geschichte des Oberrheins (1800–2000)*. Cologne: Böhlau, 2016.

Berrey, Stephen A. *The Jim Crow Routine: Everyday Performances of Race, Civil Rights, and Segregation in Mississippi*. Chapel Hill: University of North Carolina Press, 2015.

Bess, Michael K. "Routes of Conflict: Building Roads and Shaping the Nation in Mexico, 1941–1952." *Journal of Transport History* 35.1 (2014): 78–96.

Bickenbach, Matthias, and Michael Stolzke. *Die Geschwindigkeitsfabrik: Eine fragmentarische Kulturgeschichte des Autounfalls*. Berlin: Kadmos, 2014.

Bijsterveld, Karin, ed. *Soundscapes of the Urban Past: Staged Sound as Mediated Cultural Heritage*. Bielefeld: transcript, 2013.

Bijsterveld, Karin, Eefje Cleophas, Stefan Krebs, and Gijs Mom, eds. *Sound and Safe: A History of Listening Behind the Wheel*. Oxford: Oxford University Press, 2014.

Binder, Beate. *Elektrifizierung als Vision: Zur Symbolgeschichte einer Technik im Alltag*. Tübingen: Tubinger Vereinigung fur Volkskunde, 1999.

Birkenhof, Tobias. *Die Auswirkungen von Großveranstaltungen auf die langfristige ökonomische Entwicklung von Schwellenländern: Eine wachstumstheoretische und empirische Untersuchung am Beispiel der Olympischen Sommerspiele in Beijing*. Berlin: Logos, 2009.

Bjorkman, Lisa. *Pipe Politics, Contested Waters: Embedded Infrastructures of Millennial Mumbai*. Durham, NC: Duke University Press, 2015.

Black, Ernest W. *Cursus Publicus: The Infrastructure of Government in Roman Britain*. Oxford: Tempus Reparatum, 1995.

Blackbourn, David. *The Conquest of Nature: Water, Landscape and the Making of Modern Germany*. London: Pimlico, 2007.

Blass, Björn. "Räume der Entsorgung: Werte, Praktiken und Topographien von Müll (1880–1930)." PhD diss., Freie Universität Berlin, 2020.

Bleakley, Fred A. "Rural County Balks at Joining Global Village." *Wall Street Journal*, January 4, 1996.

Blum, Andrew. *Kabelsalat: Wie ich einem kaputten Kabel folgte und das Innere des Internets entdeckte*. Munich: Knaus, 2012.

Blümel, Wolfgang. "Empirische Erfassung der Präferenzen für öffentliche Güter: Die Debatte über den 'Freifahrer.'" *Zeitschrift für Wirtschaftspolitik* 37 (1988): 65–71.

Blumenberg, Hans. "Dogmatische und rationale Analyse von Motivationen technischen Fortschritts." *Zeitschrift für Kulturphilosophie* 7.2 (2013): 407–422.

Blumenberg, Hans. *Shipwreck with Spectator: Paradigm of a Metaphor for Existence.* Translated by Steven Rendall. Cambridge, MA: MIT Press, 1997.

Bobrick, Benson. *Labyrinths of Iron: Subways in History, Myth, Art, Technology, and War.* New York: Henry Holt, 1994.

Bollerey, Franziska, and Axel Föhl. "Unter dem Pflaster der Strasse: Facetten subterraner Netze." In *Technikgeschichte im Ruhrgebiet*, edited by Manfred Rasch and Dietmar Bleidick, 438–473. Essen: Klartext, 2004.

Böndel, Dirk, Anna Döpfner, and Alfred Gottwaldt, eds. *Ich diente nur der Technik: Sieben Karrieren zwischen 1940 und 1950.* Berlin: Nicolai, 1995.

Borscheid, Peter. *Das Tempo-Virus. Eine Kulturgeschichte der Beschleunigung.* Frankfurt am Main: Campus, 2004.

Bowker, Geoffrey C., and Susan Leigh Star. *Sorting Things Out: Classification and Its Consequences.* Cambridge, MA: MIT Press, 1999.

Brandt, Nat. *Chicago Death Trap: The Iroquois Theatre Fire of 1903*, Carbondale: Southern Illinois University Press, 2006.

Brantz, Dorothee. "Slaughter in the City: The Establishment of Public Abattoirs in Paris and Berlin, 1780–1914." PhD diss., University of Chicago, 2003.

Brendel, Benjamin. "Konvergente Konstruktionen: Eine Kulturgeschichte des Staudammbaus in den USA, Spanien und Ägypten 1933–1971." PhD diss., Giessen, 2017.

Briegleb, Till. "Vor allem obskur: Was der Begriff Infrastruktur heute alles bedeutet." *Süddeutsche Zeitung*, May 27, 2013.

Brock, Thomas. "Archäologie an der Pipeline." *Süddeutsche Zeitung*, August 23, 2013.

Bröckling, Ulrich. *Gute Hirten führen sanft. Über Menschenregierungskünste.* Berlin: Suhrkamp, 2017.

Brüggemeier, Franz-Josef. *Das unendliche Meer der Lüfte: Luftverschmutzung, Industrialisierung und Risikobewusstsein im 19. Jahrhundert.* Essen: Klartext, 1996.

Bublitz, Hannelore, Irina Kaldrack, Theo Röhle, and Mirna Zeman, eds. *Automatismen—Selbst-Technologien.* Munich: Fink, 2013.

Buggeln, Marc, and Inge Marszolek, eds. *Bunker: Kriegsort, Zuflucht, Erinnerungsraum.* Frankfurt am Main: Campus, 2008.

Büker, Dieter. *Mensch—Kultur—Abwasser. Von der Annehmlichkeit für wenige zur Existenzfrage der Gesellschaft: Der Umgang des Menschen mit Abwässern. Ein kulturhistorischer Längsschnitt von den Anfängen bis zum Beginn des 20. Jahrhunderts.* Essen: Die blaue Eule, 2000.

Burckhardt, Martin. "Der heisse Krieger." *Frankfurter Allgemeine Zeitung*, June 20, 2016.

Burger, Reiner. "In der Mitte entsteht ein Fluss." *Frankfurter Allgemeine Zeitung*, August 12, 2015.

Buruma, Ian. "The Worst Railroad Job." *New York Review of Books* 61.18 (2014): 19–21.

Busch, Lawrence. *Standards: Recipes for Reality*. Cambridge, MA: MIT Press, 2011.

Büschel, Hubertus. *Hilfe zur Selbsthilfe: Deutsche Entwicklungsarbeit in Afrika 1960–1975*. Frankfurt am Main: Campus, 2014.

Büsing, Friedrich Wilhelm. *Die Städtereinigung*. Stuttgart: Bergstraesser, 1897.

Butler, Richard W. "The Concept of a Tourist Area Cycle of Evolution: Implications for Management of Resources." *Canadian Geographer* 24.1 (1980): 5–12.

Caro, Robert A. *The Power Broker: Robert Moses and the Fall of New York*. New York: Knopf, 1974.

Carse, Ashley. *Beyond the Big Ditch: Politics, Ecology, and Infrastructure at the Panama Canal*. Cambridge, MA: MIT Press, 2014.

Caruso, Clelia. "Zur Einübung in die Kulturtechnik des Telefonierens." PhD diss., Universität des Saarlandes, 2010.

Castells, Manuel. "Space of Flows—Raum der Ströme. Eine Theorie des Raums in der Informationsgesellschaft." In *Stadt-Welt: Über die Globalisierung städtischer Milieus*, edited by Peter Noller, Walter Prigge, and Klaus Ronneberger, 120–133. Frankfurt am Main: Campus, 1994.

Chattopadhyay, Swati. *Unlearning the City: Infrastructure in a New Optical Field*. Minneapolis: University of Minnesota Press, 2012.

Choate, Pat, and Susan Walter. *America in Ruins: Beyond the Public Works Pork Barrel*. Washington, DC: Council of State Planning Agencies, 1981.

Chrimes, Mike, ed. *The Civil Engineering of Canals and Railways before 1850*. Aldershot, UK: Ashgate, 1997.

Cioc, Marc. *The Rhine: An Eco-Biography, 1815–2000*. Seattle: University of Washington Press, 2002.

Civilization Forum, July 26/27, 2002. https://www.civforum.de/showthread.php?6208-Zerst%F6rung-der-Infrastruktur!!!.

Clendinning, Anne. *Demons of Domesticity: Women and the English Gas Industry, 1889–1939*. Aldershot, UK: Ashgate, 2004.

Cless, Karlheinz. *Menschen am Brunnen: Ethnologische Perspektiven zum Umgang mit Wasser*. Bielefeld: transcript, 2014.

Corbin, Alain. *The Foul and the Fragrant: Odor and the Social Imagination*. Translated by Miriam L. Kochan. Cambridge, MA: Harvard University Press, 1986.

Corbin, Alain. *Les Cloches de la terre: paysage sonore et culture sensible dans les campagnes au XIXe siècle*. Paris: Flammarion, 2013.

Corn, Joseph. *User Unfriendly: Consumer Struggles with Personal Technologies, from Clocks and Sewing Machines to Cars and Computers*. Baltimore: Johns Hopkins University Press, 2011.

Cottrell, Frederick W. *Energy and Society: The Relationship between Energy, Social Change, and Economic Development*. New York: McGraw-Hill, 1955.

Coutard, Olivier, and Jonathan Rutherford. "Beyond the Networked City: An Introduction." In *Beyond the Networked City: Infrastructure Reconfigurations and Urban Change in the North and South*, edited by Olivier Coutard and Jonathan Rutherford, 1–25. London: Routledge, 2016.

Cowan, Ruth Schwartz. *More Work for Mother: The Ironies of Household Technology from the Open Hearth to the Microwave*. New York: Basic Books, 1983.

Coward, Martin. "Network-Centric Violence: Critical Infrastructure and the Urbanization of Security." *Security Dialogue* 40.4–5 (2009): 399–418.

Creveld, Martin van. *Supplying War: Logistics from Wallenstein to Patton*. Cambridge: Cambridge University Press, 2007.

"CST: Nun heulen sie wieder," *Die Zeit*, November 27, 2014.

Cvetkovski, Roland. *Modernisierung durch Beschleunigung. Raum und Mobilität im Zarenreich*. Frankfurt am Main: Campus, 2006.

Cwiertnia, Laura. "Wer lebt im Dunkeln?" *Die Zeit*, November 30, 2017.

Cyriax, Oliver. "Planning Officers." In *The Penguin Encyclopedia of Crime*. London: Penguin, 1996.

Dalakoglou, Dimitris. "The Road: An Ethnography of the Albanian-Greek Cross-border Motorway." *American Ethnologist* 37.1 (2010): 132–149.

Dalakoglou, Dimitris. "The Road from Capitalism to Capitalism: Infrastructures of (Post)Socialism in Albania." *Mobilities* 7.4 (2012): 571–586.

Daley, Robert. *The World Beneath the City*. Philadelphia: Lippincott, 1959.

Dalkmann, Holger et al., eds. *Wege von der nachholenden zur nachhaltigen Entwicklung: Infrastrukturen und deren Transfer im Zeitalter der Globalisierung*. Wuppertal: Wuppertal Institut für Klima, Umwelt, Energie, 2004.

Daniel, Ute. "Der Krimkrieg 1853–1856 und die Entstehungskontexte medialer Kriegsberichterstattung." In *Augenzeugen: Kriegsberichterstattung vom 18. zum 21. Jahrhundert*, edited by Ute Daniel, 40–67. Göttingen: Vandenhoeck & Ruprecht, 2006.

"Das mobile Chaos." *Der Spiegel*, July 13, 1998.

Dath, Dietmar. "Das schreibende Flussgespenst." *Frankfurter Allgemeine Zeitung*, March 3, 2005.

Davies, Archie. "The Coloniality of Infrastructure: Engineering, Landscape and Modernity in Recife." *Environment and Planning D: Society and Space* 39.4 (2021): 740–757.

Davis, Clarence B., and Kenneth E. Wilburn, eds. *Railway Imperialism*. New York: Greenwood, 1991.

Davis, Timothy. *National Park Roads: A Legacy in the American Landscape*. Charlottesville: University of Virginia Press, 2016.

Davison, Graeme. *Car Wars: How the Car Won Our Hearts and Conquered Our Cities*. Crows Nest: Allen & Unwin, 2004.

Decker, Hanna. "Grossprojekte werden zu optimistisch geplant." *Frankfurter Allgemeine Zeitung*, October 8, 2015.

Deckers, Daniel. "Unter den Wolken." *Frankfurter Allgemeine Zeitung*, April 5, 2013.

Dehn, Paul. *Weltwirtschaftliche Neubildungen*. Berlin: Allgemeiner Verein für Deutsche Literatur, 1904.

Demangeon, Albert, and Lucien Febvre. *Le Rhin: Problèmes d'histoire et d'économie*. Paris: Colin, 1935.

Denton, Sally. *The Profiteers: Bechtel and the Men Who Built the World*. New York: Simon & Schuster, 2016.

Depenbrock, Johannes. *Die Stellung der Kommunen in der Versorgungswirtschaft: Ein Beitrag zum Wirtschaftsverfassungsrecht*. Munich: Beck, 1981.

Devanthéry, Ariane. "Schweizerreisen." *Historisches Lexikon der Schweiz*, May 1, 2015. https://hls-dhs-dss.ch/de/articles/024575/2015-01-05/.

Diacon, Todd. *Stringing Together a Nation: Cândido Mariano da Silva Rondon and the Construction of a Modern Brazil 1906–1930*. Durham, NC: Duke University Press, 2004.

Dickens, Charles. *Hard Times*. New York: Penguin, 1995.

"Die Bundesrepublik—ein unterentwickeltes Land." *Der Spiegel*, August 30, 1961.

"Die Charlottenburger Drei." *TrenntMagazin*. https://trenntmagazin.de/recycling-die-charlottenburger-drei/#.VAR4Kih4bII; retrieved December 17, 2017.

Diefendorf, Jeffry M. *In the Wake of War: The Reconstruction of German Cities after World War II*. Oxford: Oxford University Press, 1993.

Dienel, Hans-Liudger, and Barbara Schmucki, eds. *Mobilität für alle: Die Geschichte des öffentlichen Personennahverkehrs in der Stadt zwischen technischem Fortschritt und sozialer Pflicht*. Stuttgart: Steiner, 1997.

Diesel, Eugen. *Völkerschicksal und Technik*. Stuttgart: Cotta, 1930.

Dietze, Carola. *Die Erfindung des Terrorismus in Europa, Russland und den USA 1858–1866*. Hamburg: Hamburger, 2016.

Dijk, Han van, and Roche Mongbo. "Report of the Workshop 'Urban Public Services.'" *Bulletin de l'APAD*, 23–24 (2002). https://journals.openedition.org/apad/155.

Dijstelbloem, Huub. *Borders as Infrastructure: The Technopolitics of Border Control.* Cambridge, MA: MIT Press, 2021.

Diogo, Maria Paula, and Dirk van Laak. *Europeans Globalizing: Mapping, Exploiting, Exchanging.* New York: Palgrave Macmillan, 2016.

Disco, Nil, and Eda Kranakis, eds. *Cosmopolitan Commons: Sharing Resources and Risks across Borders.* Cambridge, MA: MIT Press, 2013.

Dittmann, Frank. "Technik versus Konflikt: Wie Datennetze den Eisernen Vorhang durchdrangen." *Osteuropa* 59.10 (2009): 101–119.

Dix, Artur. *Deutschland auf den Hochstrassen des Weltwirtschaftsverkehrs.* Jena: Fischer, 1901.

Dix, Artur. "Wirtschaftsstruktur und Geopolitik." *Volkswirtschaftliche Blätter* 26.7–9 (1927): 465–484.

Doll, Nikolaus et al. "Das unfassbare deutsche Infrastruktur-Desaster." *Die Welt*, May 12, 2013.

Dombrowsky, Wolf R. et al. "4 Menschen und Katastrophen." In *BBK: Notfall- und Katastrophenpharmazie I: Bevölkerungsschutz und Medizinische Notfallversorgung*, 249–288. Bonn: Bundesamt für Bevölkerungsschutz und Katastrophenhilfe, 2009.

Dombrowsky, Wolf R. *Schutz kritischer Infrastrukturen als Grundproblem einer modernen Gesellschaft* (unpublished manuscript).

Domke, Christine. *Die Betextung des öffentlichen Raumes: Eine Studie zur Spezifik von Meso-Kommunikation am Beispiel von Bahnhöfen, Innenstädten und Flughäfen.* Heidelberg: Winter, 2014.

Dommann, Monika. "Alles fliesst: Soll die Geschichte nomadischer werden?" *Geschichte und Gesellschaft* 42 (2016): 516–534.

Dommann, Monika. *Autoren und Apparate: Die Geschichte des Copyrights im Medienwandel.* Frankfurt am Main: Fischer, 2014.

Dossmann, Axel. *Begrenzte Mobilität: Eine Kulturgeschichte der Autobahnen in der DDR.* Essen: Klartext, 2003.

Dossmann, Axel. "Erbschaft unterm Hakenkreuz." *Die Tageszeitung* (*taz*), December 19, 2006.

Dossmann, Axel. "Wie wir die Autobahnen lieben lernten." *Sowi: Sozialwissenschaftliche Informationen* 4 (1996): 235–242.

Dossmann, Axel, Jan Wenzel, and Kai Wenzel. *Architektur auf Zeit: Baracken, Pavillons, Container.* Berlin: b-books, 2006.

Dost, Paul. *Einmal Dritter Pasewalk? Wer reist in welcher Wagenklasse? Psychologische Soziologie der Wagenklassen*. Krefeld: Röhr, 1977.

Douglas, Susan J. *Listening In: Radio and the American Imagination*. Minneapolis: University of Minnesota Press, 2004.

Drew, Elizabeth. "A Country Breaking Down." *New York Review of Books*, February 25, 2016. https://www.nybooks.com/articles/2016/02/25/infrastructure-country-breaking-down.

Duranton, Gilles, and Matthew A. Turner. "The Fundamental Law of Road Congestion: Evidence from US Cities." *American Economic Review* 101 (2011): 2616–2652.

Durth, Werner, and Niels Gutschow. *Träume in Trümmern: Planungen zum Wiederaufbau zerstörter Städte im Westen Deutschlands 1940–1950*. Wiesbaden: Vieweg, 1988.

Dworog, Sabine. "Der Wandel von Staatsaufgaben: Die Konflikte um den Ausbau des Frankfurter Flughafens im Kontext gesellschaftlicher Umbrüche (1960er bis 1980er)." PhD diss., Giessen, 2007.

Dworog, Sabine. "Luftverkehrsinfrastruktur: Zur Rolle des Staates bei der Integration eines Flughafens in seine Umgebung." *Saeculum* 58.1 (2007): 115–149.

Easterling, Keller. *Extrastatecraft: The Power of Infrastructure Space*. London: Verso, 2014.

Eckermann, Johann Peter. *Conversations with Goethe: In the Last Years of His Life*. Translated by Ritchie Robertson. New York: Penguin, 2022.

Eckert, Andreas. "Exportschlager Wohlfahrtsstaat? Europäische Sozialstaatlichkeit und Kolonialismus in Afrika nach dem Zweiten Weltkrieg." *Geschichte und Gesellschaft* 32.4 (2006): 467–488.

Eckert, Astrid M. "Geteilt, aber nicht unverbunden: Grenzgewässer als deutsch-deutsches Umweltproblem." *Vierteljahrshefte für Zeitgeschichte* 62.1 (2014): 69–99.

Edgerton, David. "Creole Technologies and Global Histories: Rethinking How Things Travel in Space and Time," *Journal of History of Science and Technology* 1.1 (2007): 3–31.

Edgerton, David. *The Shock of the Old: Technology and Global History since 1900*. London: Profile, 2019.

Edwards, Paul N. "Infrastructure and Modernity: Scales of Force, Time, and Social Organization in the History of Sociotechnical Systems." In *Modernity and Technology*, edited by Thomas J. Misa, Philip Brey, and Andrew Feenberg, 185–225. Cambridge, MA: MIT Press, 2002.

Edwards, Paul N. "Meteorology as Infrastructural Globalism." *Osiris* 21 (2006): 229–250.

Edwards, Paul N., Geoffrey C. Bowker, Steven J. Jackson, and Robin Williams. "Introduction: An Agenda for Infrastructure Studies." *Journal of the Association for Information Systems* 10.5 (2009): 364–374.

"E5: Terror von Blech und Blut." *Der Spiegel*, August 25, 1975.

Elhacham, Emily, Liad Ben-Uri, Jonathan Grozovski, Yinon M. Bar-On, and Ron Milo. "Global Human-Made Mass Exceeds All Living Biomass." *Nature* 588 (2020): 442–444. https://doi.org/10.1038/s41586-020-3010-5.

Elsberg, Marc. *Blackout: Morgen ist es zu spät*. Munich: Blanvalet, 2012.

Elyachar, Julia. "Phatic Labor, Infrastructure, and the Question of Empowerment in Cairo." *American Ethnologist* 37.3 (2010): 452–464.

Emmerson, Andrew. *Discovering Subterranean London*. Oxford, UK: Shire, 2009.

Engartner, Tim. "11 000 Jahre Verspätung." *Die Zeit*, March 11, 2009.

Engell, Lorenz, Bernhard Siegert, and Joseph Vogl, eds. *Gefahrensinn*. Munich: Fink, 2009.

Engels, Jens Ivo. "Kanalregionen im Frankreich der Sattelzeit: Elemente für die Erforschung der Raumwirkungen von Infrastrukturen." *Francia* 37 (2010): 149–165.

Engels, Jens Ivo. "Machtfragen: Aktuelle Entwicklungen und Perspektiven der Infrastrukturgeschichte." *Neue Politische Literatur* 55.1 (2010): 51–70.

Engels, Jens Ivo, and Alfred Nordmann, eds. *Was heisst Kritikalität? Zu einem Schlüsselbegriff der Debatte um Kritische Infrastrukturen*. Bielefeld: transcript, 2018.

Engels, Jens Ivo, and Gerrit Jasper Schenk. "Macht der Infrastrukturen—Infrastrukturen der Macht: Überlegungen zu einem Forschungsfeld." In *Wasserinfrastrukturen und Macht: Politisch-soziale Dimensionen technischer Systeme von der Antike bis zur Gegenwart*, edited by Birte Förster and Martin Bauch, 22–58. Munich: Oldenbourg, 2014.

Ensmenger, Nathan. "The Environmental History of Computing." *Technology and Culture* 59 (4S) (2018): S7–S33. https://doi.org/10.1353/tech.2018.0148.

Erath, Marianne. "Hallo, ist da das Fräulein vom Amt?" *Emma* July/August 2009. https://www.emma.de/artikel/frauenberuf-hallo-ist-da-das-fraeulein-vom-amt-264018.

Ernstberger, Nico. *Die Versorgung Roms mit Wasser, Waren und Energie: Die Infrastruktur der Kaiserzeit*. Hamburg: Disserta, 2014.

Essbach, Wolfgang. "Zur Anthropologie artifizieller Umwelt." In *Ökohistorische Reflexionen: Mensch und Umwelt zwischen Steinzeit und Silicon Valley*, edited by Kurt W. Alt and Natascha Rauschenberger, 171–195. Freiburg im Breisgau: Rombach, 2001.

Etzemüller, Thomas. "Romantischer Rhein—eiserner Rhein: Ein Fluss als *imaginary landscape* der Moderne." *Historische Zeitschrift* 295.2 (2012): 390–424.

European Commission. "Ecosystem Services and Green Infrastructure." https://ec.europa.eu/environment/nature/ecosystems/index_en.htm; retrieved December 18, 2017.

Evans, Sandra. *Sowjetisch wohnen: Eine Literatur- und Kulturgeschichte der Kommunalka*. Bielefeld: transcript, 2011.

Ewald, François. *The Birth of Solidarity: The History of the French Welfare State*. Translated by Melinda Cooper. Durham, NC: Duke University Press, 2020.

Fack, Dietmar. *Automobil, Verkehr und Erziehung: Motorisierung und Sozialisation zwischen Beschleunigung und Anpassung 1885–1945*. Opladen: VS Verlag für Sozialwissenschaften, 2000.

Fasse Dich kurz . . . Telefonieren in der DDR." Directed by Jörg Mischke, MDR documentary, 2007.

Fässler, Peter E. *Durch den "Eisernen Vorhang": Die deutsch-deutschen Wirtschaftsbeziehungen 1949–1969*. Cologne: Böhlau, 2006.

Feest, David, and Lutz Häfner, eds. *Die Zukunft der Rückständigkeit: Chancen—Formen—Mehrwert*. Cologne: Böhlau, 2016.

Feldman, Bernard J. "What to Say about the Tacoma Narrows Bridge to Your Introductory Class." *Physics Teacher* 41 (2003): 92–96.

Feldmann, Ekke. *Bauordnungen und Baupolizei: Zur Entwicklungsgeschichte zwischen 1850 und 1950*. Frankfurt am Main: Lang, 2011.

Fenelon, James. *Four Hours of Fury: The Untold Story of World War II's Largest Airborne Invasion and the Final Push into Nazi Germany*. New York: Scribner, 2019.

Ferrell, Robert H. "The Mukden Incident: September 18–19, 1931." *Journal of Modern History* 27.1 (1955): 66–72.

Feuerbach, Leonie. "Die Toilette der Zukunft." *Frankfurter Allgemeine Zeitung*, November 4, 2014.

Fickers, Andreas. "Visibly Audible: The Radio Dial as Mediating Interface." In *Oxford Handbook of Sound Studies*, edited by Karin Bijsterveld and Trevor Pinch, 411–439. Oxford: Oxford University Press, 2012.

Fisch, Michael. "Tokyo's Commuter Train Suicides and the Society of Emergence." *Cultural Anthropology* 28.2 (2013): 320–343.

Fischermann, Thomas et al. "Fussgänger—Angst bis in die Zehenspitzen." *Die Zeit*, February 11, 2016.

Fisher, Claude S. *America Calling: A Social History of the Telephone to 1940*. Berkeley: University of California Press, 1992.

Fleischhack, Julia. *Eine Welt im Datenrausch: Computeranlagen und Datenmengen als gesellschaftliche Herausforderung in der Bundesrepublik Deutschland (1965–1975)*. Zurich, Switzerland: Chronos, 2016.

Fletcher, George P. *A Crime of Self-defense: Bernhard Goetz and the Law on Trial*. Chicago: University of Chicago Press, 1990.

Flichy, Patrice. *Une histoire de la communication moderne*. Paris: La Découverte, 2004.

Flint, Anthony. *Wrestling with Moses: How Jane Jacobs Took On New York's Master Builder and Transformed the American City*. New York: Random House, 2011.

Flitner, Michael, Julia Lossau, and Anna-Lisa Müller, eds. *Infrastrukturen der Stadt*. Wiesbaden: Springer, 2017.

Flückiger, Daniel. *Strassen für alle: Infrastrukturpolitik im Kanton Bern 1790–1850*. Baden: Hier & Jetzt, 2011.

Flyvbjerg, Bent, Nils Bruzelius, and Werner Rothengatter. *Megaprojects and Risk: An Anatomy of Ambition*. Cambridge: Cambridge University Press, 2003.

Forest de Belidor, Bernard. *Architectura Hydraulica oder die Kunst, das Gewässer des Meeres und der Flüsse zum Vorteil der Verteidigung, der Festungen, des Handels und des Ackerbaus anzuwenden*. Augsburg: Kletts, 1766.

Förster, Birte, and Martin Bauch, eds. *Wasserinfrastrukturen und Macht von der Antike bis zur Gegenwart*. Munich: Oldenbourg, 2015.

Förster, Kim. "Sicherheitsschleuse." *Arch+* 191/192 (2009): 102.

Forsthoff, Ernst. *Die Daseinsvorsorge und die Kommunen: Vortrag anlässlich der Jahrestagung des Verbands Kommunaler Unternehmen e. V. am 16. 12. 1957 in Köln*. Cologne: Sigillum, 1958.

Forsthoff, Ernst. *Die Verwaltung als Leistungsträger*. Stuttgart: Kohlhammer, 1938.

Foschepoth, Josef. *Überwachtes Deutschland: Post- und Telefonüberwachung in der alten Bundesrepublik*. Göttingen: Vandenhoeck & Ruprecht, 2012.

Foucault, Michel. *Security, Territory, Population: Lectures at the Collège de France 1977—1978*. Translated by Graham Burchell. New York: Palgrave Macmillan, 2009.

Frank, Anne. *The Diary of a Young Girl*. New York: Bantam, 1995.

Frank, Susanne, and Matthew Gandy, eds. *Hydropolis: Wasser und die Stadt der Moderne*. Frankfurt am Main: Campus, 2006.

Fraunholz, Uwe. *Motorphobia: Anti-automobiler Protest in Kaiserreich und Weimarer Republik*. Göttingen: Vandenhoeck & Ruprecht, 2002.

Frey, Manuel. *Der reinliche Bürger: Entstehung und Verbreitung bürgerlicher Tugenden in Deutschland, 1780–1860*. Göttingen: Vandenhoeck & Ruprecht, 1997.

Fried, Ferdinand. *Umsturz der Gesellschaft*. Stuttgart: Union, 1950.

Fritz-Vannahme, Joachim. "Der Anthropologe im Dschungel der Grossstadt: Ein Porträt von Marc Augé." *Die Zeit*, August 23, 1996.

Fuchs, Christian, Stephan Lebert, and Daniel Müller. "Die unheimliche Firma." *Die Zeit*, July 16, 2015.

Galbraith, John Kenneth. *The Economic Effects of the Federal Public Works Expenditures 1933–1938*. Washington, DC: Government Printing Office, 1940.

Galison, Peter. *Einstein's Clocks, Poincaré's Maps: Empires of Time*. New York: Norton, 2003.

Gandy, Matthew. *The Fabric of Space: Water, Modernity, and Urban Imagination*. Cambridge, MA: MIT Press, 2014.

Gandy, Matthew. *Recycling and the Politics of Urban Waste*. New York: Routledge, 1994.

Garrow, David J., ed. *The Montgomery Bus Boycott and the Women Who Started It: The Memoir of Jo Ann Gibson Robinson*. Knoxville: University of Tennessee Press, 1987.

Gaslicht-Kultur eV. 2021. http://www.gaslicht-kultur.de.

Gässgen, Florian. "Mr WLAN," *Der Stern* 2 (2016): 112.

Gatejel, Luminita. *Warten, Hoffen und endlich Fahren: Auto und Sozialismus in der Sowjetunion, in Rumänien und der DDR (1956–1989/91)*. Frankfurt am Main: Campus, 2014.

Gedat, Gustav-Adolf. *Was wird aus diesem Afrika? Erlebter Kampf um einen Erdteil*. Stuttgart: Steinkopf, 1938.

Geistbeck, Michael. *Weltverkehr. Die Entwicklung von Schifffahrt, Eisenbahn, Post und Telegraphie bis zum Ende des 19. Jahrhunderts*. Hildesheim: Gerstenberg, 1986.

Geisthövel, Alexa, "Das Auto." In *Orte der Moderne: Erfahrungswelten des 19. und 20. Jahrhunderts*, edited by Alexa Geisthövel and Habbo Knoch, 37–46. Frankfurt am Main: Campus, 2005.

Genscher, Hans-Dietrich. "Infrastruktur als öffentliche Aufgabe: Vortrag am 8. Juni 1970," *BMI-Dokumentation* 30 (1970): 8.

Geppert, Alexander C. T. *Fleeting Cities: Imperial Expositions in Fin-de-Siècle Europe*. New York: Palgrave Macmillan, 2010.

Gerber, Sophie. *Küche, Kühlschrank, Kilowatt: Zur Geschichte des privaten Energiekonsums in Deutschland, 1945–1990*. Bielefeld: transcript, 2014.

Gerovitch, Slava. *From Newspeak to Cyberspeak: A History of Soviet Cybernetics*. Cambridge, MA: MIT Press, 2002.

Gestwa, Klaus, and Johannes Grützenmacher. "Infrastrukturen." In *Handbuch der Geschichte Russlands, Bd. 5.2: Vom Ende des Zweiten Weltkriegs bis zum Zusammenbruch der Sowjetunion*, edited by Stefan Plaggenborg, 1089–1152. Stuttgart: Hiersemann, 2003.

Giddens, Anthony. *The Constitution of Society: Outline of the Theory of Structuration*. Cambridge, UK: Polity 1984.

Giedion, Siegfried. *Mechanization Takes Command: A Contribution to Anonymous History*. Minneapolis: University of Minnesota Press, 2013.

Giessmann, Sebastian. *Die Verbundenheit der Dinge: Eine Kulturgeschichte der Netze und Netzwerke*. Berlin: Kadmos, 2014.

Giessmann, Sebastian. *Netze und Netzwerke: Archäologie einer Kulturtechnik, 1740–1840*, Bielefeld: transcript, 2006.

Giessmann, Sebastian. "Synchronisation im Diagramm: Henry C. Beck und die Londoner Tube Map von 1933." In *Kulturtechniken der Synchronisation*, edited by Christian Kassung and Thomas Macho, 339–366. Munich: Fink, 2013.

Gilman, Nils. *Mandarins of the Future: Modernization Theory in Cold War America*. Baltimore: Johns Hopkins University Press, 2004.

Glaser, Marie Antoinette, ed. *Baustelle: Metamorphosen in der Stadt*. Baden: Müller, 2008.

Glaser, Peter. "Der blaue Planet." *Süddeutsche Zeitung*, January 30/31, 2016.

Gleichmann, Peter R. "Die Verhäuslichung körperlicher Verrichtungen," in *Materialien zu Norbert Elias' Zivilisationstheorie*, edited by Peter R. Gleichmann, Johan Goudsblom, and Hermann Korte, 254–278. Frankfurt am Main: Suhrkamp, 1979.

Glitz, Philipp. "Generation Wisch und Klick Studie zur Smartphone-Sucht." *Tagesschau .de*, September 30, 2015. www.tagesschau.de/inland/smartphone-sucht-101.html.

Goerdeler, Carl D. "Fetisch einer Minderheit." *Die Zeit*, May 27, 1994.

Goetsch, Monika. "Wie im Traum." *Süddeutsche Zeitung*, October 18/19, 2015.

Gold, John R., and Margaret M. Gold, eds. *Olympic Cities: City Agendas, Planning, and the World's Games, 1896–2016*. London: Routledge, 2011.

Gold, John R., and George Revill. "Landscapes of Defence." *Landscape Research* 24 (1999): 229–239.

Gómez-Ibáñez, José A. *Regulating Infrastructure: Monopoly, Contracts, and Discretion*. Cambridge, MA: Harvard University Press, 2003.

Gooday, Graeme. *Domesticating Electricity: Technology, Uncertainty and Gender, 1880–1914*. Pittsburgh, PA: University of Pittsburgh Press, 2008.

Göschel, Albrecht, Heidrun Kunert-Schroth, and Klaus Mittag. *Revision der sozialen Infrastruktur: Ein Rückblick auf die 80er Jahre*. Berlin: Institut für Urbanistik, 1992.

Goschler, Constantin. "Infrastruktur-Liberalismus: Rudolf Virchow als Berliner Kommunalpolitiker." *Jahrbuch zur Liberalismus-Forschung* 18 (2006): 45–58.

Gottwaldt, Alfred. "Der Bahnhof." In *Orte der Moderne: Erfahrungswelten des 19. und 20. Jahrhunderts*, edited by Alexa Geisthövel and Habbo Knoch, 17–26. Frankfurt am Main: Campus, 2005.

Graham, Stephen, ed. *Cities, War and Terrorism: Towards an Urban Geopolitics*. Oxford: Blackwell, 2004.

Graham, Stephen, ed. *Disrupted Cities: When Infrastructure Fails*. New York: Routledge, 2010.

Graham, Stephen, and Simon Marvin. *Splintering Urbanism: Networked Infrastructures, Technological Mobilities and the Urban Condition*. London: Routledge, 2001.

Graham, Stephen, and Colin McFarlane, eds. *Infrastructural Lives: Urban Infrastructure in Context*. New York: Routledge, 2015.

Grehsin, Malte. "Deutschland hat angefangen." *Frankfurter Allgemeine Zeitung*, January 4, 2014.

Grot, Klaus. *So war's damals: Dienstchronik eines Pionieroffiziers im Kalten Krieg 1954–1991*. Berlin: Miles, 2014.

Gruber, Eckart. *Fünfuhr-Tee im Adlon: Menschen und Hotels*. Berlin: Fannei & Walz, 1994.

Grützmacher, Johannes. *Die Baikal-Amur-Magistrale: Vom stalinistischen Lager zum Mobilisierungsprojekt unter Breznev*. Munich: Oldenbourg, 2012.

Gugerli, David. *Redeströme: Zur Elektrifizierung der Schweiz 1880–1914*. Zurich, Switzerland: Chronos, 1996.

Guldi, Jo. *Roads to Power: Britain Invents the Infrastructure State*. Cambridge, MA: Harvard University Press, 2012.

Gyr, Ueli. "Geschichte des Tourismus: Strukturen auf dem Weg zur Moderne," Europäische Geschichte Online (EGO), December 3, 2010. http://ieg-ego.eu/de/threads/europa-unterwegs/tourismus/ueli-gyr-geschichte-des-tourismus.

Habermann, Gerd. *Der Wohlfahrtsstaat: Die Geschichte eines Irrwegs*. Frankfurt am Main: Ullstein, 1994.

Habib, Irfan. "Science, Technological Education and Industrialization: Contours of a Bhadralok Debate." In *Technology and the Raj: Western Technology and Technical Transfers to India: 1700–1947*, edited by Robin MacLeod and David Kumar, 235–249. New Delhi: Sage, 1995.

Hachtmann, Rüdiger. "Tourismus und Tourismusgeschichte." *Docupedia-Zeitgeschichte*, December 22, 2010. https://www.docupedia.de/zg/Tourismus_und_Tourismusgeschichte.

Haefeli-Meylan, Traugott. *Die Entstehung der Briefmarke und ihre weltweite Verbreitung*. Heidelberg: Decker, 1985.

Haff, P. K. "Technology as a Geological Phenomenon: Implications for Human Well-Being." *Geological Society, London, Special Publications* 395 (2014): 301–309. https://doi.org/10.1144/SP395.4.

Hagemann, Anke. "Filter, Ventile und Schleusen: Die Architektur der Zugangsregulierung." In *Kontrollierte Urbanität: Zur Neoliberalisierung städtischer Sicherheitspolitik*, edited by Volker Eick, Jens Sambale, and Eric Töpfer, 301–328. Bielefeld: transcript, 2007.

Hagen, Kirsten von. *Telefonfiktionen*. Munich: Fink, 2015.

Hagenah, Evelyn. *Staatsaufgaben*. Frankfurt am Main: Suhrkamp, 1996.

Hahn, Thomas. "Teures Pflaster." *Süddeutsche Zeitung*, August 8/9, 2015.

Hahn, Thomas. "Tunnel und Damm als Medien des Weltverkehrs: Populäre Kommunikation in der modernen Raumrevolution." In *Topographien der Literatur: Deutsche Literatur im transnationalen Kontext*, edited by Hartmut Böhme, 479–500. Stuttgart: Metzler, 2005.

Hahnemann, Andy. *Texturen des Globalen: Geopolitik und populäre Literatur in der Zwischenkriegszeit 1918–1939*. Heidelberg: Winter, 2010.

Halliday, Stephen. *The Great Stink of London: Sir Joseph Bazalgette and the Cleansing of the Victorian Metropolis*. Stroud: Sutton, 2001.

Hampe, Erich, ed. *Der Zivile Luftschutz im Zweiten Weltkrieg. Dokumentation und Erfahrungsberichte über Aufbau und Einsatz*. Frankfurt am Main: Bernard & Graefe, 1963.

Han, Byung-Chul. *The Transparency Society*. Translated by Erik Butler. Stanford, CA: Stanford University Press, 2015.

Haneke, Georg, and Claudia Stahl. *Infrastruktur und sozialer Wandel: Sozialanthropologische Beobachtungen entlang einer Strasse im Norden Kenias*. Saarbrücken: Breitenbach, 1994.

Hannig, Nicolai. "Die Suche nach Prävention: Naturgefahren im 19. und 20. Jahrhundert." *Historische Zeitschrift* 300.1 (2015): 33–65.

Hansen, Peo, and Stefan Jonsson. *Eurafrica: The Untold History of European Integration and Colonialism*. London: Bloomsbury, 2014.

Hardin, Garrett. "The Tragedy of the Commons." *Science* 162 (1968): 1243–1248.

Harvey, Penelope. "Cementing Relations: The Materiality of Roads and Public Spaces in Provincial Peru." *Social Analysis* 54.2 (2010): 28–46.

Harvey, Penelope, and Hannah Knox. "The Enchantments of Infrastructure." *Mobilities* 7.4 (2012): 521–536.

Harvey, Penelope, and Hannah Knox. *Roads: An Anthropology of Infrastructure and Expertise*. Ithaca, NY: Cornell University Press, 2015.

Hascher, Michael. "Maut als Mittel der Strassenfinanzierung zwischen öffentlichen und privaten Interessen, 1908–2008." In *Die moderne Strasse: Planung, Bau und Verkehr vom 18. bis zum 20. Jahrhundert*, edited by Hans-Liudger Dienel and Hans-Ulrich Schiedt, 147–169. Frankfurt am Main: Campus, 2010.

Haumann, Sebastian. *"Schade, dass Beton nicht brennt . . .": Planung, Partizipation und Protest in Philadelphia und Köln 1940–1990*. Stuttgart: Steiner, 2011.

Haus, Florian C. *Zugang zu Netzen und Infrastruktureinrichtungen*. Cologne: Heymann, 2022.

Hayes, Brian. *Infrastructure: A Guide to the Industrial Landscape*. New York: Norton, 2014.

Hayes, Hammond Vinton. *Public Utilities, Their Fair Present Value and Return*. New York: Van Nostrand, 1915.

Haywood, Richard Mowbray. *The Beginning of Railway Development in Russia in the Reign of Nicholas I, 1835–1842*. Durham, NC: Duke University Press, 1969.

"Nicht neben einer Frau." *Frankfurter Allgemeine Zeitung*, March 2, 2016.

Headrick, Daniel R. *The Tentacles of Progress: Technology Transfer in the Age of Imperialism, 1850–1940*. Oxford: Oxford University Press, 1988.

Headrick, Daniel R. *The Tools of Empire: Technology and European Imperialism in the Nineteenth Century*. Oxford: Oxford University Press, 1981.

Heckmann-Strohkark, Ingrid. "Der Traum von einer europäischen Gemeinschaft: Die Internationalen Autobahnkongresse 1931 und 1932." In *Die Schweizer Autobahn*, edited by Martin Heller and Andreas Volk, 32–45. Zurich, Switzerland: Museum für Gestaltung, 1999.

Heiber, Helmut, ed. *Joseph Goebbels: Reden, Bd. 2: 1939–1945*. Düsseldorf: Droste, 1972.

Heidenreich, Elisabeth. *Fliessräume: Die Vernetzung von Natur, Raum und Gesellschaft seit dem 19. Jahrhundert*. Frankfurt am Main: Campus, 2004.

Heidenreich, Elisabeth, "Natur und Kultur heute: verwickelt in technische Fliessräume." In *Hydropolis: Wasser und die Stadt der Moderne*, edited by Susanne Frank and Matthew Gandy, 57–73. Frankfurt am Main: Campus, 2006.

Hein, Christoph. "Eine Billion Dollar für Brücken, Strassen, Flughäfen." *Frankfurter Allgemeine Sonntagszeitung*, July 2, 2006.

Heine, Eike-Christian. *Die Geschichte des Nord-Ostsee-Kanals*. Berlin: Kadmos, 2015.

Heller, Piotr. "Besonders schön ist's in der Rhön." *Frankfurter Allgemeine Sonntagszeitung*, January 6, 2013.

Hellmann, Ullrich. *Künstliche Kälte: Die Geschichte der Kühlung im Haushalt*. Giessen: Anabas, 1990.

Hellpach, Willy. *Nervosität und Kultur*. Berlin: Johannes Räde, 1902.

Hempel, Leon, Susanne Krasmann, and Ulrich Bröckling, eds. *Sichtbarkeitsregime: Überwachung, Sicherheit und Privatheit im 21. Jahrhundert*. Wiesbaden: VS Verlag für Sozialwissenschaften, 2011.

Henkel, Dietrich. "Soziale Infrastruktur: Anpassung oder Rückbau?" *Bauwelt* 76.24 (1985): 124–128.

Hennig, Christoph. *Reiselust: Touristen, Tourismus und Urlaubskultur*. Frankfurt am Main: Suhrkamp, 1999.

Hermes, Georg. *Staatliche Infrastrukturverantwortung: Rechtliche Grundstrukturen netzgebundener Transport- und Übertragungssysteme zwischen Daseinsvorsorge und Wettbewerbsregulierung am Beispiel der leitungsgebundenen Energieversorgung in Europa*. Tübingen: Mohr, 1998.

Hessler, Martina. "Die Einführung elektrischer Haushaltsgeräte in der Zwischenkriegszeit. Der Angebotspush der Produzenten und die Reaktion der Konsumentinnen." *Technikgeschichte* 65 (1998): 297–311.

Hessler, Martina. *Kulturgeschichte der Technik*. Frankfurt am Main: Campus, 2012.

Hessler, Martina. *"Mrs. Modern Woman": Zur Sozial- und Kulturgeschichte der Haushaltstechnisierung*. Frankfurt am Main: Campus, 2011.

Hetherington, Kregg, ed. *Infrastructure, Environment and Life in the Anthropocene*. Durham, NC: Duke University Press, 2019.

Hilberg, Raul. *Sonderzüge nach Auschwitz*. Mainz: Dumjahn, 1981.

Hobart, Alice Tisdale. *Oil for the Lamps of China*. Indianapolis, IN: Bobbs-Merrill, 1933.

Hochschild, Adam. *Schatten über dem Kongo: Die Geschichte eines großen, fast vergessenen Menschheitsverbrechens*. Reinbek: Rowohlt, 2000.

Hoeft, Christoph, Sören Messinger-Zimmer, and Julia Zilles, eds. *Bürgerprotest in Zeiten der Energiewende: Lokale Konflikte um Windkraft, Stromtrassen und Fracking*. Bielefeld: transcript, 2017.

Hoffmann-Axthelm, Dieter. "Stadttechnik und Stadtkultur." *Stadtbauwelt* 103 (1989): 1666–1673.

Högselius, Per, Arne Kaijser, and Erik van der Vleuten. *Europe's Infrastructure Transition: Economy, War, Nature*. New York: Palgrave, 2015.

Hohensee, Jens. *Der erste Ölpreisschock 1973/74: Die politischen und gesellschaftlichen Auswirkungen der arabischen Erdölpolitik auf die Bundesrepublik Deutschland und Westeuropa*. Stuttgart: Steiner, 1996.

Höhne, Stefan. *New York City Subway: Die Erfindung des urbanen Passagiers*. Cologne: Böhlau, 2017.

Höhne, Stefan. *Riding the New York Subway: The Invention of the Modern Passenger*. Cambridge, MA: MIT Press, 2021.

Höhne, Stefan. "Token Tactics: Artefaktpolitiken in der New Yorker City Subway." In *Versorgung und Entsorgung der Moderne: Logistiken und Infrastrukturen der 1920er und 1930er Jahre*, edited by Wiebke Porombka, Heinz Reif, and Erhard Schütz, 73–88. Frankfurt am Main: Peter Lang, 2011.

Höhne, Stefan. "Vereinzelungsanlagen: Die Genese des Drehkreuzes aus dem Geist automatischer Kontrolle." *Technikgeschichte* 83.2 (2016): 103–124.

Holtorf, Christian. *Der erste Draht zur neuen Welt: Die Verlegung des transatlantischen Telegrafenkabels*. Göttingen: Wallstein, 2013.

Hombach, Marion. "Kaiserreich in den Tropen." *GeoEpoche* 71 (2015): 84–97.

Hörl, Johannes, and Dietmar Schöndorfer, eds. *Die Grossglockner Hochalpenstrasse: Erbe und Auftrag*. Vienna: Böhlau, 2015.

Hubmayr, Gerald. *Schwarzfahren: Die Kunst des tariffreien Netzgleitens*. Cologne: Böhlau, 2002.

Hughes, Thomas P. *American Genesis: A Century of Invention and Technological Enthusiasm, 1870–1970*. New York: Viking, 1989.

Hughes, Thomas P. *Networks of Power: Electrification in Western Society, 1880–1930*. Baltimore: Johns Hopkins University Press, 1983.

Hughes, Thomas P. *Rescuing Prometheus: Four Monumental Projects That Changed the Modern World*. New York: Vintage, 1998.

Hugh-Jones, Edward M., ed. *The Push-Button World: Automation Today*. Norman: University of Oklahoma Press, 1956.

Humphrey, Caroline. "Ideology in Infrastructure: Architecture and Soviet Imagination." *Journal of the Royal Anthropological Institute* 11.1 (2005): 39–58.

Huurdeman, Anton A. *The Worldwide History of Telecommunications*. Hoboken, NJ: Wiley, 2003.

Illich, Ivan. *H_2O und die Wasser des Vergessens*. Reinbek: Rowohlt, 1994.

Illinger, Patrick. "Koloss der Meere." *Süddeutsche Zeitung*, February 7/8, 2015.

"Immer mehr Städter erzwingen hohe Investitionen in die Infrastruktur." *Frankfurter Allgemeine Zeitung*, November 28, 2015.

"Infrastruktur." *Frankfurter Allgemeine Zeitung*, March 1, 1952.

Isaacman, Allen. "Displaced People, Displaced Energy, and Displaced Memories: The Case of Cahora Bassa, 1970–2004." *International Journal of African Historical Studies* 38.2 (2005): 201–238.

Jackson, Robert H., and Carl G. Rosberg. *Personal Rule in Black Africa: Prince, Autocrat, Prophet, Tyrant*. Berkeley: University of California Press, 1982.

Janác, Jirí. *European Coasts of Bohemia: Negotiating the Danube-Oder-Elbe Canal in a Troubled Twentieth Century*. Amsterdam: Amsterdam University Press, 2012.

Jellinghaus, Lorenz. *Zwischen Daseinsvorsorge und Infrastruktur: Zum Funktionswandel von Verwaltungswissenschaften und Verwaltungsrecht in der zweiten Hälfte des 19. Jahrhunderts*. Frankfurt am Main: Klostermann, 2006.

The Jewish Peril: Protocols of the Learned Elders of Zion. London: Eyre & Spottiswoode, 1920.

Jochimsen, Reimut. *Theorie der Infrastruktur: Grundlagen der marktwirtschaftlichen Entwicklung*. Tübingen: Mohr, 1966.

Jochmann, Werner, ed. *Adolf Hitler: Monologe im Führerhauptquartier 1941–1944. Die Aufzeichnungen Heinrich Heims*. Hamburg: Knaus, 1980.

Joerges, Bernward. "Die Brücken des Robert Moses: Stille Post in der Stadt- und Techniksoziologie," *Leviathan* 27.1 (1999): 43–63.

John, Richard R. "Private Enterprise, Public Good? Communications Deregulation as a National Political Issue, 1839–1851." In *Beyond the Founders. New Approaches to the Political History of the Early American Republic*, edited by Jeffrey L. Pasley, Andrew W. Robertson, and David Waldstreicher, 328–354. Chapel Hill: University of North Carolina Press, 2004.

John, Richard R. "Telecommunications." *Enterprise and Society* 9 (2008): 507–520.

Jones, Christopher F. "A Landscape of Energy Abundance: Anthracite Coal Canals and the Roots of American Fossil Fuel Dependence, 1820–1860." *Environmental History* 15 (2010): 449–484.

Jordan, John M. "'Society Improved the Way You Can Improve a Dynamo': Charles P. Steinmetz and the Politics of Efficiency." *Technology and Culture* 30.1 (1989): 57–82.

Jordans, Angelika. "Ein Stück London in der Wüste." *Frankfurter Allgemeine Zeitung*, January 20, 1994.

Jost, Sarah, and Gabriela Wachter, eds. *Die verschwundene Arbeit in Fotografien aus Berliner Sammlungen und Archiven*. Berlin: Parthas, 2008.

Jungclaussen, John F. "Hilflos im Wasser: Regengüsse legen offen, wie marode die britische Infrastruktur ist. Geflickt wird mit Geld aus China." *Die Zeit*, February 20, 2014.

Junkelmann, Marcus. "Die Eisenbahn im Krieg: Militärische Theorien und Kriegsgeschehen bis zum Ausbruch des Ersten Weltkriegs." In *Zug der Zeit—Zeit der Züge: Deutsche Eisenbahn 1835–1985*, vol. 1, edited by Harm-Hindreich Brandt, 232–245. Munich: Orbis, 1989.

Kahrstedt, Oskar. *Internationale Bekämpfung der Arbeitslosigkeit durch Erschliessung überseeischer Gebiete: Zugleich ein Beitrag zum Problem der Vergrösserung des Welthandelsvolumens*. Berlin: Hobbing, 1931.

Kaika, Maria. "City of Flows: Der Wandel der symbolischen Bedeutung technischer Infrastrukturen in der Moderne." In *Infrastrukturnetze und Raumentwicklung*, edited by Timothy Moss, Matthias Naumann, and Markus Wissen, 87–111. Munich: Oekom, 2008.

Kalb, Christine. *Weltausstellungen im Wandel der Zeit und ihre infrastrukturellen Auswirkungen auf Stadt und Region*. Frankfurt am Main: Lang, 1994.

Kamleithner, Christa. "Neue Mischungsverhältnisse: Zum Gebrauch von Infrastrukturen." In *Infrastrukturen des Urbanen: Soundscapes, Landscapes, Netscapes*, edited by Nathalie Bredella and Chris Dähne, 253–274. Bielefeld: transcript, 2013.

Karaisl von Karais, Franz Freiherr, and Helmut Stabina. *Deutsche Strassenfibel*. Leipzig: Staackmann, 1937.

Kaschka, Ralph. *Auf dem falschen Gleis: Infrastrukturpolitik und -entwicklung der DDR am Beispiel der Deutschen Reichsbahn 1949–1989*. Frankfurt am Main: Campus, 2011.

Kathke, Torsten. *Wires That Bind: Nation, Region, and Technology in the Southwestern United States, 1854–1920*. Bielefeld: transcript, 2017.

Katz, Bernard S. "Infrastructure Ills: Or If You Can't Use It Correctly, Don't." *Advertising Age*, 60.17 (1989): 23.

Kaufmann, Jean-Claude. *Kochende Leidenschaft: Soziologie vom Kochen und Essen*. Konstanz: UVK, 2006.

Kaufmann, Stefan. *Kommunikationstechnik und Kriegführung 1815–1945: Stufen telemedialer Rüstung*. Munich: Fink, 1996.

Kaufmann, Stefan. "Landschaft beschriften: Zur Logik des American Grid Systems." In *Ordnungen der Landschaft. Natur und Raum technisch und symbolisch entwerfen*, edited by Stefan Kaufmann, 73–94. Würzburg: Ergon, 2002.

Kaul, Chandrika. *Communications, Media and the Imperial Experience: Britain and India in the Twentieth Century*. New York: Palgrave Macmillan, 2014.

Keane, Stephen. *Disaster Movies: The Cinema of Catastrophe*. London: Wallflower, 2006.

Kellermann, Bernhard. *Der Tunnel*. Frankfurt am Main: Suhrkamp, 1952.

Kenny, Nicolas. "City Glow: Streetlights, Emotions, and Nocturnal Life, 1880s–1910s." *Journal of Urban History* 43.1 (2017): 91–114.

Khannah, Parag. *Connectography: Mapping the Future of Global Civilization*. New York: Random House, 2016.

Kiaulehn, Walther. *Berlin: Schicksal einer Weltstadt*. Munich: Beck, 1997.

Kil, Wolfgang. *Luxus der Leere: Vom schwierigen Rückzug aus der Wachstumswelt: Eine Streitschrift*. Wuppertal: Müller & Busmann, 2004.

Kilgannon, Corey. "The Truth about Alligators in the Sewers of New York." *New York Times*, February 26, 2020.

Kill, Susanne, Christopher Kopper, and Jan-Henrik Peters. *Die Reichsbahn und der Strafvollzug in der DDR: Häftlingszwangsarbeit und Gefangenentransport in der SED-Diktatur*. Essen: Klartext, 2016.

Klausmeier, Axel. *Hinter der Mauer: Zur militärischen und baulichen Infrastruktur des Grenzkommandos Mitte*. Berlin: Links, 2012.

Klee, Margot. *Lebensadern des Imperiums, Strassen im Römischen Weltreich*. Stuttgart: Theiss, 2010.

Klenke, Dietmar. *Bundesdeutsche Verkehrspolitik und Motorisierung: Konfliktträchtige Weichenstellungen in den Jahren des Wiederaufstiegs*. Stuttgart: Steiner, 1993.

Kline, Ronald R. *The Cybernetics Moment: Or Why We Call Our Age the Information Age*. Baltimore: Johns Hopkins University Press, 2015.

Klose, Alexander. *Das Container-Prinzip: Wie eine Box unser Leben verändert*. Hamburg: Mare, 2009.

Knapp, Gottfried. "Orte der Verdrängung: Was nicht geliebt wird, muss auch nicht schön sein." *Baumeister: Zeitschrift für Architektur* 3 (1998): 38–41.

Knoch, Habbo. "Das Grandhotel." In *Orte der Moderne: Erfahrungswelten des 19. und 20. Jahrhunderts*, edited by Alexa Geisthövel and Habbo Knoch, 131–140. Frankfurt am Main: Campus, 2005.

Knoch, Habbo. *Grandhotels. Luxusräume und Gesellschaftswandel in New York, London und Berlin um 1900*, Göttingen: Wallstein, 2016.

Knoch, Habbo. "Life on Stage: Grand Hotels as Urban Interzones around 1900." In *Creative Urban Milieus: Historical Perspectives on Culture, Economy, and the City*, edited by Martina Hesslier and Clemens Zimmermann, 137–158. Frankfurt am Main: Campus, 2008.

Köckritz, Angela. "Der Stromdieb." *Die Zeit*, April 12, 2017.

Kohl, Helmut. "Infrastrukturreform als Innenpolitik—Möglichkeiten, Grenzen, Prioritäten." *Bergedorfer Gesprächskreis zu Fragen der freien industriellen Gesellschaft* 38 (1971).

Kolb, Anne. *Infrastruktur und Herrschaftsorganisation im Imperium Romanum*. Berlin: De Gruyter, 2014.

Kolonko, Petra. "Seide und Zement." *Frankfurter Allgemeine Zeitung*, May 13, 2017.

König, Wolfgang. *Bahnen und Berge: Verkehrstechnik, Tourismus und Naturschutz in den Schweizer Alpen 1870–1939*. Frankfurt am Main: Campus, 2000.

Korff, Gottfried. "Mentalität und Kommunikation in der Grossstadt: Berliner Notizen zur 'inneren' Urbanisierung." In *Grossstadt: Aspekte empirischer Kulturforschung*, edited by Theodor Kohlmann and Hermann Bausinger, 343–361. Berlin: Staatliche Museen Preussicher Kulturbesitz, 1985.

Koselleck, Reinhart. *Zeitschichten: Studien zur Historik*. Frankfurt am Main: Suhrkamp, 2000.

Köster, Roman. "Abschied von der 'verlorenen Verpackung': Das Recycling von Hausmüll in Westdeutschland 1945–1990." *Technikgeschichte* 81 (2014): 33–60.

Kotowski, Timo. "Teure Luftnummern in der Provinz." *Frankfurter Allgemeine Zeitung*, July 30, 2014.

Kowalczuk, Ilko-Sascha, and Arno Polzin, eds. *Fasse Dich kurz! Der grenzüberschreitende Telefonverkehr der Opposition in den 1980er Jahren und das Ministerium für Staatssicherheit*. Göttingen: Vandenhoeck & Ruprecht, 2015.

Koziolek, Helmut. "Umfassende Intensivierung der Infrastruktur." *Sitzungsberichte der Akademie der Wissenschaften der DDR* 2 (1986): 12.

Krabbe, Wolfgang. "Die Entfaltung der kommunalen Leistungsverwaltung in deutschen Städten des späten 19. Jahrhunderts." In *Urbanisierung im 19. und 20. Jahrhundert*, edited by Hans-Jürgen Teuteberg, 373–391. Cologne: Böhlau, 1983.

Krajewski, Markus. "Fehler-Planungen: Zur Geschichte und Theorie der industriellen Obsoleszenz." *Technikgeschichte* 81.1 (2014): 91–114.

Krajewski, Markus. *The Server: A Media History from the Present to the Baroque*. Translated by Ilinca Iurascu. New Haven, CT: Yale University Press, 2018.

Krebs, Stefan, Gabriele Schabacher, and Heike Weber, eds. *Kulturen des Reparierens*. Bielefeld: transcript, 2018.

Krechel, Ursula. *Landgericht*. Salzburg: Jung & Jung, 2012.

Kreiss, Christian. *Geplanter Verschleiss: Wie die Industrie uns zu immer mehr und immer schnellerem Konsum antreibt—und wie wir uns dagegen wehren können*. Berlin: Europa, 2014.

Kremer, Bernd. *Die Kunst zu überleben: Zivilverteidigung in der Bundesrepublik*. Munich: Odang 1966.

Kuchenbuch, Ludolf. "Abfall: Eine Stichwortgeschichte." In *Kultur und Alltag*, edited by Hans-Georg Soeffner, 155–170. Göttingen: Schwartz, 1988.

Küffner, Georg. "Bröckelnde Brücken." *Frankfurter Allgemeine Zeitung*, March 5, 2013.

Kühl, Uwe. *Der Munizipalsozialismus in Europa*. Munich: Oldenbourg, 2001.

Kühne, Thomas. "Massenmotorisierung und Verkehrspolitik im 20. Jahrhundert: Technikgeschichte als politische Sozial- und Kulturgeschichte." *Neue Politische Literatur* 41.2 (1996): 196–229.

Kuhr-Korolev, Corinna. *Gezähmte Helden: die Formierung der Sowjetjugend 1917–1932*. Essen: Klartext, 2005.

Küster, Hansjörg. "Das Nichts am Ende des Tunnels." *Frankfurter Allgemeine Zeitung*, March 8, 1999.

Laak, Dirk van. "Der Begriff 'Infrastruktur' und was er vor seiner Erfindung besagte." *Archiv für Begriffsgeschichte* 41 (1999): 280–299.

Laak, Dirk van. "From the Conservative Revolution to Technocratic Conservatism." In *German Ideologies since 1945: Studies in the Political Thought and Culture of the Bonn Republic*, edited by Jan-Werner Müller, 147–160. New York: Palgrave Macmillan, 2003.

Laak, Dirk van. *Imperiale Infrastruktur: Deutsche Planungen für eine Erschliessung Afrikas, 1880–1960*. Paderborn: Schöningh, 2004.

Laak, Dirk van. "Infrastruktur und Macht." In *Umwelt und Herrschaft in der Geschichte / Environnement et pouvoir: une approche historique*, edited by François Duceppe-Lamarre and Jens Ivo Engels, 105–111. Munich: Oldenbourg, 2008.

Laak, Dirk van. "Mythos 'Hessenplan': Aufstieg und Wandel einer Landesplanung nach dem Zweiten Weltkrieg." In *Raumplanung nach 1945: Kontinuitäten und*

Neuanfänge in der Bundesrepublik Deutschland, edited by Wendelin Strubelt and Detlef Briesen, 127–149. Frankfurt am Main: Campus, 2015.

Laak, Dirk van. "Technokratie im Europa des 20. Jahrhunderts—eine einflussreiche 'Hintergrundideologie.'" In *Theorien und Experimente der Moderne: Europas Gesellschaften im 20. Jahrhundert*, edited by Lutz Raphael, 101–128. Cologne: Böhlau, 2012.

Laak, Dirk van. "Vom Lebensraum zum Leitungsweg: Die Stadtstrasse als soziale Arena." In *Infrastrukturen der Stadt*, edited by Michael Flitner, Julia Lossau, and Anna-Lisa Müller, 145–162. Wiesbaden: Springer, 2017.

Laak, Dirk van. "Zukunft konkret: Zeithistorische Anmerkungen zum Handeln der praktisch Planenden." In *Ermöglichen und Verhindern: Vom Umgang mit Kontingenz*, edited by Markus Bernhardt, Stefan Brakensiek, and Benjamin Scheller, 191–208. Frankfurt am Main: Campus, 2016.

Laak, Dirk van. *Weisse Elefanten: Anspruch und Scheitern technischer Grossprojekte im 20. Jahrhundert*. Stuttgart: Deutsche Verlagsanstalt, 1999.

Laak, Dirk van, and Dirk Rose, eds. *Schreibtischtäter: Begriff, Geschichte, Typologie*. Göttingen: Wallstein, 2018.

Lachmann, Günther. *Von Not nach Elend: Eine Reise durch deutsche Landschaften und Geisterstädte von morgen*. Munich: Piper, 2008.

Lackner, Helmut. "Ein 'blutiges Geschäft.' Kommunale Vieh- und Schlachthöfe im Urbanisierungsprozess des 19. Jahrhunderts: Ein Beitrag zur Geschichte der städtischen Infrastruktur." *Technikgeschichte* 71.2 (2004): 89–138.

Ladischensky, Dimitri. "Geliebt, gehasst, gequält: Mythos und Realität des blinden Passagiers." *Mare* 60 (2007): 66–71.

Laer, Hellmut von. "Gleicht Berlin bei Nacht bald Bielefeld?" *Frankfurter Allgemeine Zeitung*, June 25, 2014.

Lagendijk, Vincent. *Electrifying Europe: The Power of Europe in the Construction of Electricity Networks*. Amsterdam: Amsterdam University Press, 2008.

Lahiri-Choudhury, Deep Kanta. *Telegraphic Imperialism: Crisis and Panic in the Indian Empire, c. 1830–1920*. New York: Palgrave Macmillan, 2010.

Lampland, Martha, and Susan Leigh Star. *Standards and Their Stories: How Quantifying, Classifying and Formalizing Practices Shape Everyday Life*. Ithaca, NY: Cornell University Press, 2009.

Lanjouw, Peter. "Infrastructure: A Ladder for the Poor." *Finance & Development* 32.1 (1995): 33–35.

Larkin, Brian. "The Politics and Poetics of Infrastructure." *Annual Review of Anthropology* 42 (2013): 327–343.

Larkin, Brian. *Signal and Noise: Media, Infrastructure, and Urban Culture in Nigeria*. Durham, NC: Duke University Press, 2008.

Larson, John Lauritz. *Internal Improvement: National Public Works and the Promise of Popular Government in the Early United States*. Chapel Hill: University of North Carolina Press, 2001.

Latham, Alan, and Peter R. H. Wood. "Inhabiting Infrastructure: Exploring the Interactional Spaces of Urban Cycling." *Environment and Planning* 47 (2015): 300–319.

Lazarova, Irina. *"Hier spricht Lenin": Das Telefon in der russischen Literatur der 1920er und 30er Jahre*. Cologne: Böhlau, 2010.

Le Goff, Jacques. *The Birth of Europe*. Translated by Janet Lloyd. Malden, MA: Blackwell, 2014.

Lea, Tess, and Paul Pholeros. "This Is Not a Pipe: The Treacheries of Indigenous Housing." *Public Culture* 22.1 (2010): 187–209.

Lee, Benjamin, and Edward LiPuma. "Cultures of Circulation: The Imaginations of Modernity." *Public Culture* 14.1 (2002): 191–213.

Lee, Paula Young, ed. *Meat, Modernity, and the Rise of the Slaughterhouse*. Durham: University of New Hampshire Press, 2008.

Leendertz, Ariane. *Ordnung schaffen: Deutsche Raumplanung im 20. Jahrhundert*. Göttingen: Wallstein, 2008.

Leggewie, Claus. "Wenn unsere Infrastruktur unter die Räder kommt." *NDR-Kultur: Gedanken zur Zeit*, December 15, 2013. https://www.yumpu.com/de/document/view/24819951/wenn-unsere-infrastruktur-unter-dierader-kommt-von-ndr; retrieved December 17, 2017.

Leighninger, Robert D., Jr. "Cultural Infrastructure: The Legacy of New Deal Public Space." *Journal of Architectural Education* 49.4 (1996): 226–236.

Lenger, Friedrich. *Metropolen der Moderne: Eine europäische Stadtgeschichte*. Munich: Beck, 2013.

Lenz, Siegfried. *The Lightship*. Translated by Michael Bullock. London: Faber & Faber, 2010.

LeRose, Chris. "The Collapse of the Silver Bridge." *West Virginia Historical Society Quarterly* 15.4 (2001).

Lessing, Theodor. *Der Antirüpel: Monatsblätter zum Kampf gegen Lärm, Rohheit und Unkultur im deutschen Wirtschafts-, Handels- und Verkehrsleben* (1908/09–1911).

Lessing, Theodor. *Der Lärm: Eine Kampfschrift gegen die Geräusche unseres Lebens*. Wiesbaden: Bergmann, 1909.

Lethen, Helmut. *Cool Conduct: The Culture of Distance in Weimar Germany*. Translated by Don Reneau. Berkeley: University of California Press, 2002.

Lichtenfels, Karl Leopold von. *Lexikon des Überlebens: Handbuch für Krisenzeiten*. Cologne: Anaconda, 2006.

Liebert, Nicola. "Die Lektion British Rail." *Edition Le Monde diplomatique: Ausverkauft. Wie das Gemeinwohl zur Privatsache wird* 6 (2009): 4.

Lindenberger, Thomas. "'Streikexzesse': Zur öffentlichen Un-Ordnung im Kaiserreich." *Deutsches Historisches Museum.* https://www.dhm.de/ausstellungen/streik/html/kaiserzeit1.html; retrieved December 18, 2017.

Ling, Rich. *New Tech, New Ties: How Mobile Communication Is Reshaping Social Cohesion.* Cambridge, MA: MIT Press, 2008.

Lintsen, Harry W., ed. *Geschiedenis van de Techniek in Nederland: De wording van een moderne samenleving 1800–1890,* 6 vols. Zutphen: Stichting Historie der Techniek, 1992–1995.

Little, Richard G. "Managing the Risk of Cascading Failure in Complex Urban Infrastructures." In *Disrupted Cities: When Infrastructure Fails,* edited by Stephen Graham, 27–40. London: Routledge, 2010.

Lobe, Adrian. "Sie setzen die Massstäbe der digitalen Welt." *Frankfurter Allgemeine Zeitung,* August 12, 2015.

Löffler, Juliane. "So gross kann Dein ******* nicht sein!" *Der Freitag,* December 19, 2014. https://www.Freitag.de/autoren/juloeffl/so-gross-kann-dein-nicht-sein.

Löffler, Petra. *Verteilte Aufmerksamkeit: Eine Mediengeschichte der Zerstreuung.* Zurich, Switzerland: Diaphanes, 2014.

Lommers, Suzanne. *Europe—on Air: Interwar Projects for Radio Broadcasting.* Amsterdam: Amsterdam University Press, 2012.

London, Jack. *War of the Classes.* New York: Regent Press, 1905.

Longen, Nicole K. "Fronarbeiten zur Finanzierung von Infrastruktur: Der Ausbau des Strassennetzes im Kurtrierer Raum 1716–1841." In *Die moderne Strasse: Planung, Bau und Verkehr vom 18. bis zum 20. Jahrhundert,* edited by Hans-Liudger Dienel and Hans-Ulrich Schiedt, 23–48. Frankfurt am Main: Campus, 2010.

Lorenz, Maren. *Vandalismus als Alltagsphänomen.* Hamburg: Hamburger, 2009.

Loske, Reinhard. "Die politische Ökologie der Infrastrukturen." In *Die Zukunft der Infrastrukturen: Intelligente Netzwerke für eine nachhaltige Entwicklung,* edited by Reinhard Loske and Roland Schaeffer, 23–44. Marburg: Metropolis, 2005.

Lotz, Christian. "Nachhaltigkeit neuskalieren: Internationale forstwissenschaftliche Kongresse und Debatten um die Ressourcenversorgung der Zukunft im Nord- und Ostseeraum (1870–1914)." PhD diss., Giessen, 2015.

Lotz, Wolfgang, and Gerd R. Ueberschär. *Die Deutsche Reichspost 1933–1945: Eine politische Verwaltungsgeschichte,* 2 vols. Berlin: Nicolai, 1999.

Löwe, Liane. "'A Nation Built on Transport': Das Auto und die US-amerikanische Gesellschaft." In *Building America: Die Erschaffung einer neuen Welt,* edited by Anke Köth, Anna Minta, and Andreas Schwarting, 247–270. Dresden: Thelem, 2005.

Löwenthal, Richard. "Bonn und Weimar: Zwei deutsche Demokratien." In *Politische Weichenstellungen im Nachkriegsdeutschland 1945–1953*, edited by Heinrich August Winkler, 9–25. Göttingen: Vandenhoeck & Ruprecht, 1979.

Lübbe, Hermann. "Technik in Politik und Ideologie." In *Technik wozu und wohin?* edited by Hardi Fischer, 131–140. Zurich, Switzerland: Artemis, 1981.

Lübken, Uwe. "'Der grosse Brückentod': Überschwemmungen als infrastrukturelle Konflikte im 19. und 20. Jahrhundert." *Saeculum* 58.1 (2007): 89–114.

Makropoulos, Michael. "Die infrastrukturelle Konstruktion der 'Volksgemeinschaft': Aspekte des Autobahnbaus im nationalsozialistischen Deutschland." In *Vernunft—Entwicklung—Leben: Schlüsselbegriffe der Moderne. Festschrift für Wolfgang Essbach*, edited by Ulrich Bröckling, Axel T. Paul, and Stefan Kaufmann, 185–203. Munich: Fink, 2004.

Malkoutzis, Nick. "How the 2004 Olympics Triggered Greece's Decline," *Bloomberg Business*, August 2, 2012.

Malm, Andreas. *Fossil Capital: The Rise of Steam-Power and the Roots of Global Warming*. London: Verso, 2016.

Malter, Bettina. "Hier ruhen unsere Milliarden." *Die Zeit*, July 3, 2014.

Mann, Michael. "Infrastructural Power Revisited." *Studies in Comparative International Development* 43.3–4 (2008): 355–365.

Mann, Michael. *The Sources of Social Power*. Vol. 2, *The Rise of Classes and Nation States 1760–1914*. Cambridge: Cambridge University Press, 1993.

Mann, Michael. *Wiring the Nation: Telecommunication, Newspaper-Reportage and Nation Building in British India, 1850–1930*. Oxford: Oxford University Press, 2017.

Marfeld, Alexander Friedrich. *Am Anfang und am Ende—der Passagier*. Berlin: Safari, 1967.

Markmiller, Anton. *Die Erziehung des Negers zur Arbeit: Wie die koloniale Pädagogik afrikanische Gesellschaften in die Abhängigkeit führte*. Berlin: Reimer, 1995.

Marland, Gregg, and Alvin M. Weinberg, "Longevity of Infrastructure." In *Cities and Their Vital Systems: Infrastructure—Past, Present, and Future*, edited by Jesse H. Ausubel and Robert Herman, 312–332. Washington, DC: National Academy Press, 1988.

Marx, Karl. *Selected Writings*, edited by David McLellan. Oxford: Oxford University Press, 2000.

Marx, Karl, and Friedrich Engels. *Collected Works*, vol. 46. New York: International Publishers, 1992.

Masquelier, Adeline. "Road Mythographies: Space, Mobility, and the Historical Imagination in Postcolonial Niger." *American Ethnologist* 29.4 (2002): 829–856.

Mattelard, Armand. *Networking the World, 1794–2000*. Minneapolis: University of Minnesota Press, 2000.

Mauch, Christoph, and Thomas Zeller, eds. *The World Beyond the Windshield: Roads and Landscapes in the United States and Europe*. Athens: Ohio University Press, 2008.

Mauelshagen, Franz. "'Anthropozän': Plädoyer für eine Klimageschichte des 19. und 20. Jahrhunderts." *Zeithistorische Forschungen / Studies in Contemporary History* 9 (2012): 131–137.

Mayntz, Renate. "Zur Entwicklung technischer Infrastruktur-Systeme." In *Differenzierung und Verselbstständigung*, edited by Renate Mayntz, Bernd Rosewitz, Uwe Schimank, and Rudolf Stichweh, 233–259. Frankfurt am Main: Campus, 1988.

Mazuri, Ali. "Nkrumah: The Leninist Czar." *Transition* 26 (1966): 8–17.

Mbembe, Achille. *On the Postcolony*. Berkeley: University of California Press, 2001.

McKay, John P. *Tramways and Trolleys: The Rise of Urban Mass Transport in Europe*. Princeton, NJ: Princeton University Press, 1976.

McLuhan, Marshall. *The Gutenberg Galaxy: The Making of Typographic Man*. Toronto: University of Toronto Press, 2017.

McNeill, John R. *Something New Under the Sun: An Environmental History of the Twentieth-Century World*. New York: Norton, 2001.

McShane, Clay. "The Origins and Globalization of Traffic Control Signals." *Journal of Urban History* 25.3 (1999): 379–404.

Mehnert, Volker. "Bier und Bratwurst statt Ruhm und Ehre." *Frankfurter Allgemeine Zeitung*, November 20, 2014.

Mehnert, Volker. "Schnell sind nur die Schmuggler." *Frankfurter Allgemeine Zeitung*, May 20, 2009.

Meier, Christian. *Das Gebot zu vergessen und die Unabweisbarkeit des Erinnerns: Vom öffentlichen Umgang mit schlimmer Vergangenheit*. Berlin: Siedler, 2010.

Meiske, Martin. *Die Geburt des Geoengineering: Grossbauprojekte in der Frühphase des Anthropozäns*. Göttingen: Wallstein, 2021.

Melnick, Mimi. *Manhole Covers*. Cambridge, MA: MIT Press, 1994.

Melosi, Martin V. *The Sanitary City: Environmental Services in Urban America from the Colonial Times to the Present*. Pittsburgh, PA: University of Pittsburgh Press, 2008.

Meneghello, Laura. "Die Rohrpost als Metainfrastruktur." PhD diss., Siegen, 2017.

Menzler-Trott, Eckart. "Vom Telefon zum Call Center: Eine kleine Geschichte des Telefonierens." *Telepolis*, May 21, 1998. https://www.heise.de/tp/artikel/2/2335/1.html.

Merki, Christoph Maria. "Vom 'Herrenfahrer' zum 'Balkanraser': Zur Geschichte des Automobilismus in der Schweiz." *Schweizerische Zeitschrift für Geschichte* 56.1 (2006): 46–56.

Merritt, Walter Gordon. *Strikes and Public Utilities: A Remedy*. New York: American Anti-Boycott Association, 1917.

Metzger, Martina. *Bewältigung, Auswirkungen und Nachwirkungen des Bombenkrieges in Berlin und London 1940–1955: Zerstörung und Wiederaufbau zweier europäischer Hauptstädte*. Stuttgart: Steiner, 2013.

Michels, Claudia. "Worte, die ausgrenzen." *Frankfurter Rundschau*, June 6, 2014.

Mierzejewski, Alfred C. *The Most Valuable Asset of the Reich: A History of the German National Railway. Vol. 1: 1920–1932*. Chapel Hill: University of North Carolina Press, 1999.

Miller, Arthur. *Death of a Salesman: Certain Private Conversations in Two Acts and a Requiem*. New York: Penguin, 1998.

Miller, Michael B. *Europe and the Maritime World: A Twentieth-Century History*. Cambridge: Cambridge University Press, 2012.

Millward, Robert. *Private and Public Enterprise in Europe: Energy, Telecommunication and Transport, 1830–1990*. Cambridge: Cambridge University Press, 2005.

Minicucci, Stephen. "Internal Improvements and the Union, 1790–1860." *Studies in American Political Development* 18.2 (2004): 160–185.

Mishra, Pankaj. *From the Ruins of Empire: The Revolt against the West and the Remaking of Asia*. New York: Penguin, 2013.

Mitchell, Allan. *The Great Train Race: Railways and the Franco-German Rivalry, 1815–1914*. New York: Berghahn, 2000.

Moesle, Marianne. "Der Schein und das Bewusstsein." *Süddeutsche Zeitung Magazin* 4 (2014): 8–13.

Mom, Gijs. *Atlantic Automobilism. Emergence and Persistence of the Car, 1895–1940*. New York: Berghahn, 2015.

Momberg, Robert. *Theorie und Politik der Infrastruktur unter Berücksichtigung institutionen- und politökonomischer Einflussfaktoren*. Frankfurt am Main: Lang, 2000.

Monbiot, George. *Captive State: The Corporate Takeover of Britain*. London: Pan, 2001.

Monnier, Alain. *Givrée*. Paris: Flammarion, 2009.

Monstadt, Jochen. "Der räumliche Wandel der Stromversorgung und die Auswirkungen auf die Raum- und Infrastrukturplanung." In *Infrastrukturnetze und Raumentwicklung: Zwischen Universalisierung und Differenzierung*, edited by Timothy Moss, Matthias Naumann, and Markus Wissen, 187–224. Munich: Oekom, 2008.

Morat, Daniel. "'Automobile gehen über mich hin': Urbane Dispositive akustischer Innervation um 1900." In *Unlaute: Noise / Geräusch in Kultur, Medien und Wissenschaften seit 1900*, edited by Sylvia Mieszkowski and Sigrid Nieberle, 127–148. Bielefeld: transcript, 2017.

Morat, Daniel, ed. *Sounds of Modern History: Auditory Cultures in 19th and 20th Century Europe*. New York: Berghahn, 2014.

Moretti, Franco, and Dominique Prestre. "Bankspeak: The Language of World Bank Reports." *New Left Review* 92 (2015): 75–99.

Mori, Takahito. *Elektrifizierung als Urbanisierungsprozess: Frankfurt am Main, 1886–1933*. Darmstadt: Hessisches Wirtschaftsarchiv, 2014.

Morse, Dan. "Still Called by Faith to the Booth: As Pay Phones Vanish, Amish and Mennonites Build Their Own." *Washington Post*, September 3, 2006.

Morton, Margaret. *The Tunnel: The Underground Homeless of New York City*. New Haven, CT: Yale University Press, 1995.

Morus, Iwan Rhys. "'The Nervous System of Britain': Space, Time and the Electric Telegraph in the Victorian Age." *British Journal for the History of Science* 33.4 (2000): 455–475.

Möser, Kurt. "Zwischen Systemopposition und Systemteilnahme: Sicherheit und Risiko im motorisierten Strassenverkehr 1890–1930." In *Geschichte der Strassenverkehrssicherheit im Wechselspiel zwischen Fahrzeug, Fahrbahn und Mensch*, edited by Armin Hermann and Harry Niemann, 159–167. Bielefeld: transcript, 2006.

Moss, Timothy. "'Cold spots' of Urban Infrastructure: Shrinking Processes in Eastern Germany and the Modern Infrastructural Ideal." *International Journal of Urban and Regional Research* 32 (2008): 436–451.

Moss, Timothy. "Divided City, Divided Infrastructures: Securing Energy and Water Services in Postwar Berlin." *Journal of Urban History* 35 (2009): 923–942.

Moss, Timothy. *Remaking Berlin: A History of the City through Infrastructure, 1920–2020*. Cambridge, MA: MIT Press, 2020.

Mukerji, Chandra. *Impossible Engineering: Technology and Territoriality on the Canal du Midi*. Princeton, NJ: Princeton University Press, 2009.

Mukerji, Chandra. "Intelligent Uses of Engineering and the Legitimacy of State Power." *Technology and Culture* 44.4 (2003): 655–676.

Mukhopadhyay, Aparajita. *Imperial Technology and "Native" Agency: A Social History of Railways in Colonial India, 1850–1920*. London: Routledge, 2018.

Mullen, Richard, and James Munson. *The Smell of the Continent: The British Discover Europe*. London: Pan Books, 2010.

Müllender, Bernd. "Ode an die Pappendecke." *Die Tageszeitung* (taz Magazin), July 7, 2001.

Müller von Blumencron, Matthias, "Die Emotionsmaschine." *Frankfurter Allgemeine Zeitung*, February 5, 2016.

Müller, Simone M. *Wiring the World: The Social and Cultural Creation of Global Telegraph Networks*. New York: Columbia University Press, 2016.

Mumford, Lewis. *The City in History: Its Origins, Its Transformations, and Its Prospects*. New York: Harvest, 1989.

Mumford, Lewis. *The Myth of the Machine*. New York: Harcourt Brace Jovanovich, 1970.

Münch, Ragnhild. *Gesundheitswesen im 18. und 19. Jahrhundert: Das Berliner Beispiel*. Berlin: Akademie, 1995.

Münkler, Stefan, and Alexander Roesler, eds. *Telefonbuch: Beiträge zu einer Kulturgeschichte des Telefons*. Frankfurt am Main: Suhrkamp, 2000.

Murakami, Haruki. *Underground: The Tokyo Gas Attack and the Japanese Psyche*. Translated by Alfred Birnbaum and Philip Gabriel. New York: Vintage, 2000.

Murphy, Craig, and JoAnn Yates. *ISO, the International Organization for Standardization: Global Governance through Voluntary Consensus*. London: Routledge, 2008.

Musil, Robert. "Denkmale." In *Gesammelte Werke*, vol. 7, 506–509. Reinbek: Rowohlt, 1978.

Musil, Robert. "Geschwindigkeit ist eine Hexerei (28. Mai 1927)." In *Gesammelte Werke*, vol. 2, 685. Reinbek: Rowohlt 1978.

Nanz, Tobias. "Das 'Rote Telefon': Ein hybrides Objekt des Kalten Krieges." *Zeitschrift für Literaturwissenschaft und* Linguistik 43.173 (2014): 135–148.

Nanz, Tobias. *Grenzverkehr: Eine Mediengeschichte der Diplomatie*. Berlin: Diaphanes, 2010.

"Netzanschluss ist Menschenrecht." *Frankfurter Allgemeine Zeitung*, October 31, 2011.

Neu, Claudia, ed. *Daseinsvorsorge: Eine gesellschaftswissenschaftliche Annäherung*. Wiesbaden: Verlag für Sozialwissenschaften, 2009.

Neubaur, Paul. "Die Usambarabahn in Deutsch-Ostafrika." *Velhagen und Klasings Monatshefte* 17.11 (1903): 575–582.

Nickles, David Paul. "Telegraph Diplomats: The United States' Relations with France in 1848 and 1870." *Technology and Culture* 40.1 (1999): 1–25.

Nickles, David Paul. *Under the Wire. How the Telegraph Changed Diplomacy*. Cambridge, MA: Harvard University Press, 2003.

Niethammer, Lutz. "'Normalisierung' im Westen: Erinnerungsspuren in die 50er Jahre." In *Ist der Nationalsozialismus Geschichte? Zu Historisierung und Historikerstreit*, edited by Dan Diner, 153–184. Frankfurt am Main: Fischer, 1987.

Nolte, Amina. "Political Infrastructure and the Politics of Infrastructure: The Jerusalem Light Rail." *City* 20.3 (2016): 441–454.

Nordmann, Hans. *Die Frühgeschichte der Eisenbahn*. Berlin: Akademie, 1947.

Norris, Frank. *The Octopus: A Story of California*. New York: Penguin, 1994.

North, Michael, ed. *Kommunikationsrevolutionen: Die neuen Medien des 16. Und 19. Jahrhunderts*. Cologne: Böhlau, 2001.

Nowak, Kai. "Teaching Self-Control: Road Safety and Traffic Education in Postwar Germany." *Historical Social Research / Historische Sozialforschung* 41.1 (2016): 135–153.

Nübel, Christoph. *Durchhalten und Überleben an der Westfront: Raum und Körper im Ersten Weltkrieg*. Paderborn: Schöningh, 2014.

Nurske, Ragnar. *Problems of Capital Formation in Underdeveloped Countries*. Oxford: Oxford University Press, 1953.

Nye, David E. *Technology Matters: Questions to Live With*. Cambridge, MA: MIT Press, 2007.

Nye, David E. *When the Lights Went Out: A History of Blackouts in America*. Cambridge, MA: MIT Press, 2010.

Oberhuber, Nadine. "Auf Sand gebaut." *Die Zeit*, October 4, 2007.

Obertreis, Julia. *Imperial Desert Dreams: Cotton Growing and Irrigation in Central East Asia, 1860–1991*. Göttingen: Vandenhoeck & Ruprecht, 2017.

Obertreis, Julia. "Karrieren, Patronage und 'Infrastrukturpoesie': Dimensionen der Infrastrukturgeschichte am Beispiel des russländischen und sowjetischen Zentralasien." In *Wasserinfrastrukturen und Macht von der Antike bis zur Gegenwart*, edited by Birte Förster and Martin Bauch, 232–265. Munich: De Gruyter Oldenbourg, 2015.

Ogle, Vanessa. "Whose Time Is It? The Pluralization of Time and the Global Condition, 1870s to 1940s." *American Historical Review* 120.5 (2013): 1376–1402.

Ohlsen, Manfred. *Der Eisenbahnkönig Bethel Henry Strousberg. Eine preussische Gründerkarriere*. Berlin: Verlag der Nation, 1987.

Oldenziel, Ruth. "Islands: The United States as a Networked Empire." In *Entangled Geographies. Empire and Technopolitics in the Global Cold War*, edited by Gabrielle Hecht, 13–41. Cambridge, MA: MIT Press, 2011.

Oltmer, Jochen. "Unentbehrliche Arbeitskräfte: Kriegsgefangene in Deutschland 1914–1918." In *Kriegsgefangene im Europa des Ersten Weltkriegs*, edited by Jochen Oltmer, 67–96. Paderborn: Schöningh, 2006.

Orland, Barbara. "Milchpropaganda vor und nach dem Ersten Weltkrieg: Konvergenzen zwischen Wissenschaft, Wirtschaft und Ernährungsreform." In *Technikgeschichte im Ruhrgebiet—Technikgeschichte für das Ruhrgebiet*, edited by Manfred Rasch and Dietmar Bleidick, 909–933. Essen: Klartext, 2004.

Ortlepp, Anke. "Airport Security and the Limits of Mobility: The Case of the United States." *Traverse: Zeitschrift für Geschichte / Révue Historique* 16.1 (2009): 75–88.

Ortlepp, Anke. *Jim Crow Terminals: The Desegregation of American Airports*. Athens: University of Georgia Press, 2017.

Osswald, Albert. *Eine Zeit vergeht.* Giessen: Ferber, 1993.

Osterhammel, Jürgen. *Die Verwandlung der Welt.* Munich: Beck, 2009.

Osterhammel, Jürgen. "Grenzen und Brücken." In *Die Flughöhe der Adler: Historische Essays zur globalen Gegenwart.* Munich: Beck, 2017.

Osterhammel, Jürgen, and Niels P. Petersson. *Geschichte der Globalisierung. Dimensionen, Prozesse, Epochen.* Munich: Beck, 2007.

Ostrom, Elinor. *Institutional Incentives and Sustainable Development: Infrastructure Policies in Perspective.* Boulder, CO: Westview, 1993.

Ostrom, Elinor, and Silke Hilfreich, eds. *Was mehr wird, wenn wir teilen: Vom gesellschaftlichen Wert der Gemeingüter.* Munich: Oekom, 2011.

Ostrom, Vincent. "Adam Smith und öffentliche Güter." In *Markt, Staat und Solidarität bei Adam Smith,* edited by Franz-Xaver Kaufmann and Hans-Günter Krüsselberg, 135–157. Frankfurt am Main: Campus, 1984.

Packer, Jeremy. "Disciplining Mobility." In *Foucault, Cultural Studies, and Governmentality,* edited by Jack Z. Bratich, Jeremy Packer, and Cameron McCarthy, 135–164. Albany: State University of New York Press, 2003.

Palla, Rudi. *Verschwundene Arbeit: Das Buch der untergegangenen Berufe.* Vienna: Brandstätter, 2014.

Pape, Lutz, and Hans-Jürgen Weinert. *Bottichwaschmaschine und Haustelegraph: Anfänge der Elektrotechnik im Haushalt.* Braunschweig: Westermann, 1993.

Park, Jinhee. "Von der Müllkippe zur Abfallwirtschaft: Die Entwicklung der Hausmüllentsorgung in Berlin (West) von 1945 bis 1990." PhD diss., Technische Universität Berlin, 2004.

Patalong, Frank. "Untersee-Kabel: Die fragilen Lebensadern des Internets." *Der Spiegel,* February 2, 2015. https://www.spiegel.de/netzwelt/web/untersee-kabel-die-fragilen-lebensadern-des-internets-a-1015809.html

Patel, Kiran Klaus. *Soldaten der Arbeit: Arbeitsdienste in Deutschland und den USA, 1933–1939/1942.* Göttingen: Vandenhoeck & Ruprecht, 2003.

Paul, Gerhard. *BilderMACHT: Studien zur Visual History des 20. und 21. Jahrhunderts.* Göttingen: Wallstein, 2013.

Payer, Peter, ed. *Sauberes Wien: Stadtreinigung und Abfallbeseitigung seit 1945.* Vienna: Holzhausen, 2006.

Payer, Peter. *Der Gestank von Wien: Über Kanalgase, Totendünste und andere üble Geruchskulissen.* Vienna: Döcker, 1997.

Payer, Peter. *Die synchronisierte Stadt: Öffentliche Uhren und Zeitwahrnehmung.* Vienna: Holzhausen, 2015.

Payer, Peter. *Unentbehrliche Requisiten der Grossstadt: Eine Kulturgeschichte der öffentlichen Bedürfnisanstalten von Wien*. Vienna: Löcker, 2000.

Peil, Corinna. *Mobilkommunikation in Japan: Zur kulturellen Infrastruktur der Handy-Aneignung*. Bielefeld: transcript, 2011.

Petermann, Thomas, Harald Bradke, Arne Lüllmann, Maik Poetzsch, and Ulrich Riehm. *Gefährdung und Verletzbarkeit moderner Gesellschaften, am Beispiel eines grossräumigen Ausfalls der Stromversorgung*. Berlin: Büro für Technikfolgen-Abschätzung beim Deutschen Bundestag, 2010.

Peters, Benjamin. *How Not to Network a Nation: The Uneasy History of the Soviet Internet*. Cambridge, MA: MIT Press, 2016.

Petroski, Henry. *The Road Taken: The History and Future of America's Infrastructure*. New York: Bloomsbury, 2016.

Petroski, Henry. *Success through Failure: The Paradox of Design*. Princeton, NJ: Princeton University Press, 2006.

Petroski, Henry. *To Engineer Is Human: The Role of Failure in Successful Design*. New York: St. Martin's, 1992.

Petroski, Henry. *To Forgive Design: Understanding Failure*. Cambridge, MA: Harvard University Press, 2012.

Pfeiffer, Eduard A. *Technik der Stadt: Vom Schaffen der technischen Betriebe im lebendigen Kreislauf der grossen Siedlungen*. Stuttgart: Franckh, 1937.

Pfister, Christian. "Das '1950er Syndrom': Die umweltgeschichtliche Epochenschwelle zwischen Industriegesellschaft und Konsumgesellschaft." In *Das 1950er Syndrom: Der Weg in die Konsumgesellschaft*, edited by Christian Pfister, 51–95. Bern: Haupt, 1995.

Pinch, Trevor. "The Invisible Technologies of Goffman's Sociology from the Merry-Go-Round to the Internet." *Technology and Culture* 51.2 (2010): 409–424.

Piper, Nikolaus. "Im Zeichen des Windhunds." *Süddeutsche Zeitung*, October 18/19, 2014.

Pitzke, Marc. "Blackout von 1977: New Yorks dunkelste Nacht." *Der Spiegel*, July 13, 2007.

Plant, Sadie. *On the Mobile: The Effects of Mobile Telephones on Social and Individual Life*. Schaumburg, IL: Motorola, 2001.

Platthaus, Andreas. "Schriftsteller, ans Telefon!" *Frankfurter Allgemeine Zeitung*, January 31, 2014.

Pleiner, Tom. *Überplanung von Infrastruktur*. Tübingen: Mohr Siebeck, 2016.

Polanyi, Michael. *Implizites Wissen*. Frankfurt am Main: Suhrkamp, 1985.

Pond, Oscar Lewis. *Municipal Control of Public Utilities: A Study of the Attitude of Our Courts toward an Increase of the Sphere of Municipal Activity*. New York: Columbia University Press, 1906.

Ponte, Alessandra. "Müllschlucker: Die Domestizierung des Abfalls." *Arch+* 191/192 (2009): 78–83.

Porombka, Wiebke. *Medialität urbaner Infrastrukturen: Der öffentliche Nahverkehr, 1870–1933*. Bielefeld: transcript, 2013.

Porter, Theodore M. "Objectivity and Authority: How French Engineers Reduced Public Utility to Numbers." *Poetics Today* 12.2 (1991): 245–265.

Porter, Theodore M. *Trust in Numbers: The Pursuit of Objectivity in Science and Public Life*. Princeton, NJ: Princeton University Press, 1995.

Potthast, Jörg. *Die Bodenhaftung der Netzwerkgesellschaft: Eine Ethnografie von Pannen an Grossflughäfen*. Bielefeld: transcript, 2007.

Potthast, Jörg. "Politische Soziologie der Zugänge: Das Beispiel der Flughafensicherheit." In *Sichtbarkeitsregime: Überwachung, Sicherheit und Privatheit im 21. Jahrhundert*, edited by Leon Hempel, Susanne Krasmann, and Ulrich Bröckling, 223–241. Wiesbaden: VS Verlag für Sozialwissenschaften 2011.

Pozharliev, Lyubomir. "Collectivity vs. Connectivity: Highway Peripheralization in Former Yugoslavia (1940s–1980s)." *Journal of Transport History* 37.2 (2016): 194–213.

Pozharliev, Lyubomir. *The Road to Socialism: Transport Infrastructure in Socialist Bulgaria and Yugoslavia (1945–1989)*. Göttingen: Vandenhoeck & Ruprecht, 2023.

Prashad, Vijay. *The Darker Nations: A People's History of the Third World*. New York: New Press, 2007.

Prausnitz, Wilhelm, ed. *Atlas und Lehrbuch der Hygiene mit besonderer Berücksichtigung der Städte-Hygiene*. Munich: Lehmann, 1909.

Puffert, Douglas J. "The Standardization of the Track Gauge on North American Railways, 1830–1890." *Journal of Economic History* 60.4 (2000): 933–960.

Putzier, Konrad. "Warum bauen die Schweizer so viele Bunker?" *Die Welt*, July 12, 2011.

Quester, Peter. "Höchstens haltbar bis . . . Wie das 20. Jahrhundert das eingebaute Kaputtgehen erfand." *TrenntMagazin*. https://trenntmagazin.de/obsolescence-hochstens-haltbar-bis/#.VijmprxFvFa; retrieved December 17, 2017.

Radkau, Joachim. *Das Zeitalter der Nervosität: Deutschland zwischen Bismarck und Hitler*. Munich: Hanser, 1998.

Radkau, Joachim. *Die Ära der Ökologie. Eine Weltgeschichte*. Munich: Beck, 2011.

Radkau, Joachim. *Geschichte der Zukunft: Prognosen, Visionen, Irrungen in Deutschland von 1945 bis heute*. Munich: Hanser, 2017.

Radkau, Joachim. *Natur und Macht: Eine Weltgeschichte der Umwelt*. Munich: Beck, 2002.

Radunski, Michael. "Schöne Grüsse aus dem Central Telegraph Office." *Frankfurter Allgemeine Zeitung*, July 13, 2013.

Rafi, Muhammad Masood, Sarosh Hashmat Lodi, and Nadeem Manzoor Hasan. "Corruption in Public Infrastructure Service and Delivery: The Karachi Case Study." *Public Works Management & Policy* 17.4 (2012): 370–387.

Rainer, George. *Understanding Infrastructure: A Guide for Architects and Planners*. New York: Wiley, 1990.

Randelhoff, Martin. "Resiliente Infrastrukturen und Städte: Kritikalität und Interdependenzen." *Zukunft Mobilität*, December 17, 2013. https://www.zukunft-mobilitaet.net/40882/analyse/resilienz-infrastruktur-stadt-wirtschaft-zukunft-resiliente-infrastrukturen.

Randow, Gero von. "Der Angriff auf die Moderne." *Frankfurter Allgemeine Zeitung*, September 19, 2001.

Randzio, Ernst. *Unterirdischer Städtebau, besonders mit Beispielen aus Gross-Berlin*. Bremen: Dorn, 1951.

Rankin, William. "Infrastructure and the International Governance of Economic Development, 1950–1965." In *Internationalization of Infrastructures: Proceedings of the 12th International Conference on the Economics of Infrastructures*, edited by Jean-François Auger, Jan Jaap Bouma, and Rolf Künneke, 61–75. Delft, Netherlands: Economics of Infrastructures, 2009.

Rao, Sudhir. "A Rail Passenger Called Gandhi . . . Railways Played a Role in the Making of the Mahatma." *Governance Now*, January 30, 2012. https://www.governancenow.com/views/columns/rail-passenger-called-gandhi.

Ratzel, Friedrich. *Erdenmacht und Völkerschicksal: Eine Auswahl aus seinen Werken*. Stuttgart: Kroner, 1940.

Rayward, W. Boyd. *The Universe of Information: The Work of Paul Otlet*. Moscow: International Federation for Documentation, 1975.

Rebske, Ernst. *Lampen, Laternen, Leuchten: Eine Historie der Beleuchtung*. Stuttgart: Franckh, 1962.

Reents, Edo. "Freudig begrüssen wir den zuckenden Schein." *Frankfurter Allgemeine Zeitung*, March 1, 2005.

Reh, Sascha. *Gegen die Zeit*. Frankfurt am Main: Schöffling, 2015.

Reichle, Jörg. "Wo wir sind, ist oben." *Süddeutsche Zeitung*. January 1/2, 2016.

Richter, Steffen. *Infrastruktur: Ein Schlüsselkonzept der Moderne und die deutsche Literatur 1848–1914*. Berlin: Matthes & Seitz, 2018.

Richter, Steffen. "Tunnelblicke: Zur literarischen Repräsentation von Machtverhältnissen anhand einer schweizerischen Infrastruktureinrichtung." In *Macht in der Deutschschweizer Literatur*, edited by Gonçalo Vilas-Boas and Teresa Oliviera, 267–286. Berlin: Frank & Timme, 2012.

Richthofen, Ferdinand von. "Land Communication between Europe and China." *Geographical Magazine* 1 (1874): 144–146.

Riehl, Wilhelm Heinrich. *Die Naturgeschichte des Volkes, Erster Band: Land und Leute.* Stuttgart: Cotta, 1925.

Riepl, Wolfgang. *Das Nachrichtenwesen des Altertums mit besonderer Rücksicht auf die Römer*. Leipzig: Teubner, 1913.

Rivière, Philippe. "Der Staat als Maschine: Das Kybernetik-Experiment in Allendes Chile." *Le Monde Diplomatique*, November 12, 2010.

Robinson, Gustavus H. "The Public Utility Concept in American Law." *Harvard Law Review* 41.3 (1928): 277–308.

Rodenberg, Julius. "Sonntag vor dem Landberger Tor (1880)." *Bilder aus dem Berliner Leben*, vol. 1. Berlin: Paetel, 1891.

Rodogno, David, Bernhard Struck, and Jakob Vogel, eds. *Shaping the Transnational Sphere: Experts, Networks, and Issues from the 1840s to the 1930s*. New York: Berghahn, 2014.

Rogers, Dennis, and Bruce O'Neill. "Infrastructural Violence: Introduction to the Special Issue." *Ethnography* 13.4 (2012): 401–412.

Röhl, Klaus Rainer. *Nähe zum Gegner: Kommunisten und Nationalsozialisten im Berliner BVG-Streik von 1932*. Frankfurt am Main: Campus, 1994.

Röhling, K. "Die Fahrgeldhinterziehung auf Strassenbahnen und ihre Bekämpfung," *Zeitschrift für Kleinbahnen* 23 (1916): 415.

Röhr, Andreas. *Die Bajkal-Amur-Magistrale: Geschichte eines sibirischen Raumerschliessungsprojektes*. Berlin: Logos, 2012.

Rolt, Lionel T. C. *From Sea to Sea: The Canal du Midi*. Columbus: Ohio University Press, 1973.

Rommerdt, Carl Christian. *Handbuch der Land- und Wasserbaukunst: Für angehende und ausübende Forstmänner, Kameralisten und Oekonomen*. Erfurt: Henning, 1828.

Rosa, Hartmut. *Beschleunigung: Die Veränderung der Zeitstrukturen in der Moderne*. Frankfurt am Main, Suhrkamp, 2005.

Rose, Dirk. "'Im Vorort': Zu einem Romanfragment Detlev von Liliencrons." *Zeitschrift für Germanistik* 16.3 (2006): 638–639.

Rose, Dirk. "Peripherie und Perspektive: Infrastrukturgeschichtliche Überlegungen zu Liliencron und Bölsche." *Zeitschrift für Germanistik* 15.2 (2005): 311–326.

Roskothen, Johannes. *Verkehr: Zu einer poetischen Theorie der Moderne*. Munich: Fink, 2003.

Roth, Ralf. *Städte im europäischen Raum: Verkehr, Kommunikation und Urbanität im 19. und 20. Jahrhundert*. Stuttgart: Steiner, 2009.

Roth, Ralf. "Wie wurden die Eisenbahnen der Welt finanziert? Einige Vergleiche." *Zeitschrift für Weltgeschichte* 10.1 (2009): 55–80.

Rüb, Matthias. "Kontinentale Durchbrüche." *Frankfurter Allgemeine Zeitung*, May 20, 2015.

Rubin, Eli. *Synthetic Socialism: Plastic and Dictatorship in the German Democratic Republic*. Chapel Hill: University of North Carolina Press, 2008.

Rudkowski, Lena. *Der Streik in der Daseinsvorsorge*. Munich: Beck, 2010.

Rüfner, Wolfgang. "§ 80 Daseinsvorsorge und soziale Sicherheit." In *Handbuch des Staatsrechts der Bundesrepublik Deutschland, Bd. III: Das Handeln des Staates*, edited by Josef Isensee and Paul Kirchhof, 1038–1084. Heidelberg: Müller, 1996.

Rügemer, Werner. *Colonia Corrupta: Globalisierung, Privatisierung und Korruption im Schatten des Kölner Klüngels*. Münster: Westfälisches Dampfboot, 2010.

Rügemer, Werner. *Cross Border Leasing: Ein Lehrstück zur globalen Enteignung der Städte*. Münster: Westfälisches Dampfboot, 2005.

Rügemer, Werner. *"Heuschrecken" im öffentlichen Raum: Public Private Partnership—Anatomie eines globalen Finanzinstruments* (Bielefeld: transcript, 2012).

Rutenberg, Jürgen von. "Zurück in die Gegenwart." *Zeit-Magazin* 29 (2015): 12–20.

Sachs, Wolfgang. *Die Liebe zum Automobil: Ein Rückblick in die Geschichte unserer Wünsche*. Reinbek: Rowohlt, 1984.

Safranski, Rüdiger. *Zeit: Was sie mit uns macht und was wir aus ihr machen*. Munich: Hanser, 2015.

Salewski, Michael. "Technik als Vision der Zukunft um die Jahrhundertwende." In *Moderne Zeiten: Technik und Zeitgeist im 19. und 20. Jahrhundert*, edited by Michael Salewski and Ilona Stölken-Fitschen, 77–91. Stuttgart: Steiner, 1994.

Sammons, Jeffrey L., ed. *Die Protokolle der Weisen von Zion: die Grundlage des modernen Antisemitismus—eine Fälschung : Text und Kommentar*. Göttingen: Wallstein, 2018.

Sampson, Tony D. *Virality: Contagion Theory in the Age of Networks*. Minneapolis: University of Minnesota Press, 2012.

Sanford, Kristen L., Joel A. Tarr, and Sue McNeil. "Crisis Perception and Policy Outcomes: Comparison between Environmental and Infrastructure Crises." *Journal of Infrastructure Systems* 1.4 (1995): 195–203.

Sassen, Saskia. *Global Networks, Linked Cities*. London: Routledge, 2002.

Saul, Klaus. "'Kein Zeitalter seit Erschaffung der Welt hat so viel und so ungeheuerlichen Lärm gemacht': Lärmquellen, Lärmbekämpfung und Antilärmbewegung im

Deutschen Kaiserreich." In *Umweltgeschichte: Methoden, Themen, Potentiale*, edited by Günter Bayerl, Norman Fuchsloch, and Torsten Mayer, 187–218. Münster: Waxmann, 1996.

Saupe, Achim. "'Human Security' and the Challenge of Automobile and Road Traffic Safety: A Cultural Historical Perspective." *Historical Social Research* 35.4 (2010): 102–121.

Schaeffer, Roland. "Produktive Nachhaltigkeit: Infrastrukturinnovationen als politische Strategie." In *Die Zukunft der Infrastrukturen: Intelligente Netzwerke für eine nachhaltige Entwicklung*, edited by Reinhard Loske and Roland Schaeffer, 45–69. Marburg: Metropolis, 2005.

Schäfers, Bernhard. "Zur Genesis und zum Stellenwert von Partizipationsforderungen im Infrastrukturbereich." *Raumforschung und Raumordnung* 32.1 (1974): 1–6.

Scharnigg, Max. "Mein halbes digitales Leben." *Süddeutsche Zeitung*. August 8/9, 2015.

Schattenberg, Susanne. *Stalins Ingenieure: Lebenswelten zwischen Technik und Terror in den 1930er Jahren*. Munich: Oldenbourg, 2002.

Scheiner, Joachim. "Stellt den McDrive doch unter Denkmalschutz!" *Frankfurter Allgemeine Zeitung*, May 28, 2013.

Scheiner, Joachim, and Christian Holz-Rau, eds. *Räumliche Mobilität und Lebenslauf: Studien zu Mobilitätsbiografien und Mobilitätssozialisation*. Wiesbaden: Springer, 2015.

Schenk, Frithjof Benjamin. *Russlands Fahrt in die Moderne: Mobilität und sozialer Raum im Eisenbahnzeitalter*. Stuttgart: Steiner, 2014.

Schenkel, Werner. "Entstehung, Entsorgung und Wiederverwertung von Müll—ein globales Problem." In *Technik und Natur*, edited by Werner Nachtigall and Charlotte Schönbeck, 483–520. Düsseldorf: VDI, 1994.

Scherer, Inge. *Grenzen des Streikrechts in den Arbeitsbereichen der Daseinsvorsorge: Schutz der Bürger vor gezielter Schädigung durch Streiks*. Berlin: Duncker & Humblot, 2000.

Schickel, Joachim. *Guerrilleros, Partisanen: Theorie und Praxis*. Munich: Hanser, 1970.

Schieb, Jörg. "Kommunikation, Unterhaltung und Konsum künftig mit einem Gerät: Grossbaustelle Datenautobahn." *Frankfurter Allgemeine Zeitung*, January 9, 1995.

Schink, Hans-Christian. *Verkehrsprojekte Deutsche Einheit: Begleitbuch zur Ausstellung*. Ostfildern-Ruit: Cantz, 2004.

Schinkel, Karl Friedrich. *Anweisung zum Bau und zur Unterhaltung der Kunststrassen*. Berlin: Zentral- und Landesbibliothek, 1834.

Schipper, Frank. *Driving Europe: Building Europe on Roads in the Twentieth Century*. Amsterdam: Aksant, 2008.

Schipper, Frank. "Unravelling Hieroglyphs: Urban Traffic Signs and the League of Nations." *Métropoles* 6 (2009).

Schivelbusch, Wolfgang. *Disenchanted Night: The Industrialization of Light in the Nineteenth Century*. Translated by Angela Davies. Berkeley: University of California Press, 1995.

Schivelbusch, Wolfgang. *Entfernte Verwandtschaft: Faschismus, Nationalsozialismus, New Deal, 1933–1939*. Frankfurt am Main: Fischer, 2008.

Schivelbusch, Wolfgang. *The Railway Journey: The Industrialization of Time and Space in the Nineteenth Century*. Berkeley: University of California Press, 2014.

Schlagintweit, Robert von. *Die Santa Fe- und Südpacificbahn in Nordamerika*. Cologne: Mayer, 1884.

Schlögel, Karl. *Das sowjetische Jahrhundert: Archäologie einer untergegangenen Welt*. Munich: Beck, 2017.

Schlözer, Christiane. "Rein ins kalte Wasser." *Süddeutsche Zeitung*, May 9, 2014.

Schmid, Georg. *Die Spur und die Trasse: (Post)moderne Wegmarken in der Geschichtswissenschaft*. Cologne: Böhlau, 1988.

Schmidt, Klaus-Peter. "Alles muss raus." *Die Zeit*, June 22, 2006.

Schmidt, Robert. "Technik, Risiko und das Zusammenspiel von Habitat und Habitus." In *Kalkuliertes Risiko: Technik, Spiel und Sport an der Grenze*, edited by Gunter Gebauer et al., 78–95. Frankfurt am Main: Campus, 2006.

Schmidt, Wilhelm R. *Als Telegrafenbauer in Deutsch-Südwest*. Erfurt: Sutton, 2006.

Schmidt, Wolfgang. *Integration und Wandel: Die Infrastruktur der Streitkräfte als Faktor sozioökonomischer Modernisierung in der Bundesrepublik 1955 bis 1975*. Munich: Oldenbourg, 2007.

Schmitt, Martin. *Internet im Kalten Krieg: Eine Vorgeschichte des globalen Kommunikationsnetzes*. Bielefeld: transcript, 2016.

Schmitt, Stefan. "Anrufen wird bald das neue Rauchen." *Die Zeit*, December 10, 2015.

Schmucki, Barbara. *Der Traum vom Verkehrsfluss*. Frankfurt am Main: Campus, 2001.

Schmucki, Barbara. "The Machine in the City: Public Appropriation of the Tramway in Britain and Germany, 1870–1915." *Journal of Urban History* 38.6 (2012): 1060–1093.

Schneider, Irmela. "Radiophone Praktiken des (Selbst-)Regierens in der Weimarer Republik." In *Formationen der Mediennutzung II: Strategien der Verdatung*, edited by Irmela Schneider and Isabell Otto, 37–53. Bielefeld: transcript, 2007.

Schneider, Ursula, and Heide David. *"Das Paradies kommt wieder . . ." Zur Kulturgeschichte und Ökologie von Herd, Kühlschrank und Waschmaschine: Katalogbuch*. Hamburg: Museum der Arbeit, 1993.

Schneider, Volker, and Marc Tenbücken, eds. *Der Staat auf dem Rückzug: Die Privatisierung öffentlicher Infrastrukturen*. Frankfurt am Main: Campus, 2004.

Schneier, Bruce. "Beyond Security Theater." *New Internationalist* 427 (2009): 10–13.

Schneier, Bruce. *Beyond Fear: Thinking Sensibly about Security in an Uncertain World*. New York: Copernicus, 2006.

Schnitzler, Antina von. "Citizenship Prepaid: Water, Calculability, and Techno-Politics in South Africa." *Journal of Southern African Studies* 34.4 (2008): 899–917.

Schoepp, Sebastian. "Letzter Zug nach Nirgendwo." *Süddeutsche Zeitung*, May 17, 2010.

Schönherr, Ekkehard. "Infrastrukturen des Glücks: Eine Bild-, Raum- und Infrastrukturgeschichte Mallorcas im 19. und 20. Jahrhundert unter Berücksichtigung des Tourismus." PhD diss., Erfurt, 2014.

Schot, Johan, and Vincent Lagendijk. "Technocratic Internationalism in the Interwar Years: Building Europe on Motorways and Electricity Networks." *Journal of Modern European History* 6.2 (2008): 196–216.

Schott, Dieter. *Die Vernetzung der Stadt: Kommunale Energiepolitik, öffentlicher Nahverkehr und die "Produktion" der modernen Stadt: Darmstadt—Mannheim—Mainz 1880–1918*. Darmstadt: Wissenschaftliche Buchgesellschaft, 1999.

Schott, Dieter. *Europäische Urbanisierung (1000–2000): Eine umwelthistorische Einführung*. Cologne: Böhlau, 2014.

Schrader, Christopher. "Hauptgewinn: eine Latrine." *Süddeutsche Zeitung*, April 17, 2015.

Schrader, Christopher. "Rettender Verzicht auf Neues." *Süddeutsche Zeitung*, September 10, 2010. https://www.sueddeutsche.de/wissen/klimawandel-rettender-verzicht-auf-neues-1.998145.

Schreiber, Christian. "Der letzte Liftboy Europas." *Next floor: Das Magazin für die Kunden der Schindler-Aufzüge AG* 1 (2013): 8–10.

Schröder, Catalina. "Wenn Anlieger zu Anlegern werden." *Die Zeit*, March 11, 2013.

Schröder, Lina. *Der Rhein-(Maas-)Schelde-Kanal als geplante Infrastrukturzelle von 1946 bis 1986: Eine Studie zur Infrastruktur- und Netzwerk-Geschichte*. Münster: Waxmann, 2017.

Schueler, Judith. *Materialising Identity: The Co-construction of the Gotthard Railway and Swiss National Identity*. Amsterdam: Aksant, 2008.

Schulz, Felix Robin. *Death in East Germany 1945–1990*. Oxford: Berghahn, 2013.

Schulze, Thomas. *Infrastruktur als politische Aufgabe: Dogmengeschichtliche, methodologische und theoretische Aspekte*. Frankfurt am Main: Lang, 1993.

Schulze, Tillmann. *Bedingt abwehrbereit: Schutz kritischer Informations-Infrastrukturen in Deutschland und den USA*. Wiesbaden: Verlag für Sozialwissenschaften, 2006.

Schüttpelz, Erhard. "Körpertechniken." *Zeitschrift für Medien- und Kulturforschung* 2.1 (2010): 101–120.

Schütz, Erhard, and Eckhard Gruber. *Mythos Reichsautobahn: Bau und Inszenierung der "Strassen des Führers" 1933–1941*. Berlin: Links, 1996.

Scott, James C. *Seeing like a State: How Certain Schemes to Improve the Human Condition Have Failed*. New Haven, CT: Yale University Press, 1998.

Seesslen, Georg. *Dietmar Schönherr: Von den phantastischen Abenteuern des Raumschiffes Orion zu Wünsch Dir was. Die Biografie*. Mannheim: Reichelt & Brockmann, 2012.

Segal, Zef. "Communication and State Construction: The Postal Service in German States, 1815–1866." *Journal of Interdisciplinary History* 44.4 (2014): 453–473.

Seligo, Hans. *Afrika im Umbau*. Jena: Urania-Freidenker, 1931.

Sennett, Richard. *The Fall of Public Man*. New York: Norton, 2017.

Siebenbiedel, Christian. "Leuchttürme machen den Seehandel sicher." *Frankfurter Allgemeine Sonntagszeitung*, October 9, 2011.

Sieberg, Herward. *Colonial Development: Die Grundlegung moderner Entwicklungspolitik durch Grossbritannien, 1919–1949*. Stuttgart: Steiner, 1985.

Sieferle, Rolf-Peter. *Rückblick auf die Natur: Eine Geschichte des Menschen und seiner Umwelt*. Munich: Landtverlag, 1997.

Siegelbaum, Lewis H., ed. *The Socialist Car: Automobility in the Eastern Bloc*. Ithaca, NY: Cornell University Press, 2011.

Siegert, Bernhard. *Relais: Geschicke der Literatur als Epoche der Post, 1751–1913*. Berlin: Brinkmann & Bose, 1993.

Siemons, Mark. "Die Wurzel der Menschlichkeit." *Frankfurter Allgemeine Zeitung*, September 26, 2014.

Simmel, Georg. "The Metropolis and Mental Life." In *Simmel on Culture: Selected Writings*, edited by David Frisby and Mike Featherstone, 174–186. London: Sage, 1997.

Simone, Abdou Maliq. "People as Infrastructure: Intersecting Fragments in Johannesburg." *Public Culture* 16.3 (2004): 407–429.

Smil, Vaclav. *Creating the Twentieth Century: Technical Innovations of 1867–1914 and Their Lasting Impact*. Oxford: Oxford University Press, 2005.

Smil, Vaclav. *Energy in World History*. London: Routledge, 2020.

Smith, Adam. *The Wealth of Nations*. New York: Modern Library, 2000.

Smith, Jason Scott. *Building New Deal Liberalism: The Political Economy of Public Works, 1933–1956*. Cambridge: Cambridge University Press, 2006.

Smith-Howard, Kendra. *Pure and Modern Milk: An Environmental History since 1900.* Oxford: Oxford University Press, 2013.

Soldt, Rüdiger. "Wir fahr'n, fahr'n, fahr'n auf der Autobahn." *Frankfurter Allgemeine Zeitung,* November 29, 2014.

Sonnenberger, Franz. "'Kolonisieren heisst transportieren'—Europa und der Beginn des Eisenbahnzeitalters in Afrika." In *Zug der Zeit—Zeit der Züge: Deutsche Eisenbahn 1835–1985,* 1:228–231. Berlin: Siedler, 1985.

Speckmann, Thomas. "Von unschuldigen Zivilisten." *Frankfurter Allgemeine Zeitung,* December 2, 2009.

Sperling, Walter. *Der Aufbruch der Provinz: Die Eisenbahn und die Neuordnung der Räume im Zarenreich.* Frankfurt am Main: Campus, 2011.

Spiewak, Martin. "Bloss nicht offline sein." *Die Zeit,* November 9, 2017.

Spode, Hasso. "Romantische Zeitreise: Tourismus als Chronotopie." In *Sommerfrische: Bilder, Orte, Praktiken,* edited by Brigitta Schmidt-Lauber, 33–45. Vienna: Institut für Europäische Ethnologie, 2014.

Staat, Yvonne, "Gute Fahrt." *Frankfurter Allgemeine Sonntagszeitung,* January 4, 2015.

Stadler, Max. "Der User." In *Das Personal der Postmoderne: Inventur einer Epoche,* edited by Alban Frei and Hannes Mangold, 75–90. Bielefeld: transcript, 2015.

Stäheli, Urs. "Infrastrukturen des Kollektiven: alte Medien—neue Kollektive?" *Zeitschrift für Medien- und Kulturforschung* 12.2 (2012): 99–116.

Star, Susan Leigh. "The Ethnography of Infrastructure." *American Behavioral Scientist* 43.3 (1999): 377–391.

Starosielski, Nicole. *The Undersea Network.* Durham, NC: Duke University Press, 2015.

Stavenhagen, Willibald. *Verkehrs-, Beobachtungs- und Nachrichten-Mittel in militärischer Beleuchtung: Für Offiziere aller Waffen des Heeres und der Marine.* Göttingen: Peters, 1905.

Stefa, Elian, and Gyler Mydyti. *Concrete Mushrooms: Reusing Albania's 750,000 Bunkers.* Milan: ACTAND, 2013.

Steffens, Lincoln. *The Shame of the Cities.* New York: McClure, Phillips, 1904.

Stein, Dietrich. *Instandhaltung von Kanalisationen.* Bochum: Stein, 1999.

Steininger, Benjamin. "Pipeline: Am Puls der fossilen Moderne." In *Stoffe in Bewegung,* edited by Kijan Espahangizi and Barbara Orland, 231–244. Zurich, Switzerland: Diaphanes, 2014.

Steininger, Benjamin. *Raum-Maschine Reichsautobahn: Zur Dynamik eines bekannt / unbekannten Bauwerks.* Berlin: Kadmos, 2005.

Steinke, Ronen. "Stopp." *Süddeutsche Zeitung,* July 20/21, 2013.

Stephenson, Neal. "Mother Earth Mother Board." *Wired*, December 1, 1996. https://www.wired.com/1996/12/ffglass.

Stier, Bernhard. *Staat und Strom: Die politische Steuerung des Elektrizitätssystems in Deutschland 1890–1950*. Ubstadt-Weiher: Verlag Regionalkultur, 1999.

Stirn, Alexander. "Kehraus im Weltraum." *Süddeutsche Zeitung*, June 4, 2017. https://www.sueddeutsche.de/wissen/weltraumschrott-kehraus-im-weltraum-1.3530728.

Stokes, Raymond G., Roman Köster, and Stephen C. Sambrook. *The Business of Waste: The United Kingdom and Germany, 1945 to the Present*. Cambridge: Cambridge University Press, 2013.

Stolk, Bram van, and Cas Wouters. "Die Gemütsruhe des Wohlfahrtsstaates." In *Macht und Zivilisation: Materialien zu Norbert Elias' Zivilisationstheorie 2*, edited by Peter Gleichmann, Johan Goudsblom, and Hermann Korte, 242–260. Frankfurt am Main: Suhrkamp, 1984.

Straeten, Jonas van der. "Measuring Progress in Megawatts? Development Narratives and the Shaping of the Tanzanian Power Infrastructure 1906–1980." PhD diss., Technische Universität Berlin, 2017.

Strasser, Susan. *Waste and Want: A Social History of Trash*. New York: Henry Holt, 1999.

Strassmann, Burkhard. "Bewahren Sie Ruhe." *Die Zeit*, November 26, 2015.

Strube, Sebastian. *Euer Dorf soll schöner werden: Ländlicher Wandel, staatliche Planung und Demokratisierung in der Bundesrepublik Deutschland*. Göttingen: Vandenhoek & Ruprecht, 2013.

Swaan, Abram de. *Der sorgende Staat: Wohlfahrt, Gesundheit und Bildung in Europa und den USA der Neuzeit*. Frankfurt am Main: Campus, 1993.

Swynegedouw, Erik. *Social Power and the Urbanization of Water: Flows of Power*. Oxford: Oxford University Press, 2004.

Tantner, Anton. *Die Hausnummer: Eine Geschichte von Ordnung und Unordnung*. Marburg: Jonas, 2007.

Tatje, Claas. "Die Pendlerrepublik: Millionen Deutsche macht der Weg zur Arbeit krank und einsam. Sie und ihre Arbeitgeber sollten etwas dagegen unternehmen." *Die Zeit*, May 22, 2014.

Tatje, Claas. *Fahrtenbuch des Wahnsinns: Unterwegs in der Pendlerrepublik*. Munich: Kösel, 2014.

"Technical Relief Grows in Germany: Hardships of Industrial Conflicts Mitigated by 1,500 Local Groups." *New York Times*, December 3, 1922.

Tegtmeyer, Gundula Madeleine. "Die Müllspezialisten von Kairo." *Natürlich: Das Magazin für ganzheitliches Leben* 3 (2006): 52–59.

"Tempo 20." *Der Spiegel*, August 20, 1963.

Tenner, Edward. *Why Things Bite Back: Technology and the Revenge of Unintended Consequences*. New York: Vintage, 1996.

Tessier, Alexandre. "La grande hôtellerie parisienne face à l'innovation: entre devoir d'adoption et facteur de distinction (1855–1914)." In *Le tourisme comme facteur de transformations économiques, techniques et sociales (XIXe–XXe siècles). Tourism as a factor of economic, technical and social transformations (XIXth—XXth centuries)*, edited by Marc Gigase, Cédric Humair, and Laurent Tissot, 179–195. Neuchâtel: Alphil, 2014.

Tessier, Alexandre. *Le Grand Hôtel: L'invention du luxe hôtelier, 1862–1972*. Rennes: Presses universitaires de Rennes, 2012.

Teuscher, Andreas. *Schweiz am Meer—Pläne für den "Central-Hafen" Europas inklusive Alpenüberquerung mit Schiffen im 20. Jahrhundert*. Zurich, Switzerland: Limmat, 2014.

Teuteberg, Hans Jürgen, and Cornelius Neutsch, eds. *Vom Flügeltelegraphen zum Internet: Geschichte der modernen Telekommunikation*. Stuttgart: Steiner, 1998.

Theile, Merlind. "Die Achse des Guten." *Die Zeit*, September 4, 2015.

Thiemeyer, Guido, and Isabel Tölle. "Supranationalität im 19. Jahrhundert? Die Beispiele der Zentralkommission für die Rheinschifffahrt und des Oktroyvertrages 1804–1832." *Journal of European Integration History* 17 (2011): 177–196.

Thomas, Peter. "Hundert Jahre Einstellungssache." *Frankfurter Allgemeine Zeitung*, August 20, 2013.

Thüsing, Gregor. *Arbeitskampf in der Daseinsvorsorge*. Tübingen: Mohr Siebeck, 2013.

Tischler, Julia. *Light and Power in a Multiracial Nation: The Kariba Dam Scheme in the Central African Federation*. Basingstoke, UK: Palgrave Macmillan, 2013.

Tissot, Laurent. *Naissance d'une industrie touristique: Les Anglais et la Suisse au XIXe siècle*. Lausanne, Switzerland: Payot, 2000.

Tissot, Laurent. "Tourismus." *Historisches Lexikon der Schweiz*. www.hls-dhsdss.ch/textes/d/D14070.php; retrieved December 17, 2017.

Tomilina, Natalia G., ed. Пять колец под кремлевскими звездами. Moscow: MFD, 2011.

Torp, Cornelius. "Revolution zur See: Maritime Frachtkosten und Weltmarkt im 'langen' 19. Jahrhundert." In *Das maritime Europa: Werte—Wissen—Wirtschaft*, edited by Jürgen Elvert, Lutz Feldt, Ingo Löppenberg, and Jens Ruppenthal, 93–108. Stuttgart: Steiner, 2016.

Toth, Jennifer. *The Mole People: Life in the Tunnels Beneath New York City*. Chicago: Chicago Review, 1993.

Trevor-Roper, Hugh R., ed. *Hitler's Table Talk 1941–1944: Secret Conversations*. London: Enigma, 2000.

Trotha, Trutz von. "Was war Kolonialismus? Einige zusammenfassende Befunde zur Soziologie und Geschichte des Kolonialismus und der Kolonialherrschaft." *Saeculum* 55 (2004): 49–95.

Trovalla, Ulrika, and Eric Trovalla. "Infrastructure Turned Suprastructure: Unpredictable Materialities and Visions of a Nigerian Nation." *Journal of Material Culture* 20.1 (2015): 43–57.

Trurnit, Hanno. *Geschichte(n) hinterm Zähler: Die Beziehungen zwischen Energieversorgern und ihren Kunden*. Munich: Trurnit, 1996.

Tucholsky, Kurt. "Berlin! Berlin!" *Gesammelte Werke in zehn Bänden*, vol. 2. Reinbek: Rowohlt, 1975.

Tucholsky, Kurt. "Der Kontrollierte." *Gesammelte Werke in zehn Bänden, vol. 2*. Reinbek: Rowohlt, 1975.

Tucholsky, Kurt. *Von der Kunst, richtig zu reisen*. Münster: Coppenrath, 2015.

Tümmers, Horst Johannes. *Der Rhein: Ein europäischer Fluss und seine Geschichte*. Munich: Beck, 1994.

Tunzelmann, Alex von. "The Railway Man," *Guardian*, June 24, 2015.

Uekötter, Frank. *Die Wahrheit ist auf dem Feld: Eine Wissensgeschichte der deutschen Landwirtschaft*. Göttingen: Vandenhoek & Ruprecht, 2010.

Ullrich, Otto. "Technologie." In *Wie im Westen so auf Erden: Ein polemisches Handbuch zur Entwicklungspolitik*, edited by Wolfgang Sachs, 390–408. Reinbek: Rowohlt, 1993.

Ullrich, Volker. "Quietschen, Hupen, Fauchen." *ZeitGeschichte* 2 (2013): 24–26.

United States Strategic Bombing Survey. *The Effects of Strategic Bombing on German Morale*. 2 vols. Washington, DC: Government Printing Office, 1947.

Urry, John. *The Tourist Gaze: Leisure and Travel in Contemporary Societies*. London: Sage, 1990.

Vahrenkamp, Richard. "Logistik als Metasystem der Infrastruktur: Die Grenzen der Eisenbahn-Logistik und der Aufstieg der Lastwagen-Logistik 1900 bis 1938." In *Versorgung und Entsorgung der Moderne: Logistiken und Infrastrukturen der 1920er und 1930er Jahre*, edited by Wiebke Porombka, Heinz Reif, and Erhard Schütz, 123–150. Frankfurt am Main: Lang, 2011.

Veidt, Karl. "Als Divisionspfarrer im Argonnerwald, 1914–1917." *Hessische Quellen zum Ersten Weltkrieg*. https://www.lagis-hessen.de/de/purl/resolve/subject/qhg/id/35-4; retrieved December 17, 2017.

Vierhaus, Rudolf, ed. *Das Tagebuch der Baronin von Spitzemberg, geb. Freiin von Varnbüler: Aufzeichnungen aus der Hofgesellschaft des Hohenzollernreiches*. Göttingen: Vandenhoeck & Ruprecht, 1989.

Virilio, Paul. *Polar Inertia*. Translated by Patrick Camiller. London: Sage, 2000.

Virilio, Paul. *Speed and Politics*. Translated by Mark Polizzotti. Los Angeles: Semiotext[e], 2006.

Voeten, Teun. *Tunnel People*. Oakland, CA: PM, 2010.

Vogt, Franz. *Anstandsbüchlein für das Volk*. Donauwörth: Auer 1894.

Völkel, Markus. "Einigkeit und Freiheit. Die Eisenbahn ein Mittel nationaler Politik?" *Zug der Zeit—Zeit der Züge: Deutsche Eisenbahn 1835–1985*, vol. 1, 218–227. Berlin: Siedler, 1985.

Vorderer, Peter and Christoph Klimmt, "Das neue Normal." *Die Zeit*, January 28, 2016.

Vowinckel, Annette. *Flugzeugentführungen: Eine Kulturgeschichte*. Göttingen: Wallstein, 2011.

Waldman, Jonathan. *Rust: The Longest War*. New York: Simon & Schuster, 2015.

Ward, Christopher J. *Brezhnev's Folly: The Building of BAM and Late Soviet Socialism*. Pittsburgh, PA: University of Pittsburgh Press, 2009.

Weber, Eugen. *Peasants into Frenchmen: The Modernization of Rural France 1870–1914*. Stanford, CA: Stanford University Press, 1988.

Weber, Heike. *Das Versprechen mobiler Freiheit*. Bielefeld: transcript, 2015.

Weber, Heike. "'Entschaffen' Reste und das Ausrangieren, Zerlegen und Beseitigen des Gemachten." *Technikgeschichte* 81.1 (2014): 1–32.

Weber, Heike. "Stecken, drehen, drücken: Interfaces von Alltagstechniken und ihre Bediengesten." *Technikgeschichte* 76 (2009): 233–254.

Weber, Heike, and Ruth Oldenziel. "Recycling and Re-Use in the Twentieth Century," *Contemporary European History* 22.3 (2013): 347–370.Weisband, Marina. *Wir nennen es Politik: Ideen für eine zeitgemässe Demokratie*. Stuttgart: Klett-Cotta, 2013.

Wells, Louis T., and Eric S. Gleason. "Is Foreign Infrastructure Investment Still Risky?" *Harvard Business Review* 9–10 (1995): 44–55.

Wells, Malcolm. *Infra Structures: Life Support for the Nation's Circulatory Systems*. Brewster, MA: Underground Art Gallery, 1994.

Welzer, Harald. *Mentale Infrastrukturen—wie das Wachstum in die Welt und in die Seelen kam*. Berlin: Heinrich-Böll-Stiftung, 2011.

Welzer, Harald. "Wasser aus der Leitung." *Frankfurter Allgemeine Zeitung*, August 6, 2013.

Wenzlhuemer, Roland. *Connecting the Nineteenth-Century World: The Telegraph and Globalization*. Cambridge: Cambridge University Press, 2013.

Wenzlhuemer, Roland. *Globalgeschichte schreiben: Eine Einführung in 6 Episoden*. Munich: UVK/Lucius, 2017.

Wenzlhuemer, Roland. "Verbrechen, Verbrechensbekämpfung und Telegrafie: Kriminalhistorische Perspektiven auf die Entkoppelung von Transport und

Kommunikation im langen 19. Jahrhundert." *Historische Zeitschrift* 301.2 (2015): 347–373.

Westerman, Frank. *Ingenieure der Seele: Schriftsteller unter Stalin—eine Erkundungsreise.* Berlin: BVT, 2003.

Westermann, Andrea. *Plastik und politische Kultur in Westdeutschland.* Zurich, Switzerland: Chronos, 2007.

Whitman, Walt. *The Portable Walt Whitman.* Edited by Michael Warner. New York: Penguin, 2004.

Whitmore, Mark. "Transport and Supply during the First World War." https://www.iwm.org.uk/history/transport-and-supply-during-the-first-world-war; retrieved December 20, 2017.

Widmann, Marc. "Menschens Kinder." *Süddeutsche Zeitung,* May 4/5, 2014.

Wiedenfeld, Kurt. "Der volkswirtschaftliche Einfluss der modernen Verkehrsmittel und die deutsche Volkswirtschaftslehre des 19. Jahrhunderts." In *Die Entwicklung der deutschen Volkswirtschaftslehre im neunzehnten Jahrhundert: Festgabe für Gustav Schmoller, 2. Teil.* Leipzig: Duncker & Humblot, 1908. 1–28.

Wiegand, Ralf. "Gedenkenlos." *Süddeutsche Zeitung,* April 18/19, 2015.

Wiegel, Michaela. "Ins Wesen von Paris fahren," *Frankfurter Allgemeine Zeitung,* July 19, 2000.

"Wie Straßen die Schwächen Berns offenlegen. Gespräch mit Daniel Flückiger." *Berner Zeitung,* October 20, 2011.

Wildt, Michael. "'Der muss hinaus! Der muss hinaus!' Antisemitismus in deutschen Nord- und Ostseebädern 1920–1935." *Mittelweg* 36.4 (2001): 2–25.

Wildt, Michael. *Volk, Volksgemeinschaft, AfD.* Hamburg: Hamburger, 2017.

Wilhelm, Daniel. *Die Kommunikation infrastruktureller Grossprojekte: Die Elektrifizierung Oberschwabens durch die OEW in der ersten Hälfte des 20. Jahrhunderts.* Stuttgart: Steiner, 2014.

Wilhelmer, Lars. *Transit-Orte in der Literatur: Eisenbahn—Hotel—Hafen—Flughafen.* Bielefeld: transcript, 2015.

Willemsen, Roger. *Wer wir waren: Zukunftsrede.* Frankfurt am Main: Fischer, 2017.

Windmüller, Sonja. *Die Kehrseite der Dinge: Müll, Abfall, Wegwerfen als kulturwissenschaftliches Problem.* Münster: LIT, 2004.

Winkler, Hartmut. *Switching / Zapping: Ein Text zum Thema und ein parallel laufendes Unterhaltungsprogramm.* Darmstadt: Häusser, 1991.

Winter, Max. *Im unterirdischen Wien.* Berlin: Seemann, 1905.

Winther, Tanja. *The Impact of Electricity: Development, Desires and Dilemmas.* New York: Berghahn, 2008.

Wissen, Markus, and Matthias Naumann. "Raumdimensionen des Wandels technischer Infrastruktursysteme. Eine Einleitung." In *Infrastrukturnetze und Raumentwicklung: Zwischen Universalisierung und Differenzierung*, edited by Timothy Moss, Matthias Naumann, and Markus Wissen, eds. 17–34. Munich: Oekom, 2008.

Wolf, Birgit. *Sprache in der DDR: Ein Wörterbuch*. Berlin: De Gruyter, 2000.

Wolfrum, Edgar. *Welt im Zwiespalt: Eine andere Geschichte des 20. Jahrhunderts*. Stuttgart: Klett-Cotta, 2017.

Wong, Edward. "The Air-Raid-Shelter Apartments Under Beijing." *New York Times Magazine*, April 22, 2011.

World Bank. *World Development Report 1994: Infrastructure for Development*. Oxford: Oxford University Press, 1994.

World Wildlife Fund. "Earth Hour," March 26, 2022. https://www.worldwildlife.org/pages/earth-hour.

Worster, Donald. *Under Western Skies: Nature and History in the American West*. Oxford: Oxford University Press, 1992.

Wu, Tim. *The Master Switch: The Rise and Fall of Information Empires*. New York: Vintage, 2013.

Wu, Xijujie. *Ein Jahrhundert Licht: Eine technikethnologische Studie zur Beleuchtung im chinesischen ländlichen Alltag*. Wiesbaden: Harrassowitz, 2009.

Yacowar, Maurice. "The Bug in the Rug: Notes on the Disaster Genre." In *Film Genre Reader III*, edited by Barry Keith Grant, 277–295. Austin: University of Texas Press, 2003.

Zeiss, Heinrich. "Hygiene und Technik." *Technikgeschichte* 25 (1936): 66–73.

Zejnelovic, Marko. *Der "Highway of Nations": Eine postkoloniale Studie zum Panamakanal als imperiale Infrastruktur*. Augsburg: Institut für Spanien, Portugal- und Lateinamerikastudien, 2014.

Zelger, Sabine. *Das Pferd frisst keinen Gurkensalat: Kulturgeschichte des Telefonierens*. Cologne: Böhlau, 1997.

Zennss-Reimann, Regine. *Unterm Flügelrad: Frauen bei der Eisenbahn*. Berlin: Museum für Verkehr & Technik, 1993.

Ziblatt, Daniel. "Why Some Cities Provide More Public Goods than Others: A Subnational Comparison of the Provision of Public Goods in German Cities in 1912." *Studies in Comparative International Development* 43 (2008): 273–289.

Zons, Julia. *Casellis Pantelegraph: Geschichte eines vergessenen Mediums*. Bielefeld: transcript, 2015.

Zweig, Stefan. "Blick über die elektrische Stadt in die Zukunft hinein." *Neue Rundschau* 1 (1992): 109–114.

Zwicky, J. Friedrich. *Public Utilities*. Jena: Osterwalg, 1937.

INDEX

Advertising, *55*, 161
Aesthetics of Infrastructure, 130–131, 133–134
Africa, 35, 57, 71, 99–101, 104, 107, 124
Agriculture, 50, 57, 60, 82, 84, 114, 144, 197
Airplanes and Airports, 141, 159–161
Anti-Corruption Laws, 185
Anti-Semitism, 77–78, 165
Automobiles. *See* Cars
Autonomy, 174

Benjamin, Walter, 44
Bicycles. *See* Cycling
Bismarck, Otto Von, 37, 57
Bolivia, Water War (Cochabama), 115
Bombings, 61, 131, 167
Bridges, 16, 26, *83*, 126, 134, 151, 153–156
Bunkers, 149–150, 168
Bush, George W., 169–170

Canal Systems, 21–25, 30–31
Cars, 61–72, 84–85, 91–92, 133, 197.
Castells, Manuel, 16, 195
Cell Phones. *See* Smartphones
Chappe Semaphore, 34
China, 31, 103, 149, 180–181
Cities, 32, 38–46, 179, 195–196
Civic Engagement, 127
Civil War (US), 29, 31, 166
Cleanliness, 45–46, 105–106, 194
Colonization, 27–28, 56–57, 96–98, 168
Communism. *See* Socialism
Construction Projects, 70, 85
Consumerism, 80–82
Contracts, Fixed-Term, 115
Crimean War, 36
Cuban Revolution, 168–169
Cults, 165
Cybernetics, 88
Cycling, 158, 179

Death Railway, 122–123
Deregulation, 117, 120–121, 200
Detachment, 180–181
Digitization, 130
Disasters, 150–151
Disease, 40, 73–74, 95
Distraction, 177

Domestic Sphere, 49, 53, 79, 102
Dupuit, Jules, 13, 153

Earth Hour, 200
Economics, 13, 54, 114–118, 121, 155, 185–186, 191
Edison, Thomas Alva, 11, 50
Education, 158, 170
Efficiency, 139
Egypt, 31, 104, 110, 119–120
Electricity, 49–53, *55*, 76, 102, 163–164
Energy Sources, 22, 94, 108
Engels, Friedrich, 47–48
England, 21–23, 26, 76
Environmental Disasters, 163
Epidemics, 43
Etiquette, 157, 175
Euergetism, 138

Facebook, 195
France, 13, 23, 34, 56, 74, 128, 153
Freedom, 91–92, 164, 204

Gandhi, Mahatma, 72–73
Gas, 49–51, 53–54, 94, *132,* 194
Genscher, Hans-Dietrich, 84, 95
Germany, 2, 24–25, 33, 61–62, 70, 86–87, 92–95, 107, 116, 124–125
Giedion, Siegfried, 6
Globalization, 15, 35, 98–100, 128, 191, 194, 197
Goethe, Johann Wolfgang Von, 10–11, 33, 143
Graffiti, 179
Guevara, Ernesto "Che," 168

Habit/Habitus, 158
Hackers, 179
Haussmann, Georges Eugène, 43–44, 85
Heidegger, Martin, 199
Highways. *See* Roadways
Hitler, Adolph, 70, 139, 166
Honecker, Erich, 92, 95
Hotels, 62–67
Housing, 40, 54–55, 81, 121
Hydraulic Engineering, 126
Hygiene. *See* Cleanliness

Imperialism and Technology, 35
Indigenous People, Displacement of, 26–28
Individuality, 182
Industrialization, 47–48, 193
Information Age, 192–193
Infrastructure
  as term, 5–9, 58, 83, 119, 152, 170, 201–202
  overview, 201–205
  definitions, 12–18
  historiographical approaches, 9–12
  literature on, 86–87
International Organization for Standardization, 15, 135–136
Internet, 169–170, 189–190, 193, 199

Jacobs, Jane, 85
Jewish People, 77–78
Johnson, Lyndon B., 89, 153

Land Settlement, 26
Lawrence, Thomas E. (Lawrence of Arabia), 167
Laws, 41
Lenin, Vladimir Ilyich, 48, 89, 124
Libraries, 190, 192
Light, Electric, 102
Lighthouses, 30, 117, 148–149

Marshall Plan, 83
Marx, Karl, 33, 48
Material Culture and Memory, 79–80
Mcculloch, Robert, 26
Mcluhan, Marshall, 35
Mechanization, 49
Memory, 79–80, 144
Merkel, Angela, 190

Military, 58, 84–85, 116, 166
Milk, 191
Mindfulness, 181
Mining Industry, 13
Modernity and Modernization, 4, 12, 32–33, 35–36, 40, 61–62, 130, 198, 204
Monuments vs. Infrastructures, 134
Moses, Robert, 44, 85–86
Multitasking, 177, 181
Mumford, Lewis, 16, 134
Mussel Slough Tragedy, 26–27, 31

Nature, 16, 129–130, 163, 196
Neoliberalism, 95, 118, 156, 204
New Media, 145–146, 178
Noise, 131–133
Norris, Frank, 27, 41
Nuclear Disasters, 94, 108, 167–168

Oil, 49, 80, 94, 99, 103, 193
Olympics, 136–138
Open Access, 193

Parks, Rosa, 72
Peace, Rhetoric of, 33–34, 37
Perception, 178
Pest Control, 105, 162
Phonographs, 55
Plumbing Systems, 46
Pneumatic Infrastructure, 146, *147*
Police and Security, 116
Popular Music, 144–145
Postal System, 113–114, 147, 195
Poverty, 97
Prison and Prisoners, 60, 68
Privacy, 177
Private Ownership, 74
Private Sector, 120, 127–128
Privatization, 120–122
Productivity, 124
Propaganda, 165
Protests, 170
Public Space, 77, 124–125, 177
Public Works, 9–10, 13, 21–46, 74–75

Racism, 72, 77
Radio Technology, 62
Railways, 25, 30–34, 97, 117, 122–123, 143, 156, 161, 180
Recycling, 107, 110
Refrigeration, 79, 82
Regulation, 117, 126–127, 200
Religion and Religious Groups, 52, 71, 96, 98–99, 104, 110, 180
Resources, 103, 196–197
Revolutionary Thinking, 48
Riepl's Law, 146–147
Roadways, 15, 91, 121, 135
Rockefeller Principle, 54
Roman Empire, 8–9, 21–22, 38
Romanticism, 24
Roosevelt, Franklin D., 75
Russia, 31–33, 136–138

Safety and Security, 116, 157–160, 162–163, 169, 203
Sewers, 38, 42, 46, 150, 152. *See also* Underground Spaces
Shipping and Delivery, 191
Signs, 135, 159
Simmel, Georg, 17, 175
Smartphones, 1, 130, 181, 187. *See also* Telephones
Smith, Adam, 10, 12
Socialism, 47, 56, 88–89, 134, 203–204
Sound, 131–133
Soviet Union, 48, 56, 89–91, 136–137, 139, 141
Stalin, Joseph, 139
Standardization and Uniformization, 175
Star, Susan Leigh, 17, 134
Stephenson, Neal, 190
Subways, 165, 173–174, 176
Suez Canal (Egypt), 31, 119–120

Suicide, 173
Superstructures, 199
Supply and Demand, 194

Taxes, 116
Technocratic Ideology, 139–140
Telecommunications, 126–127, 195
Telegraphy, 34, 37, 59, 146, 148, 191
Telephones, 35, 55–56, 59, 90–91, 184, 187. *See also* Smartphones
Terrorist Attacks, 164–165, 169
Thailand-Burma Line, 122
Thoreau, Henry David, 131
Toilets, 46, 77, 194, 200
Tokyo (Japan), 173–174, 187
Tourism Industry, 64–67
Tucholsky, Kurt, 65, 175, 183
Turkey, Modernization of, 95
Twitter, 36

Underground Spaces, 61, 134, 149, 165, 168. *See also* Sewers
United States, 26–29, 74, 85, 125, 129, 152, 166
Utilities, 115
Utopia, 189–190

Vandalism, *118*
Villages, 32, 35
Vogt, Franz, 157, 164
Vulnerability, 163

Wars, 57–60, 166–167
Waste, 38, 104–107, 109–110, 131
Watches, 186
Water and Waterways, 43, 105–106, 115, 194, 197
Weber, Max, 14, 199
Westernization, 96, 99–101
Whatsapp, 195
Whitman, Walt, 176
Wiebeking, Carl Friedrich von, 24
Wilhelm, Kaiser, Ii, 37
Women's Work, 50–51, 55, 60, *184*
Work and Labour, 13, 47, 60, 71, 123–125, 151–152, 156, 160–161, 182–184
World Bank, 97, 115, 156
Wunderwald, Gustav, *3*

Infrastructures Series

Edited by Paul N. Edwards and Janet Vertesi

Paul N. Edwards, *A Vast Machine: Computer Models, Climate Data, and the Politics of Global Warming*

Lawrence M. Busch, *Standards: Recipes for Reality*

Lisa Gitelman, ed., *"Raw Data" Is an Oxymoron*

Finn Brunton, *Spam: A Shadow History of the Internet*

Nil Disco and Eda Kranakis, eds., *Cosmopolitan Commons: Sharing Resources and Risks across Borders*

Casper Bruun Jensen and Brit Ross Winthereik, *Monitoring Movements in Development Aid: Recursive Partnerships and Infrastructures*

James Leach and Lee Wilson, eds., *Subversion, Conversion, Development: Cross-Cultural Knowledge Exchange and the Politics of Design*

Olga Kuchinskaya, *The Politics of Invisibility: Public Knowledge about Radiation Health Effects after Chernobyl*

Ashley Carse, *Beyond the Big Ditch: Politics, Ecology, and Infrastructure at the Panama Canal*

Alexander Klose, translated by Charles Marcrum II, *The Container Principle: How a Box Changes the Way We Think*

Eric T. Meyer and Ralph Schroeder, *Knowledge Machines: Digital Transformations of the Sciences and Humanities*

Geoffrey C. Bowker, Stefan Timmermans, Adele E. Clarke, and Ellen Balka, eds., *Boundary Objects and Beyond: Working with Leigh Star*

Clifford Siskin, *System: The Shaping of Modern Knowledge*

Lawrence Busch, *Knowledge for Sale: The Neoliberal Takeover of Higher Education*

Bill Maurer and Lana Swartz, *Paid: Tales of Dongles, Checks, and Other Money Stuff*

Katayoun Shafiee, *Machineries of Oil: An Infrastructural History of BP in Iran*

Megan Finn, *Documenting Aftermath: Information Infrastructures in the Wake of Disasters*

Ann M. Pendleton-Jullian and John Seely Brown, *Design Unbound: Designing for Emergence in a White Water World, Volume 1: Designing for Emergence*

Ann M. Pendleton-Jullian and John Seely Brown, *Design Unbound: Designing for Emergence in a White Water World, Volume 2: Ecologies of Change*

Jordan Frith, *A Billion Little Pieces: RFID and Infrastructures of Identification*

Morgan G. Ames, *The Charisma Machine: The Life, Death, and Legacy of One Laptop per Child*

Ryan Ellis, *Letters, Power Lines, and Other Dangerous Things: The Politics of Infrastructure Security*

Mario Biagioli and Alexandra Lippman, eds, *Gaming the Metrics: Misconduct and Manipulation in Academic Research*

Malcolm McCullough, *Downtime on the Microgrid: Architecture, Electricity, and Smart City Islands*

Emmanuel Didier, translated by Priya Vari Sen, *America by the Numbers: Quantification, Democracy, and the Birth of National Statistics*

Andrés Luque-Ayala and Simon Marvin, *Urban Operating Systems; Producing the Computational City*

Michael Truscello, *Infrastructural Brutalism: Art and the Necropolitics of Infrastructure*

Christopher R. Henke and Benjamin Sims, *Repairing Infrastructures: The Maintenance of Materiality and Power*

Stefan Höhne, *New York City Subway: The Invention of the Urban Passenger*

Timothy Moss, *Conduits of Berlin: Remaking the City through Infrastructure, 1920–2020*

Claude Rosental, translated by Catherine Porter, *The Demonstration Society*

Blake Atwood, *Underground: The Secret Life of Videocassettes in Iran*

Huub Dijstelbloem, *Borders as Infrastructure: The Technopolitics of Border Control*

Dylan Mulvin, *Proxies: Standards and their Media*

Eric Monteiro, *Digital Oil: Machineries of Knowing*

Klaus Hoeyer, *Data Paradoxes: The Politics of Intensified Data Sourcing in Healthcare*

Dirk van Laak, *Lifelines of Our Society: A Global History of Infrastructure*